General Thermodynamics

D. R. OLANDER

CRC Press is an imprint of the
Taylor & Francis Group, an **informa** business

CRC Press
Taylor & Francis Group
6000 Broken Sound Parkway NW, Suite 300
Boca Raton, FL 33487-2742

Printed in the United States of America on acid-free paper
10 9 8 7 6 5 4 3 2 1

International Standard Book Number-13: 978-0-8493-7438-8 (Hardcover)

Library of Congress Cataloging-in-Publication Data

Olander, Donald R.
General thermodynamics / Donald Olander.
p. cm.
Includes bibliographical references and index.
ISBN 978-0-8493-7438-8 (alk. paper)
1. Thermodynamics. I. Title.

QC311.O47 2008
536'.7--dc22 2007016721

Visit the Taylor & Francis Web site at
http://www.taylorandfrancis.com

and the CRC Press Web site at
http://www.crcpress.com

Table of Contents

Preface

Thermodynamics touches more specialties in science and engineering than any other theory. All freshmen in these fields encounter thermodynamics in freshman chemistry and physics. In large universities, entire upper-division or graduate courses are devoted to the aspects of thermodynamics pertinent to particular disciplines. All of these courses start with the basics of classical thermodynamics, namely the first and second laws, reversibility, properties of materials and equilibrium. Thereafter the contents are the aspects of thermodynamics needed in each field:

Mechanical engineering: power cycles, refrigeration, and combustion
Chemical engineering: aqueous and organic solution properties and separation processes
Materials science and engineering: nonaqueous materials, phase diagrams and heterogeneous chemistry
Chemistry: chemical reaction equilibria and some statistical thermodynamics
Physics: statistical thermodynamics
Geophysics: high-pressure effects and high-temperature solid/liquid equilibria
Bioengineering: thermodynamics of (mainly) the human body, including ligand binding to macromolecules and the chemistry of proteins and other energy-transport species

This book is intended principally as a concise text for a broad-scope course in thermodynamics for upper division students in science and/or engineering. It covers all of the above applications in varying degrees of thoroughness; statistical thermodynamics is not included, mainly because the atomic-level approach is better covered in a separate book.

The present book and the course from which it grew are not intended to replace the specialized thermodynamics courses offered in mechanical engineering and chemical engineering. However, it should be useful for both small colleges that offer serious science courses as well as large universities with extensive teaching in science and engineering. As offered (as an engineering course) in one of the latter institutions, enrollment consisted of majors in materials science and engineering, nuclear engineering, civil and environmental engineering, bioengineering, engineering physics, and a smattering from five other engineering departments. In smaller colleges that have no engineering school, the book is intended for students in physics and chemistry departments.

The book starts with a brief review of the historical development of thermodynamics in the nineteenth century contained in the book by Von Baeyer (1998). This book provides a delightful, nontechnical description of the science of thermodynamics and its principal early contributors. The monograph by Van Ness (1983) offers a basic technical introduction to the subject.

REFERENCES

Von Baeyer H. C. 1998. *Maxwell's Demon—Why Warmth Disperses and Time Passes*. New York: Random House.
Van Ness H. C. 1983. *Understanding Thermodynamics*. Mineola, NY: Dover.

For students interested in more detailed expositions of certain portions of the subject, the following specialized textbooks are recommended:

Mechanical Engineering

Van Wylen G., R. Sonntag, and C. Borgnakke. 1994. *Fundamentals of Classical Thermodynamics*. 4th ed. or later editions. New York: Wiley.
Cengel Y. and M. Boles. 2006. *Thermodynamics*. 5th ed. New York: McGraw-Hill.

Chemical Engineering

Smith J. M., H. C. Van Ness, and M. M. Abbott. 2001. *Chemical Engineering Thermodynamics*. 6th ed. New York: McGraw-Hill.

Materials Science and Engineering

Gaskell D. 1981 or later editions. *Introduction to Metallurgical Thermodynamics*. 2nd ed. New York: McGraw-Hill.

Bioengineering

Edsall J. and H. Gutfreund. 1983. *Biothermodynamics*. New York: Wiley.
Haynie D. 2001. *Biological Thermodynamics*. Cambridge: Cambridge University Press.

Environmental engineering

Valsaraj K. 2000. *Elements of Environmental Engineering*. Chap. 4, 2nd ed. Boca Raton, FL: CRC Press.

Chemistry

Atkins P. 1978. *Physical Chemistry*, Part I, Equilibrium. New York: W. H. Freeman.

Author

Donald R. Olander has an A.B. in chemistry and a B.S. in chemical engineering from Columbia University and a Ph.D. in chemical engineering from Massachusetts Institute of Technology (MIT). During the earlier era in the field of nuclear engineering when he was at MIT, Olander wrote his doctoral dissertation on the reprocessing of spent nuclear fuels via solvent extraction. Nuclear engineering was a young discipline in the 1950s, and at most universities it was not yet a formal department.

Prior to graduate school, he spent a year studying chemistry in Paris as a Fulbright Scholar at the École du Chimie Physique, where the acids were still kept in wine bottles. In 1958, Olander joined the Chemical Engineering Department at the University of California at Berkeley. In 1961, he moved to Berkeley's new Nuclear Engineering Department, where he remains today as the James Fife Chair in Engineering.

Olander's early work was in basic surface science, analyzing the mechanisms of reactions of gases with solid surfaces using a technique called Molecular Beam Mass Spectrometry. His book, *Fundamental Aspects of Nuclear Reactor Fuel Elements* published in 1976 by the Energy Research and Development Administration (the predecessor of the Department of Energy), is a classic in the field that is still used by every nuclear materials course in the country. A second edition is forthcoming.

His later work involved theoretical studies on the enrichment of uranium by gas centrifuges and included an article in the *Scientific American* in 1978. Olander's other major contribution to science was a paper entitled "The Large Cake-Cutting Problem," in which he presented a mathematical method for cutting large cakes that insured equal servings for all. It was published in the *Journal of Irreproducible Results.*

Subsequently, he moved into the field of nuclear materials, in which his specialty is the chemistry and thermodynamics of nuclear fuels. In 1999, a collection of papers written for the occasion by his colleagues in the field of nuclear materials, former students, and postdocs was published in a special issue of the *Journal of Nuclear Materials* in his honor. In 2000, Professor Olander was elected to membership in the National Academy of Engineering. He will retire in July 2008 after fifty years at the University of California.

Acknowledgments

Thanks and sincerest appreciation goes to Professor John Prausnitz, who has been the author's thermodynamics guru for five decades.

A good part of the motivation for this book was the interaction with the students in the thermodynamics course shared between the author and Professor Andreas Glaeser. His contributions to devising, implementing and teaching a course of such an unusually wide scope from which this book arose is highly appreciated.

To my son Ben, for his dramatic design of the book's cover.

~

A day in the country is fine
But what do I do with my time?
Why I write thermodynamics
But avoid thermodynamics
'Cause entropy's just so sublime

1 Concepts and Definitions

1.1 A BRIEF HISTORY OF THERMODYNAMICS

The driving force for the development of thermodynamics was the development of the steam engine by James Watt around 1770. From 1800 to 1900, thermodynamic theory was slowly and fitfully developed. By 1900, "classical" thermodynamics was essentially complete. In time, various specialized branches of thermodynamics developed.

1.1.1 The Concept of Temperature

Qualitatively, temperature characterizes an object as hot, cold, warm, etc. Galileo (ca 1610) is said to have quantified temperature by the height of liquid in a vertical tube that extended into a pool of the liquid. The tube terminated at the top in a large air-filled glass bulb, which served as the temperature sensor. Such a device was suitable only for measuring changes in ambient air temperature. A major advance was made by the instrument-maker Fahrenheit, who in 1724 constructed the earliest version of the mercury thermometer, descendants of which exist to this day. Because of its compactness and unitary construction in a single piece of glass, Fahrenheit's thermometer could be inserted into a liquid or into orifices of the human body. By adjusting the quantity of ordinary sea salt in an ice-water mixture, the lowest temperature attained was fixed at 0° on the Fahrenheit scale.* By placing the thermometer in his mouth, he decreed a body reference temperature of 96°F. The mercury column responded linearly to other temperatures. For example, when inserted into boiling water, the mercury column rose to 2.21 times the rise at the bodily reference point of 96°F (both with respect to the 0°F position). Thus, the boiling point of pure water is 212°F. In similar fashion, the temperature of an ice–water mixture (without salt) is 32°F.

Celsius (1742) defined 0°C as the boiling point of water and 100°C as the melting point of ice. Soon thereafter, Linnaeus reversed these designations and divided the intervening scale into 100 parts, giving rise to the centigrade scale. With only minor differences, the centigrade scale morphed into the Celsius scale, which, however, has been given a much sounder fundamental basis than simply relying on the two phase changes of water.

The first of the fundamental characterizations of the temperature scale is based on the notion of the *absolute zero temperature,* introduced by Kelvin (ca 1885). This reference mark can be determined by the version of Galileo's gas thermometer shown in Figure 1.1. The vessel of volume V contains n moles of an ideal gas, which obeys the law:

* The topic of freezing-point depression, as illustrated by salt/ice/water mixtures, is treated in Section 8.6.7.

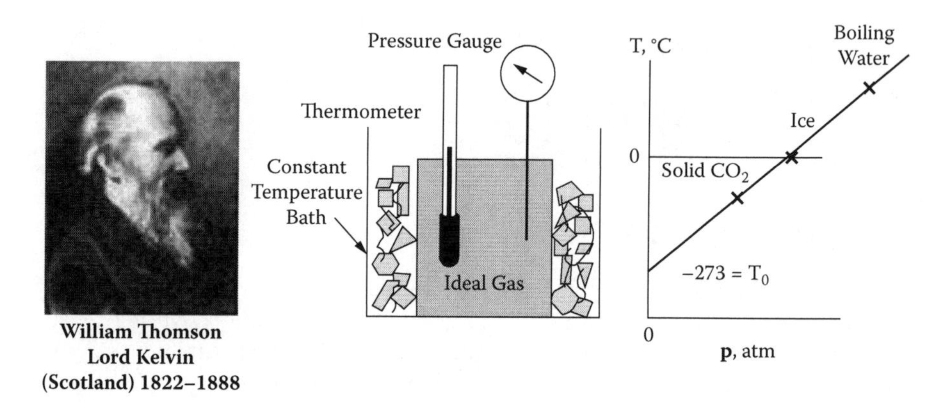

FIGURE 1.1 Kelvin and the gas thermometer.

$$pV = nRT$$

where p is the pressure of the gas in the vessel and R is a constant. Being part of what is in fact a thermodynamic relation, the temperature T in the ideal gas law must have a fundamental meaning that is not captured by the arbitrary Celsius scale. The objective of the experiment shown in Figure 1.1 is to relate the "thermodynamic" temperature in the above equation to the Celsius temperature.

In the experiment, the pressure is measured as a function of the gas temperature, which is fixed by the bath surrounding the vessel. In Figure 1.1, ice/water is represented as the thermal medium, and the temperature assigned a value of 0 on the Celsius scale. Changing the temperature by using boiling water or dry ice as the thermal reservoir and measuring the gas pressure for each produces a straight-line plot on the T-p graph shown on the right-hand side of the figure. Points on this line are independent of the gas used, to the extent that it behaves ideally. The line in Figure 1.1 is of the form $T°C = Ap + T_0$, with the experimental intercept being –273°C. A new temperature unit called the Kelvin, with a different zero than that on the Celsius scale but one degree the same in both, is defined by $T(K) = T°C + 273$.

In addition to the intercept, the slope of the line in Figure 1.1 is nR/V. With n and V known, R is 82 cm^3-atm/mole-K. Different units for pressure and volume give different values of R; however, the temperature units must remain Kelvins. R is called the universal gas constant, because it is independent of the gas used, always with the proviso that it behaves ideally. Since the units of pressure times volume are the same as force times distance or energy, R can be converted to 1.986 calories/mole-K or 8.314 Joules/mole-K.

Use of the usual melting point of water as the zero reference for the Celsius scale is not sufficient for a quantity as fundamental as temperature. For instance, ice melts at a slightly higher temperature in Denver than in New York City. This difference is due to the effect of total pressure (exerted by the air) on the melting point.* To avoid this difficulty, an international committee has recently decreed that

* Discussed in Chapter 5.

the reference state be that in which solid, liquid and gaseous water coexist in the absence of air or any other gas. Such a state defines the *triple point,* which for water is designated as 0.01°C (273.16 K), where the vapor pressure is 0.00611 atm. The specification of the triple-point temperature as 0.01°C means that melting point of ice under atmospheric pressure remains 0°C or 273.15 K.

1.1.2 Heat and Work

Work in a mechanical sense was a well-known concept very early in history. Perhaps the most fundamental form of work is that done against or by gravity in raising or lowering weights. Other devices that produce (or accept) work include springs, pistons and cylinders, moving an object against a resisting force, turbines, electrons moving in an electric potential gradient and muscle contraction. Heat, however, is none of these and cannot be classified as work.

Heat, energy and temperature are closely related but strictly distinct concepts. Briefly but sufficiently, heat is energy in transit from one body to another at a lower temperature. The key to the definition is "in transit." One cannot refer to the "heat content" of an object. A body contains energy but not heat. The latter is manifest only as energy moving in response to a temperature difference. Nor is heat equivalent to temperature; "hot," although sounding similar to heat, refers to temperature.

An earlier, and erroneous, conception of heat was prevalent at the end of the eighteenth century. Heat was viewed as a "fluid" (called caloric) that is contained within a body and moves from a body at high temperature to one at low temperature. The similarity to the flow of a real fluid due to a difference in pressure is obvious. The problem with the theory is that caloric was thought to be something contained in a finite quantity by a body. If caloric flows out of a body, there should be a limit to the extent of heat transfer. This is obviously not so. Rubbing an eraser on paper heats the paper slightly, but in no way removes anything from the inside of the eraser. What is actually happening is conversion of the work expended by the scribbler in moving the eraser over the paper into heat. About 75 years passed before the caloric idea of heat was finally rejected in favor of the temperature difference-driven energy-in-transit notion described in the preceding paragraph.

The person most responsible for the refutation of the caloric theory was an American, who took on the name Count Rumford in the service of a German nobleman. His job was fabrication of cannons for the prince, a task that involved boring out large ingots of iron. As shown in Figure 1.2, this involved horsepower delivered to the boring operation by a series of gears and pulleys. Rumford noticed that the water in which the cannon was immersed continued to heat up and boil as long as the boring operation continued. This, he reasoned, could not be explained by the caloric theory of heat, which posited a fixed quantity of the ineffable fluid in a body. He recognized the work of the horses, transmitted to the boring tool, was the source of the heat entering the water. Thus was the equivalence of heat and work first dimly recognized and heat given its current interpretation as energy leaving or entering a body.

If the flow of energy in Rumford's apparatus is followed back from the contraction of the muscles of the horses, the next transfer mechanism is utilization of the

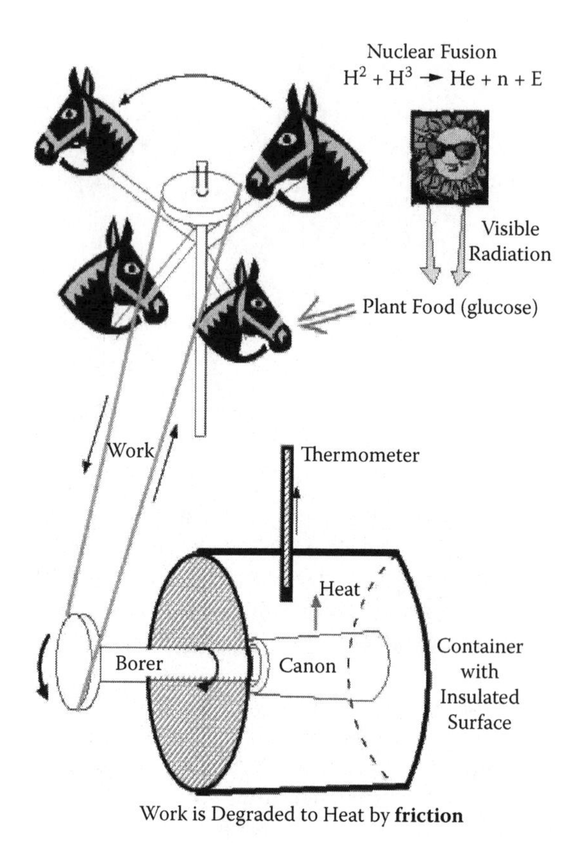

Benjamin Thompson
Count Rumford
(America) 1753–1814

FIGURE 1.2 Rumford's cannon-boring job.

chemical energy carried in the horse's blood in the molecular form of glucose. This chemical is in turn produced from carbon dioxide and water by absorption of visible light from the sun in the grass eaten by the animals. The electromagnetic radiation is a byproduct of the fusion reactions in the sun. In a sense, then, Rumford's cannon boring operation is nuclear-powered.

Although heat and work were recognized as manifestations of energy transfer between bodies, they could not be quantitatively coupled unless they could be expressed in common units. Typically, work is expressed as the product of force and distance, or Newton-meters, which equals a Joule. The conventional unit for heat, on the other hand, is the calories which is the quantity of energy that needs to be delivered to one gram of water to raise its temperature by one degree Celsius. The person most associated with solving this problem is James Joule.

Along with many serious experiments, Joule is reported to have conducted one that is both correct technically but seemingly extraordinarily difficult to render quantitative. He is supposed to have measured the temperature of the water flowing over the top of a waterfall and of the pool at the bottom (Figure 1.3). The latter is slightly higher than the former because of the conversion of potential energy to kinetic energy and finally to thermal energy (heat). Whatever the validity of this

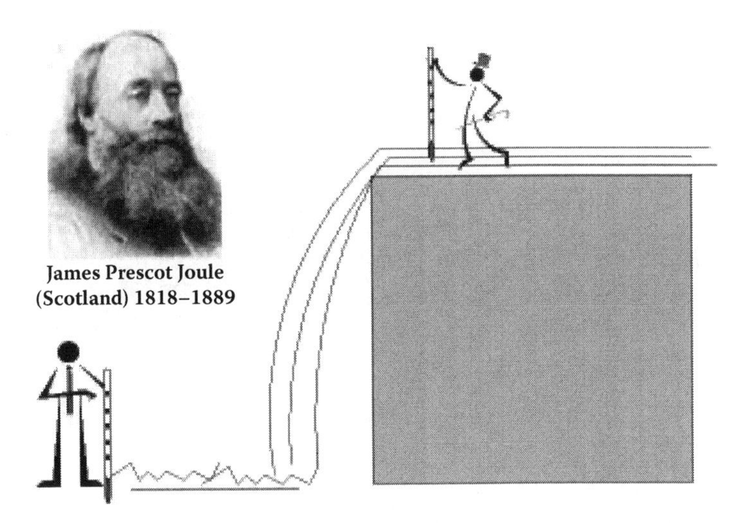

FIGURE 1.3 Joule and the waterfall experiment.

tale, the correspondence of the units of heat and work, also known as the *mechanical equivalent of heat*, was established as 4.184 Joules per calorie.*

1.1.3 Carnot, Clausius, and the Second Law

At the beginning of the nineteenth century, improvement of steam engines was a frustrating trial-and-error operation; eventually the practitioners began to realize that no matter how clever the improvement, there appeared to be a limit to the efficiency of these devices. In 1824, a young (28-year-old) military engineer named Sadi Carnot produced a fundamental analysis of any engine that converted heat to work that was to become the basis of the second law of thermodynamics.** The working fluid did not have to be steam; the theory did not depend on the source of heat or the details of the engine. The efficiency depended solely on the temperature from which the engine received heat and the temperature of the reservoir to which it rejected unusable heat. This deduction was remarkable because it was proposed long before two major features of thermodynamics with which it deals were understood. The first is the nature of heat and the second is the notion of absolute temperature.

As shown in Figure 1.4, what Carnot conceived was an idealized *cycle* that represented any continuous engine. With the steam engine as the motivation, the engine receives heat from a high-temperature reservoir (e.g., hot steam), produces mechanical work (e.g., driving a piston) and, very importantly, rejects heat to a cold reservoir (e.g., condenser water or the atmosphere). Carnot drew his far-reaching conclusions from two critical insights. First, the process in the engine had to be cyclic, in which the working fluid endlessly circulated through the device but periodically returned to its initial state. Second, and more profound, the steps through which the

* See Problem 1.8.

** "Reflections on the Motive Power of Fire and on Machines Fitted to Develop that Power."

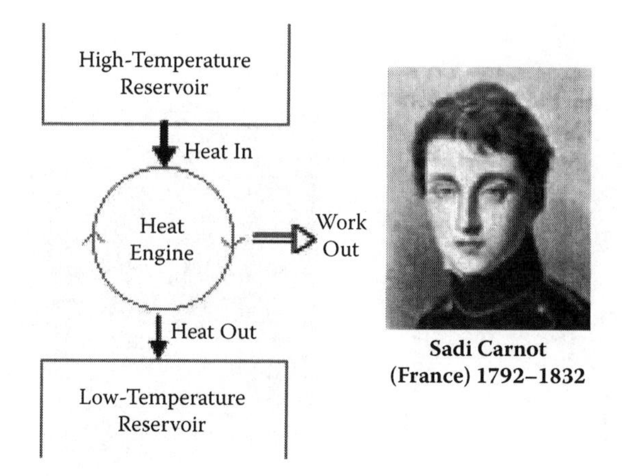

FIGURE 1.4 Carnot's heat engine.

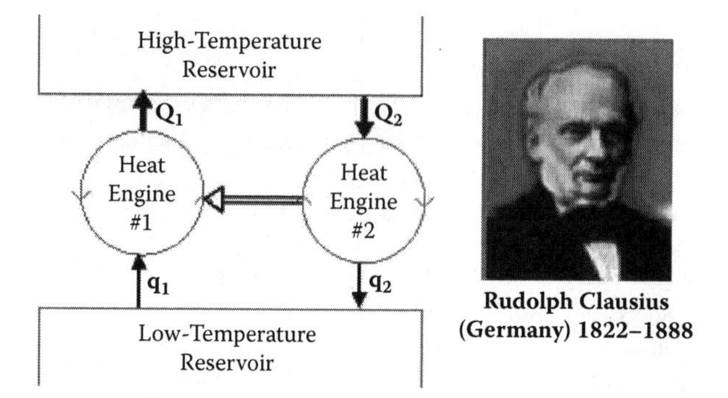

FIGURE 1.5 Clausius' demonstration of the maximum efficiency of Carnot's engine.

engine and working fluid passed had to be *reversible*. Reversibility is a concept that is difficult to define precisely, but is readily recognized when observed or described. Some of the more obvious requirements are the absence of dissipative processes such as friction between moving parts or turbulence in the working fluid. Not so obvious is the requirement that the temperature differences between the hot reservoir and the fluid in the engine at the point of receipt of the input heat flow and between the engine at its coolest point and the low-temperature reservoir accepting rejected heat had to be infinitesimally small. Seen another way, only infinitesimally small changes in the temperatures of the two thermal reservoirs are required to reverse the entire operation (that is, for the engine to consume work and act as a refrigerator).

A far-reaching consequence of the seemingly innocuous claim that mechanical work can be extracted from a reversible engine connecting two bodies at different temperatures is a qualitative statement of the second law of thermodynamics. This connection was enunciated not by Carnot but by Rudolph Clausius some two decades after Carnot's epic work. The argument, based on Figure 1.5, is as follows. A corollary of Carnot's assertion that a temperature difference must be capable of producing work is the impossibility of the reverse process (transferring heat up a temperature difference) without supplying external work. Engine no. 1 is perfectly reversible but operated as a refrigerator. The objective is to prove that no other engine operating between the same two temperatures can have a higher efficiency of converting heat to work than that of engine no. 1.

Suppose that engine no. 2 is more efficient than engine no. 1. Then less heat is required for no. 2 to produce the work required to operate no. 1, or $Q_1 > Q_2$. Taken together, the two engines are transferring $Q_1 - Q_2$ of heat from the low-temperature reservoir to the high-temperature reservoir, *without assistance from external work.* This violates the corollary to Carnot's assertion, so the supposition that engine no. 2 can have a higher efficiency than engine no. 1 is incorrect. Two conclusions are drawn from this exercise: (1) no engine can have a higher efficiency than a reversible engine when operating between the same two temperature reservoirs; (2) it is impossible to transfer heat from low to high temperature without doing work.

Conclusion (2) is a qualitative statement of the second law of thermodynamics. Conclusion (1) is the logical basis for a quantitative statement of the second law. In his time, Carnot could not have converted his analysis into quantitative form, for several reasons.

First, heat (as caloric) was thought to be conserved. Just as the amount of water passing through a waterwheel is not diminished, so the amount of caloric (heat) rejected by an engine was believed to be the same as the amount entering the engine. The work performed by the engine was at the expense of the quality of the caloric, not its quantity.

Second, the first law of thermodynamics had not been formalized. Had it been, Carnot would have recognized that the equality of the work done by the cyclical engine and the net heat it received was a consequence of this law. In fact, the notion of energy was fuzzy (as it is today).

Third, the concept of a thermodynamic temperature (with an absolute zero value) was yet to be formulated, so Carnot was constrained to asserting that the efficiency of the ideal engine increased as the hot reservoir became hotter and as the cold reservoir became colder.

Lastly, the concept of entropy and its quantitative connection to heat was decades away. Carnot would have needed this to formulate the efficiency of the ideal engine even if the thermodynamic temperature scale had been established.

1.1.4 The First Law and Energy

About the middle of the nineteenth century, it all came together. Rumford's cogitations on the boring of cannons suggested that heat and work were somehow interconvertible and that the work of the horses corresponded to the heat generated by the cannon-boring operation. The concept of temperature was well established and recognized as a different entity from heat. An essential step was the quantitative understanding of the relation of heat and work, as furthered by the work of Joule.

Assembling all of these hints into a formula for the first law is attributed to the German physicist Rudolph Clausius, who, in 1850, more or less said that in a particular process, the difference between heat and work is a property of the substance involved called energy. More specifically, the type of energy was termed "internal energy," in order to differentiate it from potential and kinetic forms of energy as treated in mechanics. A body contains a certain amount of internal energy, and when some of it leaves the body, it does so as either heat or work, or both.

There is no clear definition of internal energy. It is somehow contained in the motion of molecules of a gas, in the vibrations of atoms in a solid, the attraction between atoms in molecules and in the forces that bind the nucleus. All we really know is that it is "conserved," in the sense that it can never be created or destroyed. In a particular process, its change is the difference between heat input to and work done by the body. Richard Feynman said, "We have no knowledge of what energy is. It is an abstract thing ...." Fortunately, none of these uncertainties prevent attaching firm numerical values to the internal energy of all sorts of substances, at least to within an additive constant. In transferring from one place to another, energy takes on a variety of forms, all of them collectible under the rubric of either heat or work.

Although internal energy appears to be intimately connected to heat and work via the first law of thermodynamics, its essential quality is closer to those of temperature and pressure; all three are thermodynamic properties, or attributes of a body held in well-defined conditions.

1.1.5 Entropy and the Second Law of Thermodynamics

The idea that attributes of a body constitute its thermodynamic properties has been alluded to in the preceding section. Some properties, such as volume, pressure and even temperature, are clearly understandable and easily measurable. There is no meter for internal energy yet it is a property as surely as the preceding three. What then is a metric for a property? The most reliable seems to be the following: *in moving from one condition to a new one, a quantity is a property if its change does not depend on the path taken.* On this basis, neither heat nor work is a property. However, their difference is path-independent, and the property represented by this difference is the *internal energy*—this is just the first law.

Unlike internal energy, entropy has no connection to ordinary human experience. It is a near-totally abstract quantity. Yet it is a property on the same level as temperature, pressure, volume and internal energy. The path-independence criterion was utilized by Clausius in discovering (for that is the only word for it) *entropy.* His method was the same as that described above for internal energy, namely, to search for a combination of heat and work in processes between the same two states that is independent of the path between the two states. Clausius found that the heat divided by the absolute temperature (or more generally, the integral of the heat divided by the temperature) is the same for all process routes between fixed beginning and end states. He called this odd property entropy (symbols), and with it, quantitative forms of the first and second laws were complete.

Two decades following its discovery by Clausius, Ludwig Boltzmann derived an equation relating entropy to the degree of order of a system. Order and disorder are comprehensible to the mind, so entropy finally had an anchor in human experience.

Implicit in Boltzmann's equation is the third law of thermodynamics, which states that the entropy of crystalline solids is zero at 0 Kelvin. This is due to the

FIGURE 1.6 Statue of Boltzmann on his tomb, on which is carved his famous equation.

perfectly ordered arrangement of the atoms and the cessation of their vibration. The $\mathcal{W}$ in Boltzmann's equation (Figure 1.6) is the number of ways that a large number of indistinguishable particles can be arranged. If all atoms in the solid are in their equilibrium positions and their lowest vibrational state, only one arrangement is possible, or $\mathcal{W} = 1$. The consequence is $S = 0$.

1.1.6 Gibbs and Chemical Thermodynamics

Between 1876 and 1878, J. Willard Gibbs, a professor of mathematical physics at Yale University, published a series of three papers entitled "On the Equilibrium of Heterogeneous Substances," which was destined to stand with Carnot's "Reflections..." as an indispensable foundation of thermodynamics. The first two papers dealt with the graphical representation of thermodynamic properties and processes (movement of a system from one state to another). Particularly significant was the introduction of a graph representing the entropy and temperature of a steam engine at various stages. The conditions at the four key points of the engine's cycle, (see Figure 4.4) enabled the heat terms in Figure 1.4 to be expressed quantitatively. Along with the analogous pressure versus volume graph, which gave the net work of the four processes, Carnot's heat engine was placed on a firm quantitative basis.

Gibbs was the first to represent the thermodynamic properties of matter on three-dimensional plots. As shown in his second paper, these depicted a surface of a thermodynamic property, say pressure, as a function of two other properties, usually temperature and volume (although Gibbs' favorite was entropy as a function of internal energy and volume). Such diagrams represent the *equation of state* of a substance in a manner that remains to this day.

Gibbs Phase Rule
Gibbs Free Energy
Gibbs-Helmholz Equation
Gibbs-Duhem Relation
Chemical Potential

FIGURE 1.7 J. Willard Gibbs and some of his important contributions to the thermodynamics of chemical and nonreacting multicomponent systems.

Far and away the most important of Gibbs' contributions to thermodynamics was his third paper. In it, he developed the concept of equilibrium, not just for multiple phases of a pure substance, but for systems that contained two or more chemical species, or *components*, as Gibbs called them. He showed that a new thermodynamic property originating from a particular combination of internal energy, pressure, temperature, volume and entropy was the determining factor in expressing all types of thermodynamic equilibria. This new property is now called the *Gibbs free energy.* In essence, Gibbs succeeded in transforming the portion of physical chemistry that dealt with thermodynamics from a qualitative collection of observations to the quantitative, theoretical science it is today. Figure 1.7 lists some of the other thermodynamic terms that bear his name.

1.1.7 Historical Summary

The fundamentals of thermodynamics were essentially fully established by the beginning of the twentieth century. More precisely, this discipline is termed *classical thermodynamics* to distinguish it from the later arrival of *statistical thermodynamics* (to which Gibbs contributed). The latter variant incorporates the motion and interactions of the atoms, molecules and electrons in a system in order to calculate its macroscopic thermodynamic properties. Delving into the submicroscopic behavior of the particles of a body is beyond the ken of classical thermodynamics.

Statistical thermodynamics relies on quantum mechanics to permit statistical averaging of the energy states of individual particles, which ultimately leads to prediction of macroscopic properties. The intermediate in this process is a quantity called the *partition function*, which is directly related to a classical property called the Helmholz free energy. For example, the particles of an ideal gas do not (by definition) attract or repel each other and are very small relative to the volume available to them. With only this information, statistical thermodynamics is able to produce the ideal gas law, a relation that classical thermodynamics simply accepts as a fact of nature. Other insights provided by statistical thermodynamics include the behavior of the specific heat of solids as the temperature approaches absolute zero and the motions of atoms responsible for the internal energy of gases (translation, rotation) and of solids (vibration).

1.2 THERMODYNAMIC NOMENCLATURE

Thermodynamics can be fairly regarded as a science of relationships. It provides logical connections between a welter of seemingly unrelated properties of substances and modes of changing states. In common with all other branches of science, thermodynamics has its own terminology. Some terms are familiar to everyone (such as temperature and pressure); others are mysterious to the nonspecialist (such as entropy and reversibility); one is intermediate between these two extremes, which people have a feeling for but no one can define (energy).

- A *phase* is one of the three states of matter, solid, liquid or gas. If two or more phases coexist, they must be uniform on a molecular scale and separated by sharp interfaces.
- A *system* is the material contained inside a well-defined portion of space.
  - If the system contains a fixed quantity of matter whatever change is taking place, it is said to be *closed.* The gas in a sealed container is an example of a closed system.
  - If matter flows in and out of the system, it is termed *open.* An orifice in a tube with imaginary permeable boundaries upstream and downstream constitutes an open system.
  - A system that is uniform throughout is a *homogeneous* system; such a system consists of a single phase.
  - If the system consists of more than one phase, it is *heterogeneous.* Ice and water in a glass is an example.
- A *boundary* forms the periphery of the system.
  - It can be moveable or not; the latter is termed a *rigid* boundary.
  - It can allow heat to pass through it or be insulated; the latter characterizes an *adiabatic* boundary.
- The *properties* of a system are its attributes, or quantities, that characterize its condition. All thermodynamic properties are fixed if as few as two properties are specified. For example, specification of the temperature and pressure of a gas fixes its internal energy as well as its volume. Thermodynamics cannot predict the law that relates pressure, temperature, and energy any more than it can predict the p-V-T relation of the gas. It only requires that there be such a law.
- The *state* of a system is fixed by its properties.
- A *process* is a change in the state of a system. The route taken is called a *path.* There are many possible paths that a process may take.
- The *surroundings* are all matter outside the boundary of a system. Thermodynamically, the universe consists of a system, its boundary and the surroundings. As a practical matter, the surroundings are the immediate environs of the system.
- A system is at *internal equilibrium* if nothing about it changes with time and it does not support gradients of any property.
- *External equilibrium* prohibits differences in any thermodynamic property that the system can sense in the surroundings. For a nonadiabatic boundary,

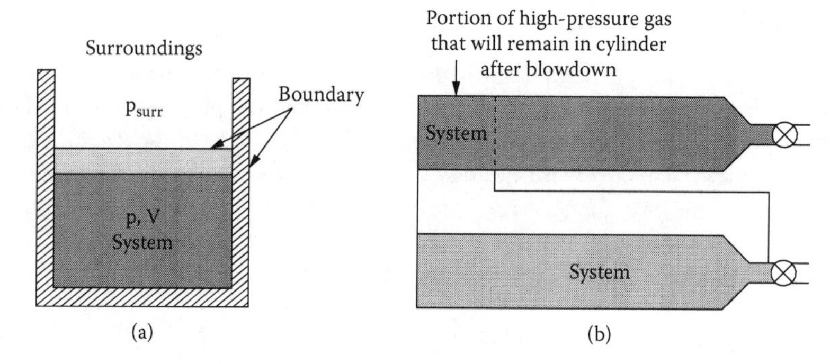

FIGURE 1.8 Two closed systems: (a) a cylinder/piston; (b) blowdown of gas in a cylinder.

temperatures must be equal; for a nonrigid boundary, pressures must be equal.

- A process is *reversible* if it proceeds with the system in internal equilibrium as well as in external equilibrium with its surroundings. In addition, there can be no dissipative events as the system changes state (e.g., friction between moving parts).
- *Component*: chemically identifiable species in a system whose composition changes during a process or which can affect a thermodynamic property (e.g., an inert gas pressurizing a liquid changes its vapor pressure). A system of one component is called a *pure substance*.

Figure 1.8(a) shows a system (gas or liquid) contained inside a cylinder with a piston. This is a very common combination used to assist thermodynamic analyses. Not all systems are as easily identifiable as the one contained in the cylinder/piston boundary.

Figure 1.8(b) shows a common high-pressure gas cylinder with a valve at one end. If the valve is opened, gas flows out. When the pressure in the cylinder equals the pressure of the surroundings, the flow ceases. In order to know the state of the gas remaining in the cylinder following this blowdown process, a system needs to be identified. The appropriate system is the gas contained at the end of the cylinder to the left of the dashed vertical line. This line, together with the cylinder wall to its left, constitutes an imaginary boundary for the system at the initial high pressure. The location of the line is chosen so that all gas within the system boundary fills the cylinder at the end of the blowdown. The process is imagined to be movement of the dashed vertical line towards the right as gas leaves through the valve at the end of the cylinder. When viewed in this manner, the system is closed because it contains the same quantity of gas throughout the blowdown process.

In addition to valves and their kissing cousins, nozzles, typical open systems are turbines, which generate electricity from a flowing gas, and pumps, which are the reverse of turbines. Pumps consume electric power to increase the pressure of the fluid passing through. Figure 1.9 shows a schematic of an open system with a rigid boundary enclosing most of the system. The imaginary meshes at the inlet and exit flow areas complete the system boundary. The system, which is whatever is contained in this composite boundary, may or may not vary (slowly) with time.

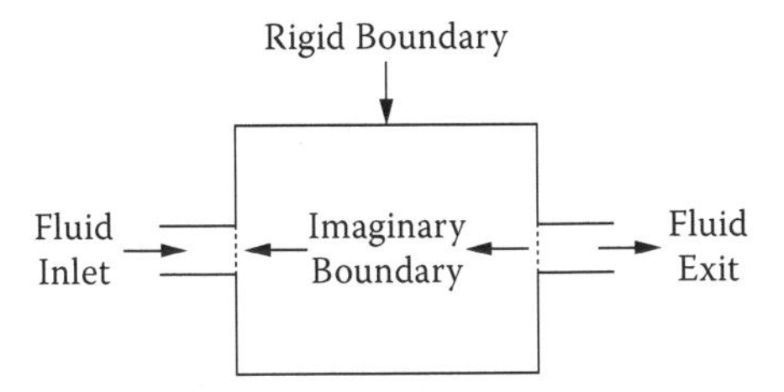

FIGURE 1.9 A generic open system.

1.3 HEAT AND WORK

Thermodynamics is also concerned with what drives a system from one state to another. For such a change to occur, the system must exchange mass and/or energy with its surroundings, the latter in the form of *heat* and *work*. These forms of energy-in-motion are manifest when they cross the boundaries, real or imaginary, that separate a system from its surroundings.

The concepts of heat and work are fundamentally different from the properties of a material. Heat and temperature are particularly prone to confusion, possibly because "hot" sounds more like the former than the latter. Something that is hot possesses an elevated temperature, but not necessarily a large quantity of heat. A weather report of "sweltering heat" is considered to be synonymous with the term "blazing hot," but thermodynamically, "heat" and "hot" are distinct concepts.

To say that a body (or system) contains a certain quantity of heat is incorrect; the body or system possesses internal energy. Heat appears as this energy crosses the system's boundary in the form of conduction, convection or radiation. We often speak of chemical reactions or nuclear radiation as heating a body. However, from a thermodynamic point of view, these are agents for increasing the internal energy of the body. The term "heat" is restricted to thermal energy crossing a system boundary driven by a temperature difference between system and surroundings.

The other form of system–surroundings interaction is *work*. This is a catch-all term for forms of energy transfer that have in common that they are not heat but are in principle completely interconvertible among themselves. The most common form of work in thermodynamic discussions is that produced by a force F acting over a distance ΔX, which represents displacement of the system boundary. This action involves a quantity of work given by $W = F\Delta X$. If ΔX is multiplied by A, the area over which the force acts, and F is divided by A, the work equation becomes $F = (F/A)(A\Delta X)$. Because F/A defines pressure p, and the product $A\Delta X$ is the volume change ΔV, the work involved can also be written as $W = p\Delta V$. This form of mechanical work done on (or by) the system is called *"pV" work.*

A useful device for expressing pV work is the piston-cylinder combination shown in Figure 1.8(a). Movement of the piston changes the volume V of the gas or liquid system contained in the cylinder. If the piston is weightless, the expansion is slow and there is no friction as the piston moves, the system pressure p is equal to the pressure of the surroundings p_{surr}. In this case there is no ambiguity in the pressure to be used in the pV work formula.

In general, however, care is needed in using the proper pressure in the pV work term. For all real expansion/compression processes, friction causes the system pressure p to differ from the pressure of the surroundings, p_{surr}. Since the surroundings are the provider or recipient of the work in the process, pV work must be calculated using p_{surr}. Section 1.9 focuses on this point in more detail.

Another common form of work is *shaft* work, which is transmitted to the surroundings by means of rotational motion rather than expansion or contraction of the system boundaries, as in the pV form. Shaft work can be performed in an open system such as the one shown in Figure 1.9. An example of shaft work is spinning of a turbine by flowing high-pressure steam in an electric power plant.

A third form of work that falls within the purview of thermodynamics is *electrical* work. This form of work is best exemplified by the ability of a battery to run a motor by means of the electrical current it generates.

1.4 CHARACTERISTICS OF SYSTEM BOUNDARIES

As noted above, boundaries (real or imaginary) separate the system from its surroundings. The characteristics of this boundary govern what forms of mass and energy can pass between system and surroundings. These features of the boundary can be divided into opposing limits of the ability to transmit mass, work, or heat.

- *Heat transmission*: Movement of energy as heat requires a difference in temperature between system and surroundings and a boundary through which heat can pass. A boundary that does not permit heat to cross is said to be *adiabatic*. Boundaries that pass heat are termed *diathermic*, although this term is rarely employed. The quantitative heat-conducting properties of the boundary are of no concern; kinetic (or rate) processes are not within the purview of thermodynamics.
- *Work transmission*: From a thermodynamic perspective, the work-transmitting properties of the boundary are not significantly more complicated than its heat-transfer ability. The single most commonly encountered type of work is done by a moveable boundary, by which the system performs pV work. Other forms of work (e.g., shaft and electrical work) can cross a system boundary whether or not it is rigid (i.e., incapable of movement, hence of performing pV work).
- *The isolated system*: Thermodynamics reserves a special name for a boundary that is both adiabatic and rigid, and is not penetrated by rotating shafts, electrical wires or other devices that could transmit non-pV forms of work. A system protected by such a boundary is called *isolated.* It would appear that a system that cannot be influenced by its surroundings is of little practical interest. This is indeed so. However, the isolated system occupies a hallowed niche in thermodynamic theory because it provides one of the simplest ways of elucidating some of its more esoteric features, such as equilibrium, spontaneity of change and entropy.
- *Mass transmission*: The mass-transmitting capabilities of a system boundary possess limits analogous to those of heat and work transmissibility.

The boundaries of the *closed* system are impervious to all matter; the material inside a closed system retains its elemental identity during passage of heat and/or work across its boundaries. However, the system's molecular composition may change by chemical reaction. In an *open* system, matter flows across inlets and outlets in the boundary (Figure 1.9). At steady state, the quantity of matter in an open system is constant. In contrast to a closed system, gradients of thermodynamic properties are permitted in open systems (e.g., the pressure decrease through a turbine).

1.5 THERMODYNAMIC PROCESSES

A thermodynamic *process* is the act of changing the state of a system. The state of the system is defined by a few properties such as temperature, pressure, etc. The process may occur spontaneously, such as the reaction of H_2 and O_2 to form H_2O, or it may be induced as a result of the interchange of heat and work with the surroundings. We are always interested in the initial and final states of a process, and often in the path followed between these two states. However, thermodynamics is blind to the rate of the process.

Thermodynamic processes are always characterized by restraints, meaning that during the change one or more of the system's properties are either held constant or are related in a known way. When a single property is fixed, the process is labeled with the prefix *iso*. Common iso processes are described below:

- *Isothermal*, in which the process occurs at constant temperature. Common isothermal processes are changes of phase: the melting of a solid to a liquid (or vice versa) or vaporization or condensation.
- *Isobaric* processes occur at constant total pressure. Heating a gas in a cylinder with a piston exposed to a constant external pressure is such a process.
- *Isochoric* processes take place at constant volume. Systems with rigid boundaries by definition can undergo only isochoric changes.
- *Isentropic* processes involve a more subtle restraint than the above three. Isentropic processes occur at constant values of the system's entropy. Such processes must be adiabatic, but the absence of heat exchange with the surroundings is not a sufficient condition. To be isentropic, the process must also occur without dissipative effects, or irreversibilities. The concept of the reversible process is discussed below and is intimately tied to the second law of thermodynamics. Although never completely attainable, nearly isentropic processes are of enormous practical importance. The passage of steam through a well-designed turbine or the flow of a liquid through an efficient pump represent processes that are close to isentropic, or its synonym, adiabatic-reversible.
- *Cyclical* processes are combinations of "one-way" processes such as those described above that return the system to its original state. The path followed by the working fluid in an electric power plant or in a refrigerator or air conditioner are common cyclical processes.

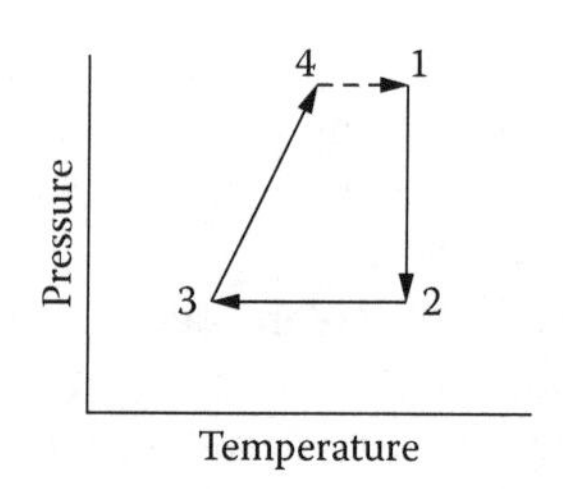

FIGURE 1.10 Four sequential processes diagrammed in pressure-temperature space.

Figure 1.10 illustrates the above processes on a *process diagram.* Its coordinates are pressure and temperature, although any other pair of thermodynamic properties would do equally well. The arrows represent processes and the numbers represent thermodynamic states. The process $1 \to 2$ is isothermal, and is followed by an isobaric process $2 \to 3$. If the substance involved in these processes is an ideal gas, the process $3 \to 4$ is isochoric.* If the figure is closed by the dashed arrow, a cyclic process is produced. It is not necessary that the paths connecting the thermodynamic states (numbered corners) be straight lines. Nor is a sequence of connected processes necessary; the simplest process connects two states (points on a process diagram).

1.6 THERMODYNAMIC PROPERTIES

Even pure substances possess a large number of quantitative properties, including electrical, magnetic, optical, mechanical, transport and thermodynamic. Only the last category is of interest here.

Thermodynamic properties are sometimes called *state functions* because they depend only on the state or condition of the system. They do not depend on the process or the path by which the particular state was achieved. For example, water vapor at a specified pressure and temperature is the same whether created by evaporating liquid water or by reacting H_2 and O_2.

1.6.1 FUNDAMENTAL VERSUS AUXILLIARY PROPERTIES

Pure substances have only five *fundamental* or *primitive* thermodynamic properties, which are those that cannot be derived from other thermodynamic properties. These are, with their common symbols:

T = temperature
p = pressure
V = volume
U = internal energy
S = entropy

In addition, the following three auxilliary thermodynamic properties are combinations of the primitive properties.

* For an ideal gas, $p = RT/v$, so the slope of the arrow $3 \to 4$ is R/v.

$H = U + pV$ = enthalpy
$F = U - TS$ = Helmholz free energy
$G = H - TS$ = Gibbs free energy

The auxilliary properties were conceived because the particular combination of the properties they represent are natural variables describing commonly encountered processes. For example, in isobaric (constant-pressure) processes involving only pV work, the heat exchanged between the system and its surroundings is equal to the change in the system's enthalpy. In another example, when pressure and temperature are fixed, chemical equilibrium is achieved when the Gibbs free energy is a minimum. The Helmholz free energy is rarely used in engineering thermodynamics, but its importance lies in the link that it provides between microscopic and macroscopic thermodynamics.

1.6.2 Intensive versus Extensive Properties

The eight basic thermodynamic properties can be classified as *intensive* or *extensive.* Intensive means the independence of the property on the quantity of substance (or equivalently, on the size of the system). Temperature and pressure are intensive properties. All of the others are extensive: their value is proportional to the quantity of the substance in the system. This feature permits V, S, U, H, F, and G to be made intensive simply by dividing by the quantity of the substance in the system. Quantity can be measured in terms of mass, moles, or number of molecules. Taking the number of moles, n, as the measure of quantity, the intensive counterparts of V, ... G are $v = V/n$, ... $g = G/n$. The lower-case designations are reserved for intensive properties, which are also called *specific* properties. The specific volume v is the reciprocal of the density of the substance. Pressure and temperature cannot be extensive; they are uniquely intensive properties.

1.6.3 Derivative Properties

Other thermodynamic properties are defined as partial derivatives of one of the eight properties listed above. The *heat capacities* (also called *specific heats*) at constant volume and at constant pressure,

$$C_V = \left(\frac{\partial u}{\partial T}\right)_V \qquad C_P = \left(\frac{\partial h}{\partial T}\right)_p \tag{1.1}$$

physically represent the increases in internal energy and enthalpy, respectively, per degree of temperature increase. They are written as partial derivatives because of the restraints indicated by the subscripts on the derivatives. For C_V, the increase in temperature is required to occur at a fixed volume. For C_p, on the other hand, the system's pressure is maintained constant during the increase in temperature.

Two other important derivative thermodynamic properties involve the fractional changes in volume as temperature or pressure is increased. The *coefficient of thermal expansion* α and the *coefficient of compressibility* β are defined by:

$$\alpha = \frac{1}{v}\left(\frac{\partial v}{\partial T}\right)_p \qquad \beta = -\frac{1}{v}\left(\frac{\partial v}{\partial p}\right)_T \tag{1.2}$$

Because the specific volume (or density) of a substance depends on both temperature and pressure, α and β are defined as partial derivatives in order to indicate the property that is to be held constant during the increase of the other property. Both α and β are positive numbers, which accounts for the negative sign in the definition of β.

1.6.4 Absolute versus Relative Properties

Another distinction among the basic thermodynamic properties is whether or not they possess absolute values. That is, is there a state in which a particular property has a value of zero? For volume the existence of an absolute measure is obvious. Pressure and temperature also possess unequivocal states in which these properties vanish. The absolute character of these properties is best understood by considering the atomistic nature of matter, in particular of gases.

1.6.4.1 Pressure

Pressure in a gas is a consequence of the momentum transferred to a wall by rebounding of impinging molecules. Such momentum transfers appear as a force on the surface, which, when scaled to a unit surface area, is the gas pressure. A common means of reducing pressure is by changing the density of the gas using a vacuum pump. This reduces the frequency of molecular impacts on the walls and as a consequence diminishes the rate of momentum transfer to the surface. Being a force per unit area, pressure in the SI (i.e., metric) system has units of Newtons (N) per square meter, or Pascals (Pa). Normal atmospheric pressure is 10^5 Pa, or 0.1 MPa.

The origin of pressure in a liquid is quite different from that in a gas. Molecular rebounding from the system walls is not a major contributor. Imagine pressing on the top of a box filled with a liquid, as in Figure 1.11a. The force on top of the piston appears as a pressure in the liquid inside the box, evenly and isotropically distributed thoughout (hydrostatic pressure). The temperature and density of the liquid barely changed in this process. Pressure in a liquid is approximately analogous to the force on opposing walls of a box generated by connecting them with a turnbuckle* (Figure 1.11b).

1.6.4.2 Temperature

That temperature can be measured in absolute terms has been discussed in Section 1.1.1.

1.6.4.3 Volume

The absolute nature of volume is a consequence of our notions of three-dimensional space. The volume changes with pressure and temperature given by Equation (1.2) are likewise properties with absolute values.

---

* A turnbuckle is a cast-metal sleeve with a left-handed screw thread at one end and right-handed one at the other end. Twisting the sleeve causes the threaded rods to move outward.

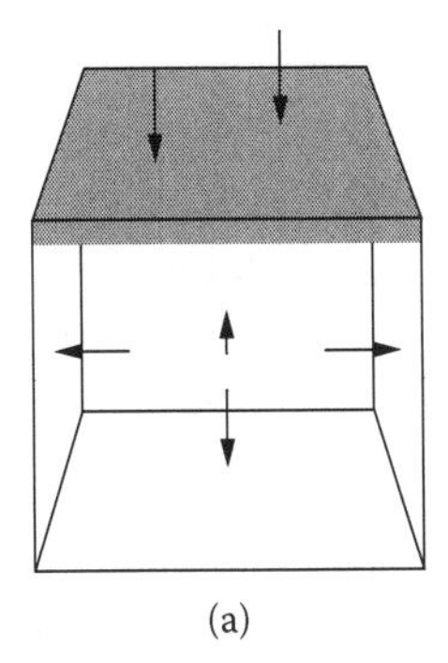

(a)

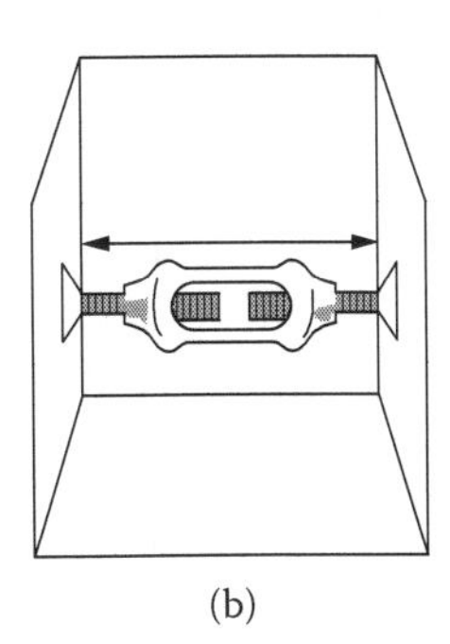

(b)

FIGURE 1.11 Turnbuckle analogy of pressure in a liquid.

1.6.4.4 Entropy

Entropy is at once the most fundamental of the thermodynamic concepts and the least connected to common experience. As reviewed in Section 1.1.5, entropy acquired meaning as a measure of the degree of order of a system. Highly ordered systems such as solids have low entropies, and conversely, systems possessing a high degree of disorder, such as gases, exhibit high entropies. The regularity of a liquid is intermediate between that of solids and gases, and as a result, the entropy of a liquid falls between that of its corresponding solid and gaseous states. By virtue of the third law of thermodynamics ($S = 0$ at 0 K), entropy has an absolute value.

1.6.4.5 Internal Energy, Enthalpy, and Free Energy

Internal energy in various forms is stored by the molecules or atoms of a substance. The energy of crystalline solids is contained partly as the interparticle potential energy that is responsible for the stability of the solid phase. Additional energy is stored as kinetic energy of the atoms or molecules that vibrate about their equilibrium positions in the crystal. The increase in vibrational energy of a solid with increasing temperature is responsible for the specific heat of this state of matter. The intermolecular (or interatomic) potential energy is independent of temperature.

The chief mode of energy storage in gases is the translational kinetic energy of the moving particles. Interparticle potential energy in gases is small; in an ideal gas, this component is by definition zero. When the interparticle interactions are not negligible, the gas behaves nonideally.

Molecular gases also hold energy in the form of vibrations and rotations of the individual molecules (atomic gases such as helium do not have contributions of this type). These motions become significant contributors to the internal energy, and hence to the heat capacity, only at high temperatures. For the most part, the specific heats of gases are due to the increase in molecular speeds with increasing temperature.

The internal energy of the liquid state is due mainly to particle vibrations, as in the solid state. Even though the structure is not as regular as that of crystalline solids, liquids possess none of the translational motion of gases. The proximity of the properties of many liquids and solids (e.g., density, internal energy) makes the term *condensed phase* a useful description when the distinction between liquid and solid is not important.

The internal energy u and its energy-like cousins h, f, and g do not have absolute values. This lack is not an impediment to thermodynamic calculations, however, which involve only changes in these properties. To facilitate computation, the internal energy (or enthalpy or free energy) is set equal to zero at an arbitrarily chosen temperature and pressure called the *reference state*. The most common reference state is room temperature (298 K) and 1 atm pressure. Other choices are possible. For example, the enthalpy rather than the internal energy of a substance may be assigned a value of zero at 298 K at 1 atm pressure. However, the reference state must be unique: u can be set equal to zero at some reference condition denoted by T_o and p_o, but h cannot also be zero in the same state. This is because $h = u + pv$, so in the reference state, the specific volume v_o is implicitly fixed by the values of T_o and p_o. Thus, if $u = 0$ at p_o and T_o, h must be $p_o v_o$ in this state.

By virtue of their definitions as derivatives, the specific heats of Equation (1.1) are independent of the reference state chosen for u or h. C_P and C_V are absolute properties.

Similarly, the volume dependencies on temperature and pressure, α and β, possess absolute values.

1.7 REVERSIBLE AND IRREVERSIBLE PROCESSES

The old English nursery rhyme provides a good description of an irreversible process:

Humpty Dumpty sat on a wall
Humpty Dumpty had a great fall
All the King's horses and all the King's men
Couldn't put Humpty Dumpty together again

Although not all processes are as dramatic as Humpty Dumpty's fall, the implication of permanent changes in the system and/or surroundings captures the essence of mechanical irreversibility. In thermodynamic processes, irreversibilities occur in a number of ways:

1. Friction between moving parts, which degrades potentially useful work into heat.
2. Rapid (pressure-unbalanced) expansion or compression of a gas. If contained in a cylinder with a piston, rapid change in gas volume is accompanied by diminishing oscillations of the piston about its final equilibrium position.
3. Heat transfer between system and surroundings through a finite temperature difference.

These three modes of irreversibility are not independent of each other. Damping of piston oscillation in No. 2 is a consequence of solid–solid friction between piston and cylinder wall (No. 1). Friction of any sort ultimately degrades kinetic energy or work to heat, a process that cannot be reversed.

Irreversibility can occur within the system, in the surroundings or in both. Within the system, *internal irreversibility* is associated with friction within a moving fluid caused by turbulence, viscous flow, or rapid changes that imbalance the uniform conditions in a gas or liquid. Expansion or compression work done by a system can be calculated by Equation (1.3) only if internal reversibility prevails. *External irreversibilities* occur in the surroundings, in the boundary between system and surroundings (e.g., friction between the moving piston and the cylinder in Figure 1.8a), or jointly between system and surroundings, as in heat exchange over a nonzero temperature difference. When irreversibilities are absent from both system and surroundings, the process is said to be *totally reversible.*

All real processes contain various degrees of irreversibility, but the concept of the perfectly reversible process is extraordinarily useful in thermodynamic analysis. Irreversible processes can be quantitatively treated by the first law only, whereas both the first and second laws can be applied to reversible processes.

A reversible process is one that can be made to go backward without any change in the system or the surroundings. To possess this feature, the reversible process must proceed through a series of infinitesimal stages in which internal equilibrium and external equilibrium are maintained. Such processes are sometimes called *quasistatic.*

Reversibility can also be identified by comparing the work done *by* (or *on*) the system with the work done *on* (or *by*) the surroundings. If the two works are identical, the process is totally reversible. The following very popular example illustrates this method of detecting irreversibility by analyzing compression of an ideal gas by adding weights to the top of a piston. The book by Van Wylen and Sonntag (Section 6.3) gives a brief qualitative version of this example. Abbott and Van Ness (1989, Section 1.5) present a slightly modified version. Van Ness (1983, 19–22) gives a more colorful qualitative description of this example using grains of sand as the small masses added to the piston.

1.7.1 Example: Compression of an Ideal Gas Using Sliding Weights

The two methods of compressing N moles of an ideal gas by reversible and irreversible processes with common initial and final states are illustrated in Figures 1.12 and 1.13. In both cases, a total mass M is added to the top of a piston in a cylinder containing an ideal gas. The piston is assumed to move freely in the cylinder, so that its motion does not introduce irreversibilities into either process. The only difference between the two methods is in the way that the mass is placed on the piston's pedestal. In Figure 1.12, the mass M is divided into a large number n of very small masses m such that $M = nm$. The small masses are slid onto the piston one at a time from shelves at different elevations. In Figure 1.13, the entire mass M is placed on the piston at once.

In both cases, the cylinders are immersed in a constant-temperature bath, which maintains the gas in Figure 1.12 at temperature T during the entire process. In Figure 1.13, only the initial and final states are sure to be at temperature T because these are equilibrium states whereas the middle one is not. The constant-temperature condition is not critical to the argument. The key issue is the manner in which the masses are added to the piston.

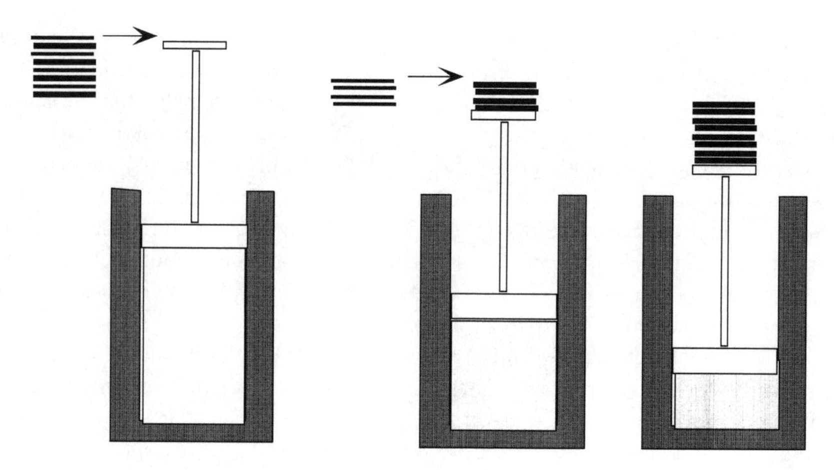

FIGURE 1.12 Reversible, isothermal compression of an ideal gas.

1.7.1.1 Reversible Compression (Figure 1.12)

Sliding a small mass onto the pedestal causes the piston to compress the gas in the cylinder very slightly. Each addition of a small mass approximates an infinitesimal equilibrium stage so that the overall process is reversible. At all points in the compression process, the conditions of external equilibrium, namely:

$$T = T_{surr}$$

and

$$p = p_{surr} + jmg/A$$

are satisfied. T_{surr} and p_{surr} are, respectively, the constant temperature and constant pressure of the surroundings (i.e., the environment in which the cylinder is placed). The number of small masses added to the piston is denoted by j $(0 \leq j \leq n)$, g is the acceleration of gravity, and A is the cross-sectional area of the piston and cylinder. The shelves holding the small masses are not equally spaced.

Because the process is reversible, the work involved in compressing the gas can be calculated in two ways. The first is the (negative) pV work done by the gas as its volume is reduced from V_o to V_f by the weight of the small masses slid onto the pedestal of the piston (work is considered to be positive if done *by* the system):

$$W_{rev} = \int_{V_o}^{V_f} p dV \tag{1.3}$$

The integral can be performed by expressing p in terms of V using the ideal gas law, $pV = nRT$. Since the process is isothermal, T is constant and:

$$W_{rev} = -NRT\int_{V_f}^{V_o}\frac{dV}{V} = -NRT\ln\left(\frac{V_o}{V_f}\right) \quad (1.3a)$$

The second method of calculating the work involved in the process depicted in Figure 1.12 is from the point of view of the surroundings. pV work is performed by the external pressure as the piston descends; in addition, there is a loss of gravitational potential ΔE_p as the small masses are slid onto the pedestal and sink with the piston. These two contributions yield:

$$W_{surr} = p_{surr}(V_o - V_f) + \Delta E_p$$

Detailed calculation* shows that $W_{rev} = W_{surr}$. This equality is a hallmark of a reversible process. Problem 1.2 analyzes the analogous problem of loading weights onto a spring rather than a piston.

The process in Figure 1.12 can be reversed by sequentially sliding small masses off the pedestal at their original elevations. When the initial state is recovered, both system and surroundings will have been restored to their original states.

1.7.1.2 Irreversible Compression (Figure 1.13)

When the entire mass M is placed on the piston, as in Figure 1.13, the system is not in equilibrium with the surroundings. The piston descends rapidly, oscillates as the gas acts as a spring and eventually settles down to the final elevation. Although W_{rev} does not apply to this situation (because the process is not internally reversible), the equation for W_{surr} does. In this case, the potential energy loss ΔE_p is easily determined from the elevation change of the large mass, which is directly related to the volume change of the gas by Δ(elevation) = $(V_o - V_f)/A$. The work done by the surroundings in this irreversible process is:

$$W_{irr} = p_{surr}(V_o - V_f) + Mg(V_o - V_f)/A$$

At the final equilibrium state, the force balance on the piston gives:

$$p_f - p_{surr} = Mg/A$$

Combining these two equations and using the ideal gas law yields:

$$W_{irr} = p_f(V_o - V_f) = p_f V_f\left(\frac{V_o}{V_f} - 1\right) = NRT\left(\frac{V_o}{V_f} - 1\right)$$

* D. Olander, "Compression of an Ideal Gas as a Classroom Example of Reversible and Irreversible Processes," International Journal of Engineering Education **16** (2000) 524.

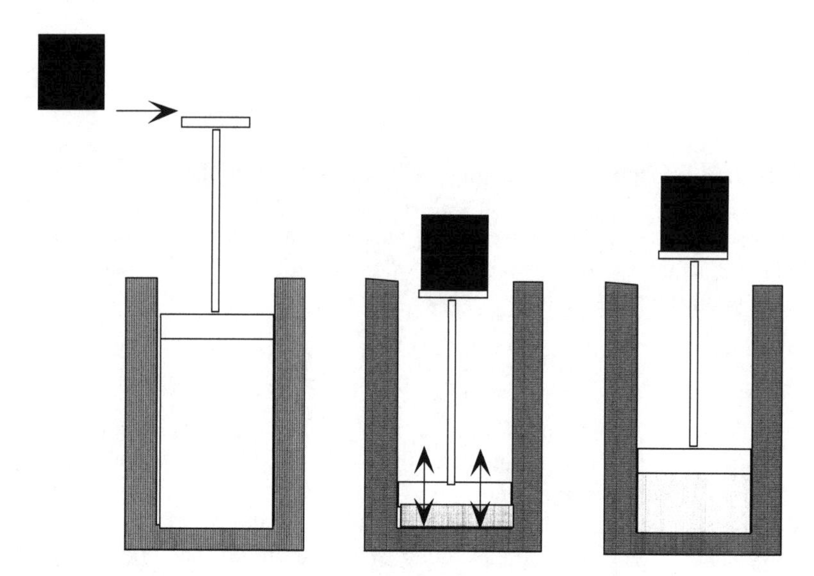

FIGURE 1.13 Irreversible, isothermal compression of an ideal gas.

Comparing this equation with W_{rev} for $V_o/V_f = 3$ (as an example) shows that the ratio of the reversible work of compression to the irreversible work is:

$$\frac{W_{rev}}{W_{irr}} = \frac{\ln 3}{3-1} = 0.55$$

That is, the surroundings need to supply only about one half as much work for the reversible process as it does for the irreversible process.

Finally, simply sliding the large mass off the piston in the final state in Figure 1.13 will return the system (gas plus piston) to its initial state but will leave the surroundings with the large mass at a lower elevation than it was initially. This is characteristic of an irreversible process.

In the above example, the work *done by the surroundings* on the system in the irreversible case is always greater than the work required were the process reversible. As a corollary, the work *done by the system* on the surroundings is always greater than can be had from the irreversible process. *The first stricture says that the effort to do a job is always greater than you think it should be and the second warns that you can never do as much as theoretically possible.* This foreshadows the second law of thermodynamics.

1.8 THE FIRST LAW OF THERMODYNAMICS

One of the most amusing and astute explanations of the first law was given by the physicist Richard Feynman in his lectures to physics students at the California Institute of Technology. Feynman's story is too long to be repeated here, but Van Ness (1983, 2–8) recounts it in full, and is well worth reading.

It is an empirical observation, never refuted, that the change in a property labeled the internal energy of a closed system resulting from addition of heat and performance of work is given by:

$$\Delta U = Q - W \tag{1.4}$$

where
$\Delta U = U(\text{final}) - U(\text{initial})$ = change in system internal energy
Q = heat added to the system
W = work done by the system

Equation (1.4) is the most common form of the first law of thermodynamics. It applies to a process that takes a system from an initial to a final state. It neglects changes in the system's potential energy and kinetic energy. The signs of the terms on the right side are important.

Q represents heat *added to* the system, so that if the process results in removal of heat, Q is negative. Similarly, work *done by* the system is considered to be positive, which accounts for the negative sign in Equation (1.4).

Actually, the left side of Equation (1.4) should be the change in *total* energy of the system, which includes kinetic and potential energies as well as internal energy. However, for commonly encountered closed systems, kinetic and potential energy changes are usually absent, so Equation (1.4) suffices. For open systems, where fluid flows across system boundaries, however, inclusion of these two terms is often necessary (see Chapter 4).

The first law can also be written for a differential slice of the process:

$$dU = \delta Q - \delta W \tag{1.5}$$

The manner of expressing the differentials in Equation (1.5) arises from the fundamentally different natures of internal energy and heat or work. The differential dU represents an infinitesimal change in the property U. The differentials δQ and δW, on the other hand, denote infinitesimal quantities of heat and work exchanged between system and surroundings. Q and W are not thermodynamic properties, and so cannot "change" the way that U can.

The differential and integral forms of the first law are directly related because:

$$\Delta U = \int dU = U(\mathit{final}) - U(\mathit{initial})$$

and

$$Q = \int \delta Q \qquad W = \int \delta W \tag{1.6}$$

Integration proceeds from the initial state to the final state. The integral of dU is the difference between the values of U in these two states, but the integrals of Q

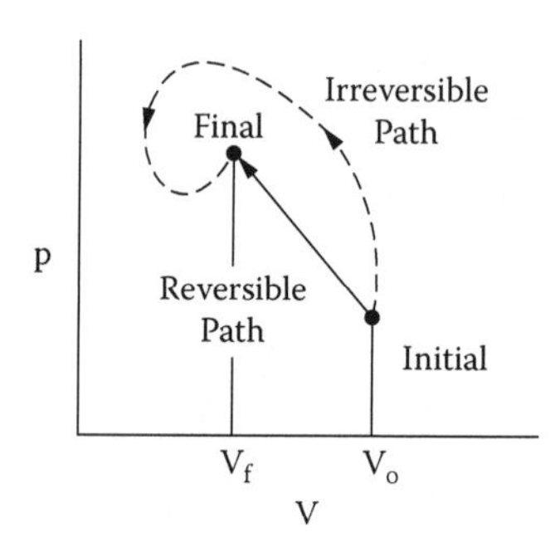

FIGURE 1.14 Reversible and irreversible compression paths.

and W cannot be similarly written because these quantities do not have "values" associated with the state of the system; they depend on the path taken in the process. "Path" means the variation of the system's state (e.g., p and V) as it moves from the initial to the final state.

Although Q and W individually depend on the path, the difference $Q - W$ is independent of the path because it represents ΔU, a property change. For instance, in the examples in the preceding sections of compressing an ideal gas reversibly (Figure 1.12) and irreversibly (Figure 1.13), $\Delta U = 0$ because the initial and final temperatures are the same and the gas is ideal. However, the example at the end of the preceding section shows that the work done by the surroundings on the system in the irreversible process is approximately twice that of the reversible process. Consequently, Equation (1.4) requires that the heat released to the environment in the irreversible process must also be about twice that released in the reversible process.

Figure 1.14 shows the paths followed by the gas inside the cylinder in Figures 1.12 and 1.13 in the two compression processes on a $p - V$ coordinate diagram. According to Equation (1.3), the area under the reversible curve is W_{rev}. The path of the irreversible compression process is shown schematically as a dashed curve; the $p - V$ trajectory cannot be deduced from thermodynamics because the system is not in equilibrium except in the initial and final states.

An important consequence of the first law expressed by Equation (1.4) is that increasing ΔU can be achieved either by adding heat to the system ($Q > 0$) with $W = 0$ or by performing work on the system (i.e., $W < 0$) in an adiabatic process ($Q = 0$). Work addition can be accomplished in a number of ways, including compression (i.e., pV work), or in the form of shaft work, typically represented by a paddle rotating in a fluid contained in a sealed insulated vessel. Friction of the moving paddle in the fluid degrades the work to heat, which causes the increase in internal energy. Irrespective of the form of work done on the system, as far as the first law is concerned, heat and work produce identical effects, and are said to be equivalent. One of the early triumphs of thermodynamics was the quantitative measurement of the "mechanical equivalent of heat" as 4.184 J/calorie by the Scottish scientist Joule in the nineteenth century.

The above discussions of the first law dealt with one-way processes, which began with an initial state and ended in a final state. A second representation of the first

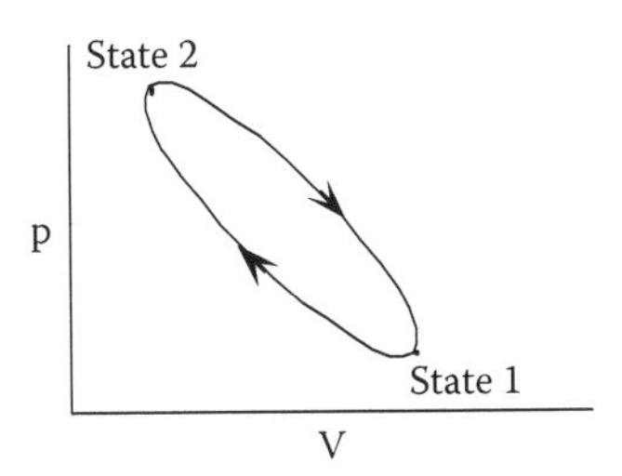

FIGURE 1.15 A two-step cyclical process between states 1 and 2.

law (and historically the first) applies to a cyclical process, such as the one shown in Figure 1.15.

Here, the system moves from state 1 to state 2 by the lower path and returns by the upper path. No matter what the shapes of these two trajectories, ΔU for the cycle $1 \rightarrow 2 \rightarrow 1$ must be zero because U is a path-independent thermodynamic property. The form of the first law for the cycle, which follows from integrating Equation (1.6) with

$$\oint dU = 0$$

is:

$$\oint \delta W = \oint \delta Q \tag{1.7}$$

where the integrals traverse the complete cycle. The cyclical work and heat depend on the path chosen but their path-integrals are always equal.

The forms of the first law given by Equations (1.4) and (1.7) apply to a system and do not involve the surroundings. However, the surroundings, which supply the Q to the system and receive W from it, must also obey the first law. The equation:

$$\Delta U + \Delta U_{surr} = 0 \tag{1.8}$$

is at once the law of conservation of energy and a form of the first law of thermodynamics. As in Equation (1.4), kinetic energy and potential energy changes have been neglected in Equation (1.8).

Because $\Delta U_{surr} = Q_{surr} - W_{surr}$, Equations (1.4) and (1.8) imply that $Q + Q_{surr} = W + W_{surr}$. This equation does not mean that $Q_{surr} = -Q$ and $W_{surr} = -W$; these individual equalities hold only for reversible processes such as the compression of the ideal gas in Figure 1.12.

Figure 1.16 shows two systems, called heat reservoirs, labeled 1 and 2, in good thermal contact through a heat-transmitting interface. Taken together, the pair constitutes an isolated system because the encasing boundary is both rigid and adiabatic.

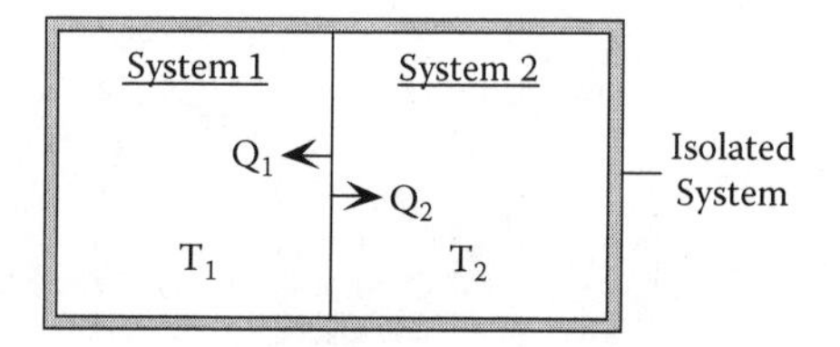

FIGURE 1.16 Heat flows between two systems separated by a diathermal interface and encased in a rigid adiabatic boundary.

No work is done by system 1 on system 2 (or vice versa) but because $T_1 \neq T_2$, heat flows from one system to the other. The direction of the arrows in the diagram follows the convention of the heat in the first law, but one of the Qs must be negative. The first law for the combined system is:

$$\Delta U_1 + \Delta U_2 = Q_1 + Q_2 = 0$$

so that $Q_1 = -Q_2$.The first law cannot predict which heat is negative; common experience and the second law require that heat flows from the hotter to the colder system, but this remains to be proven.

1.9 THE SECOND LAW OF THERMODYNAMICS

Textbooks on thermodynamics with a mechanical engineering flavor introduce the second law by considering the nature of heat and work in cyclical processes known as heat engines. Historically, this is indeed how the second law was first made quantitative and, as a byproduct, the thermodynamic property called entropy emerged. In these notes, the process is reversed. The second law in its several guises is first presented in the final forms that are close analogs of the forms of the first law in Section 1.8. Discussion of the relationship of heat engines and cyclical processes to the second law and entropy is deferred until Chapter 4.

We all have an intuitive feeling for the second law, but rarely make the explicit connection. We know, for example, that without external intervention, heat will never flow from a cold body to a hot one, steam will not spontaneously decompose into H_2 and O_2, and Humpty Dumpty cannot be put together again. The opposite of these processes cause the entropy of the universe to increase, which is precisely what the second law requires of spontaneous changes.

In the microscopic view of thermodynamics, entropy characterizes the state of disorder of a system. Because heating a system usually creates a less highly ordered state, entropy changes are closely related to heat, but are not at all associated with work. The work that raises a weight by a frictionless pulley does not affect the state of order or organization of the system or the surroundings, and so work is entropy-neutral. This qualitative notion of the connection of heat and entropy is embodied in the following quantitative statement of the second law:

$$\Delta S = \int \frac{\delta Q_{rev}}{T} = \frac{Q_{rev}}{T} \quad (\text{if } T = \text{constant}) \tag{1.9}$$

where ΔS and the integral represent changes from an initial state to a final state. The subscript "rev" means that the process must be reversible, not just in heat transfer but in the mechanical aspects as well (see Section 1.8). The last equality in Equation (1.9) is the special case of heat transferred at a constant temperature.

Entropy as a thermodynamic property was discovered well before the advent of quantum mechanics or Boltzmann's famous $S = k\ln\mathcal{W}$ where $\mathcal{W}$ is the number of molecular arrangements a substance can form subject to macroscopic restraints. Following many years of investigation in the mid-nineteenth century, the German physicist Rudolph Clausius "discovered" Equation (1.9) in much the same way that the first law was revealed; namely, by examining the heat and work exchanged between a system and its surroundings as the system changed from one state to another. There are innumerable paths that a system can take between two end states, each path involving different values of Q and W. The internal energy difference ΔU is inferred from the invariant values of the $Q - W$ for all paths. Being a thermodynamic property change, ΔU does not depend on the path taken.

In analogous fashion, Clausius showed that the integral

$$\int \delta Q / T$$

is also the same for all processes between fixed end states, with the additional proviso that the processes are reversible. The inescapable conclusion of this observation is that $\int \delta Q / T$ is the change in a thermodynamic property, which Clausius named *entropy*. A more detailed yet quite readable exposition of the method described above for uncovering the meaning of entropy is given in Van Ness (1983, 55–61). Von Baeyer (1998, Chapter 7) gives a less technical recounting of Clausius' role in this search.

Equation (1.9) is the second law analog of Equation (1.4) for the first law; both relate changes in a thermodynamic property to heat and work. In Equation (1.9), the convention for the sign of the heat is the same as that adopted for the first law: Q is positive if heat is added to the system.

Although Equation (1.4) applies to any process, Equation (1.9) is valid only for reversible changes. The complete statement of the second law that complements Equation (1.9) accounts for irreversible processes by the inequality:

$$\Delta S > \int \left(\frac{\delta Q_{irr}}{T} \right) \quad \text{or} \quad \Delta S > \left(\frac{Q_{irr}}{T} \right) \quad (\text{if } T = \text{constant}) \tag{1.10}$$

This equation does not require that ΔS be positive for any irreversible process. Q can be sufficiently negative (i.e., heat is removed from the system) to more than compensate for the entropy increase due to irreversibilities. All that Equation (1.10) requires is that in such processes, the entropy change is more positive than the heat addition divided by the temperature. That is, both sides of the inequality in Equation (1.10) may be negative, but the right side is more negative than the left side.

The second law can also be written for a differential portion of a process:

$$dS = \left(\frac{\delta Q}{T}\right)_{rev} \qquad dS > \left(\frac{\delta Q}{T}\right)_{irr} \tag{1.11}$$

These equations are the second law analogs of Equation (1.5) for the first law.

When the process involves a complete cycle, as in Figure 1.15, the system's entropy returns to its initial value so that

$$\Delta S(cycle) = \oint \delta S = 0$$

For the cycle, Equations (1.9) and (1.10) become:

$$0 \geq \oint \frac{\delta Q}{T} \tag{1.12}$$

where the equality applies if all steps in the cycle are reversible and the inequality describes a cycle with irreversibilities. The latter is called the *inequality of Clausius,* and can be regarded as another form of the second law. Equation (1.12) is the second law analog of Equation (1.7) for the first law.

The preceding discussion of the second law has dealt exclusively with the system, without regard to the entropy changes in the surroundings. Considering the system and surroundings (the "universe"), the total entropy change is:

$$\Delta S + \Delta S_{surr} \geq 0 \tag{1.13}$$

where the equality applies to reversible processes. Contrary to the energy analog given by Equation (1.8), entropy is *not* conserved in processes with irreversibilities. The logical consequence of this fact, that the universe is destined to degrade to a uniform mass of indistinguishable dust, has troubled philosophers and cosmologists for a century. However, experience, both scientific and practical, has empirically demonstrated, without exception, the correctness of Equation (1.13).

A practically important special case of Equation (1.13) is provided by changes in an isolated system. Such a system is completely cut off from its surroundings, so that no matter what transpires in the system, $\Delta S_{surr} = 0$. Equation (1.13) still applies, and takes the form:

$$\Delta S \geq 0 \quad \text{(for an isolated system)} \tag{1.14}$$

The appropriate interpretation of Equation (1.14) is that the entropy of an isolated system seeks its maximum value, which occurs at the state of equilibrium. This equation is often referred to as the *maximum entropy principle*. Two examples of its application follow. Problems 3.19 and 3.20 utilize this principle for similar situations.

1.9.1 Example: The Direction of Heat Flow

The first law cannot determine the direction of the heat flows between systems 1 and 2 in Figure 1.16. The reservoirs are infinite in size, so that exchange of a finite quantity of heat does not affect their temperatures, which are taken to be $T_1 > T_2$. We know from experience that heat will flow from system 1 to system 2, or; in terms of the directions indicated by the arrows in Figure 1.16, that Q_1 must be negative and Q_2 must be positive. The second law proof of this intuitive conclusion starts by noting that because T_1 and T_2 are not infinitesimally close, the process must be irreversible, so $\Delta S = \Delta S_1 + \Delta S_2 > 0$.

The entropy changes of the individual reservoirs, on the other hand, are given by the reversible formula Equation (1.9). The reason is that no irreversibilities occur *within* the reservoirs; changes in S of these bodies depend only on the quantity of heat transferred, and not on its origin or destination. Therefore, the above inequality becomes:

$$\frac{Q_1}{T_1} + \frac{Q_2}{T_2} > 0$$

But the first law applied to this situation requires that $Q_1 = -Q_2$, which converts the above equation to:

$$Q_2\left(-\frac{1}{T_1} + \frac{1}{T_2}\right) = Q_2\frac{T_1 - T_2}{T_1T_2} > 0$$

Since temperatures are always positive, and since the initial restriction was $T_1 > T_2$, the above equation shows that Q_2 must be positive. That is, the direction of heat flow is from the hot body to the cold body.

The above application of the second law may seem needlessly formal, but a more challenging analysis of the process depicted in Figure 1.16 is the following: If the initial state of the isolated system is $T_1 > T_2$,

- What is the final common equilibrium temperature of the two systems?
- What is the entropy increase of the isolated system when the final (equilibrium) state is attained?
- Does the final state represent the maximum possible entropy of the isolated system?

Answering these questions requires knowing how temperature affects internal energy (or enthalpy) and entropy, which are considered in Problem 3.15.

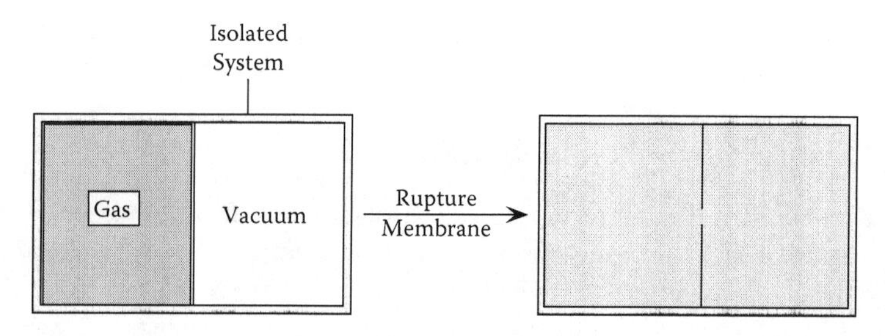

FIGURE 1.17 An ideal gas contained on one side of an isolated system by a membrane that is subsequently ruptured to allow the gas to redistribute.

1.9.2 Example: Entropy Change Resulting from a Rapid (Irreversible) Expansion of an Ideal Gas

In Figure 1.17, N moles of an ideal gas are initially partitioned into the left half of an isolated system by a membrane. The process consists of puncturing the membrane, thereby allowing the gas to occupy both halves equally.

The system is isolated, so $\Delta U = 0$, hence the gas temperature does not change. The gas, however, has doubled its volume. The entropy change occasioned by this expansion cannot be calculated for the irreversible process depicted in Figure 1.17. However, because entropy is a state function, ΔS can be calculated by constructing a reversible process (any convenient one) that brings the system between the same states. Such a reversible process is the reverse of the one shown in Figure 1.12, in which the work done on the system is given by Equation (1.3a). Because $\Delta U = 0$ for isothermal expansion of an ideal gas, the first law requires that the heat absorbed be equal to the work performed. The entropy change for the reversible isothermal expansion of an ideal gas to twice the volume is:

$$\Delta S = \frac{Q_{rev}}{T} = \frac{W_{rev}}{T} = nR\ln\left(\frac{V_f}{V_o}\right) = NR\ln 2 > 0$$

The entropy change in the irreversible process of Figure 1.17 is also $NR\ln 2$, despite the absence of heat exchange with the surroundings. It is positive, in conformance with the principle of maximum entropy for an isolated system. From a microscopic perspective, doubling the volume increases the number of quantum states available to the gas, and so increases its state of disorder. That the uniform state on the right in Figure 1.17 is the equilibrium state is intuitively obvious. It can be shown that any other distribution has a lower entropy than the uniform distribution, which leads to equal gas pressures in the two halves of the container.

Problem 3.17 shows that the entropy increase due to this irreversible expansion is not restricted to ideal gases as in the above example. Several examples of the second law applied to reversible compression of an ideal gas are given below.

Example 1: What is the entropy change of an ideal gas that is compressed reversibly and isothermally from volume V_1 to V_2? (See Figure 1.12) Explain the sign of the entropy change.

The internal energy of an ideal gas depends only on temperature; for an isothermal process, $\Delta U = 0$. The first law (Equation [1.4]) gives:

$$Q_{rev} = W_{rev} = NRT \ln(V_2 / V_1)$$

where the last equality comes from Equation (1.3a). Because the gas is compressed, $V_2/V_1 < 1$, and the heat is negative (heat is released during the process) and the work is negative (work is done on the system by the surroundings). N is the moles of gas in the cylinder.

Because the process is reversible, Equation (1.9) is the appropriate form of the second law, and the entropy change is:

$$S_2 - S_1 = Q_{rev} / T = NR \ln(V_2 / V_1)$$

The entropy change is negative in the process, and can be explained in two ways. From a macroscopic viewpoint, the loss of heat from the system implies a reduction in entropy. From a microscopic viewpoint, the reduced volume of the final state 2 means that the molecules occupy fewer translational quantum states. Hence, order is increased, and entropy is decreased.

Example 2: (from Potter and Somerton) Consider the piston-spring device shown below. Initially, the piston rests on stops and the spring at its equilibrium length (no force on piston).

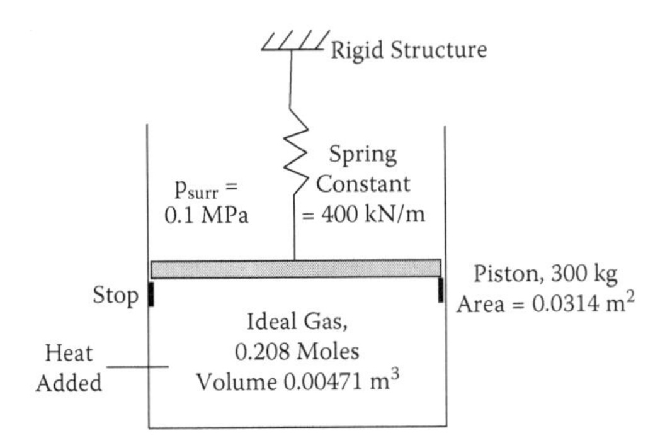

1. *Heat is added to the gas in the cylinder until the piston just leaves the stops. What is the gas temperature at this point?*

   Because the spring is still at its equilibrium length, the pressure at piston liftoff is due only to the gas pressure of the surroundings and the force due to the piston mass:

   $$p_1 = p_{surr} + Mg / A = 1.94 \times 10^5 \text{ N/m}^2$$

   where $g = 9.81$ m/s^2 is the acceleration of gravity. The temperature at piston liftoff is given by the ideal gas law:

$$T_1 = \frac{p_1 V_1}{nR} = 528 \text{ K}$$

2. *Heat is added reversibly to the gas until the final volume is* $V_2 = 0.000628 \, m^3$. *How much* pV *work is done by the system?*

For a volume V greater than V_1, the pressure is greater than p_1 due to compression of the spring. The spring compression distance is $x = (V - V_1)/A$, and the pressure is:

$$p = p_{surr} + Mg / A + kx / A = p_1 + k(V - V_o) / A^2$$

where k is the spring constant. The work done by the expanding gas is:

$$W = \int_{V_1}^{V_2} pdV = p_1(V_2 - V_1) + \frac{k}{A^2}\left[\frac{1}{2}(V_2^2 - V_1^2) - V_1(V_2 - V_1)\right] = 804 \text{ J}$$

3. *Show that the process is reversible by calculating the work done on the surroundings.*

The work done on the surroundings consists of the pV work by the rising piston against the pressure of the surroundings, W_{press}; the work done in elevating the mass of the piston, W_M; and the work done in compressing the spring, W_{sp}.

$$W_{surr} = p_{surr}(V_2 - V_1) = 157 \text{ J}$$

$$W_M = Mgx_2 = Mg(V_2 - V_1) / A = 147 \text{ J}$$

$$W_{sp} = \frac{1}{2}kx_2^2 = \frac{1}{2}k\left(\frac{V_2 - V_1}{A}\right)^2 = 500 \text{ J}$$

Adding these three work components gives 804 J, which is the same as the work done by the system calculated in part (2). The equality of the work done by the system and the work done on the surroundings is a necessary and sufficient condition for reversibility of a process.

1.10 THE FUNDAMENTAL DIFFERENTIALS

1.10.1 INTERNAL ENERGY

In a closed system, the first law for differential changes is:

$$du = \delta q - \delta w_{(pV)} - \delta w_{ext}$$

The work term has been divided up into an expansion/contraction (pV) term and the last term representing all other forms of work (such as electrical work supplied by a battery or chemical work expended as ATP induces muscle contraction). If the heat and pV work terms are reversible, they can be replaced by Tds and pdV, respectively, and the above equation becomes:

$$du = Tds - pdv - \delta w_{ext} \tag{1.15}$$

Similar differentials can be obtained from the last three energy-like properties.

1.10.2 Enthalpy

$h \equiv u + pv \rightarrow dh = du + pdv + vdp$, then use Equation (1.15):

$$dh = Tds + vdp - \delta w_{ext} \tag{1.16}$$

1.10.3 Helmholz Free Energy

$f \equiv u - Ts \rightarrow df = du - Tds - sdT$, then use Equation (1.15):

$$df = -pdv - sdT - \delta w_{ext} \tag{1.17}$$

1.10.4 Gibbs Free Energy

$g \equiv h - Ts \rightarrow dg = dh - Tds - sdT$, then use Equation (1.16):

$$dg = -vdp - sdT - \delta w_{ext} \tag{1.18}$$

1.10.5 Working Forms

If the external work term is removed, these equations become:

$$du = Tds - pdv \tag{1.15a}$$

$$dh = Tds + vdp \tag{1.16a}$$

$$df = -pdv - sdT \tag{1.17a}$$

$$dg = vdp - sdT \tag{1.18a}$$

The importance of these equations cannot be overestimated. In their present form, they are the starting points for most thermodynamic analyses of one-component substances. With an additional term, Equation (1.18a) constitutes the basis of chemical thermodynamics. From them, the several conditions of equilibrium in a closed system can be derived (see below).

Another significant aspect of these equations relates to the processes to which they apply. Their derivation was premised on the assumption of reversibility. However, the end result contains only state functions (properties), so that *they apply to irreversible processes as well.*

The importance of these four equations is reflected in their collective name. The most common is *fundamental differentials*. When rearranged, the first two

(Equations [1.15a] and [1.16a]) are termed the *Tds equations.* They are also known as the *Gibbs equations.*

1.11 EQUILIBRIUM

In the preceding sections, the term *equilibrium* has been used without definition. For thermodynamics to be of use, the state of the system under examination must be at equilibrium, which means that it has no tendency to change with time if unprovoked.* Two aspects of equilibrium are *internal* (within the system) and *external* (between the system and its surroundings).

1.11.1 INTERNAL EQUILIBRIUM

Internal equilibrium means that the matter in the system is uniform on a molecular level, or that it has no concentration, pressure or temperature gradients within it. For example, the gas in a box in which all gas molecules spontaneously occupy one half of the volume with the other half empty is not an equilibrium system. Macroscopic thermodynamics has nothing to say about such a situation. Microscopic or statistical thermodynamics shows that such a state cannot be ruled out, but is highly improbable.

In purely mechanical systems, equilibrium is represented by the state of the system that has the lowest energy. Thus, the equilibrium state of a pendulum is achieved when the weight hangs motionless in a vertical position. In thermodynamic systems, on the other hand, there are many criteria for equilibrium, depending on the constraints placed on the system. Constraining a system means fixing at least two of its properties. Depending on the constraint, equilibrium is expressed as the state in which a particular thermodynamic property is a maximum or a minimum. Examples of thermodynamic equilibrium are given below.

Consider an isolated system. Because its boundary is rigid and cannot pass matter, the system can exchange neither heat nor any form of work with its surroundings. These conditions imply constraints of fixed internal energy and volume, $dV = dU = 0$, which reduces Equation (1.15), to:

$$\delta W_{ext} = TdS_{U,V} \tag{1.19}$$

The subscripts on dS indicate that any change in S must occur at constant internal energy and volume. δW_{ext} can be thought of as *virtual* work, which could be realized if the system could communicate with its surroundings. An example is a box containing a pendulum with its arm at maximum swing. If connected by a line to a pulley and weight outside the box, external work could be performed by the pendulum in lifting the weight. Without such a line, the potential for work cannot be realized.

* Time-independence of the state of a system is not a foolproof criterion of equilibrium. A system with Q = W is at steady-state but not necessarily at equilibrium (e.g., there may be temperature gradients in the system).

If the boundary does not even permit virtual work ($\delta W_{ext} = 0$), the system is said to be in equilibrium, and Equation (1.19) reduces to:

$$dS_{u,v} = 0 \quad (1.19a)$$

In words, this brief requirement is:

At equilibrium, the entropy of an isolated system is a maximum.

Equation (1.19a) does not say whether the extremum is a maximum or a minimum, but the proof is as follows. If the system is not in equilibrium, it is still capable of doing external work. In so doing, the left-hand side of Equation (1.19) is positive, and so must be the right-hand side. Therefore, for any isolated system out of equilibrium, $dS_{u,v} > 0$, or the entropy increases until the maximum is attained and Equation (1.19a) applies.

A different equilibrium criterion applies when the process constraints are the more common ones of constant temperature and pressure. In this case, equilibrium is attained when the Gibbs free energy* of the system is a minimum. This is proven by setting $dT = dp = 0$ in Equation (1.18), resulting in:

$$\delta W_{ext} = -dG_{T,p} \quad (1.20)$$

The equilibrium criterion is obtained from the inability-to-do-work criterion by setting $\delta W_{ext} = 0$, which results in:

$$dG_{T,p} = 0 \quad (1.20a)$$

or, in words,

At equilibrium, the Gibbs free energy of a system held at constant T *and* p *is a minimum.*

That the equilibrium state is a minimum is demonstrated by considering Equation (1.20). A system at constant T and p that is out of equilibrium can still perform real external work (not just virtual work), and in so doing its free energy decreases. Therefore, for any isothermal, isobaric system out of equilibrium, $dG_{T,p} < 0$, or the free energy decreases until equilibrium is achieved. This equilibrium condition is particularly useful in dealing with chemical reactions (Chapter 9) or in assessing the stability of the various phases (vapor, liquid or solid) of a system (Chapter 5).

Figure 1.18 shows the conditions represented by Equations (1.19a) and (1.20a) in graphical form.

In the left-hand sketch, the "state of the system" might be temperature nonuniformity inside the impervious boundary. Or, it could be an unequal distribution of the material within the boundary, as illustrated in the example of Figure 1.17.

* Henceforth, the Gibbs free energy is termed "free energy."

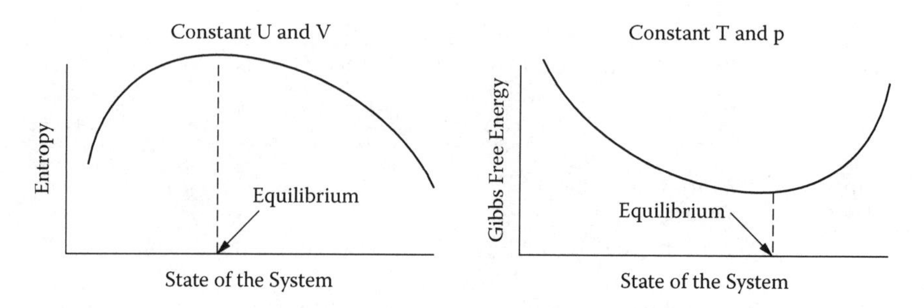

FIGURE 1.18 Equilibrium conditions for certain restraints.

Off-equilibrium states in the right-hand diagram of Figure 1.18 most often represent a reactive mixture that has not attained chemical equilibrium.

In a multiphase system, the requirements for equilibrium are the equality of temperature and pressure in each phase. Multicomponent systems entail additional equilibrium requirements related to the concentrations of the various species they contain: a property called the *chemical potential* of each species must be the same in all phases. The chemical potential of a species is related to (but is not equal to) its concentration. Detailed discussion of this topic is deferred to Chapter 7.

1.11.2 External Equilibrium

If the boundaries can transmit heat and matter and can expand and contract, the system and its surroundings must satisfy the following:

- The temperatures of the system and its surroundings must be the same.
- The pressure of the system and the surroundings must be equal.
- The chemical potential of all species to which the boundary is permeable must be the same in the system and in the surroundings.

1.12 COMPONENTS, PHASES, AND THE GIBBS PHASE RULE

Until now, the number of components in the system has not been addressed, and the examples illustrating concepts have been based on a single phase. *Components* are distinct chemical species whose quantities can be independently varied. *Phases* are regions of a system in which all properties are uniform and are distinct from other regions in the same system.

Single-phase systems are gas, liquid, or solid. Two-phase systems are combinations of these phases, and correspond to the following forward and reverse processes:

- Gas-liquid: vaporization/condensation
- Gas-solid: sublimation/condensation
- Liquid-solid: melting (fusion)/freezing (crystallization)

Somewhat less common than the above combinations of phases are two-liquid systems and two-solid systems. When two liquids do not mix, such as oil and water,

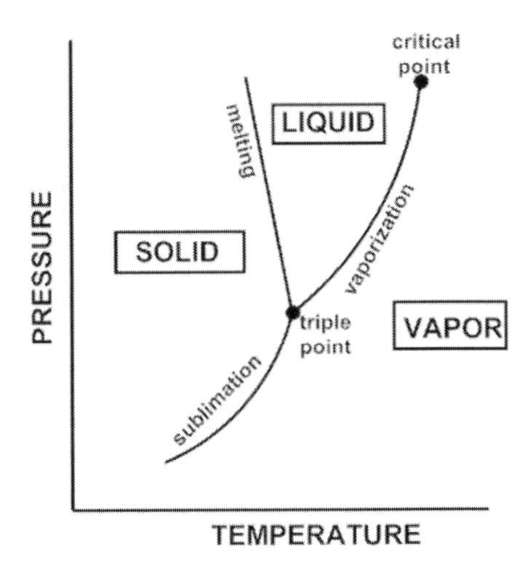

FIGURE 1.19 Pressure-temperature diagram for a pure substance.

they are said to be *immiscible*. Systems consisting of two coexisting solids are also quite common; metal and its oxide, such as rust on iron, are important in metallurgy and in the natural process of corrosion.

1.12.1 One-Component Systems

In a one-component system, coexistence of solid, liquid and vapor occurs at a unique set of conditions called the *triple point*. The triple point of water is at 0.01°C and 0.61 kPa, and so is not observed except in the laboratory where the pressure can be reduced below atmospheric pressure (100 kPa). Every pure substance (element or compound) has a vapor-liquid-solid triple point with its associated unique p–T combination. The phase field of a pure (i.e., one-component) substance in the pressure-temperature space is illustrated schematically in Figure 1.19.

This diagram contains three curves that separate areas of stability of solid, liquid and vapor phases. Single-phase states exist over wide regions of the p–T space. States along the curves represent the three common two-phase systems mentioned above. The sublimation curve extends in principle to the absolute zero temperature, and the melting line continues, in theory, to unlimited pressures. However, the vaporization curve ends at a distinct p–T combination called the *critical point*. Beyond this point, there is no difference in the properties of liquid and vapor phases, and the interface between the two no longer exists.

1.12.2 Two-Component Systems

In much of materials science, there is a need to represent the state of multicomponent, multiphase systems in which the gas phase is excluded for convenience and chemical reactions are not involved. Such representations are called *phase diagrams*. They show the regions of existence of various phases on a plot with temperature as the ordinate and composition as the abscissa. A simple phase diagram of the binary A–B

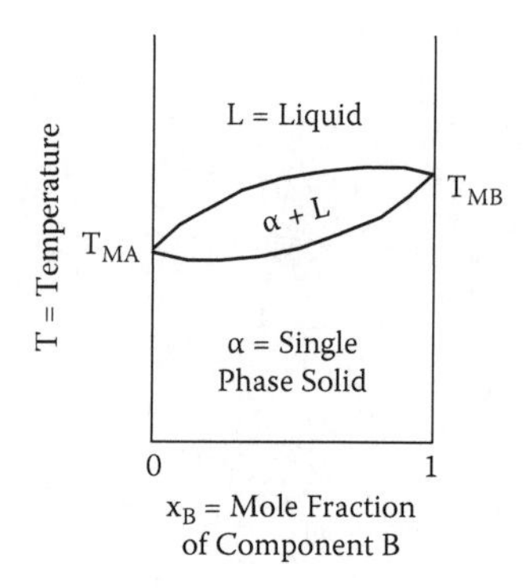

FIGURE 1.20 The phase diagram for an ideal A–B binary system.

system is shown in Figure 1.20. The ends of the lens-shaped portion of the diagram intersect at the melting points of pure A, T_{MA}, on the left, and the corresponding temperature T_{MB} for component B on the right. Below the lower curve, the system is a single-phase solid solution, in which A and B are homogeneously mixed in a specific crystal lattice structure called α. Above the upper curve, the system is a single liquid without, of course, a regular structure. In between these two curves is the two-phase region designated α + L, where the solid and the liquid coexist. Two-component systems (or phase diagrams) are usually termed *binary* systems (or phase diagrams).

1.12.3 Counting Components

Specification of the number of components in a system is less clear-cut than fixing the number of phases present. Each chemical species of fixed molecular makeup that can be mixed in arbitrary amounts in the system is considered to be a component. Composition is dictated by various measures of concentration. The most common is the *mole fraction*, or occasionally the mass fraction. If the system contains **C** components, only **C** – 1 mole fractions need be specified in order to fix unambiguously the composition. For example, a mixture of O_2 and N_2 is a two component system, but the mole fraction of only one component is needed to specify the system's composition. In addition, if the composition of the mixture does not change during the process under consideration, the binary system can be treated as a pseudo single-component system. In flowing through a valve, for example, air can be considered as a single species with thermodynamic properties that are the concentration-weighted average of the pure-species values.

If a chemical reaction occurs between the species in a system, the number of components is reduced by one. The system M + MO_2 + O_2 contains three molecular species but only two components, M and O. Another example is a gas containing H_2, O_2 and H_2O. At low temperature and in the absence of an ignition source, the hydrogen does not burn and the mixture is a true three-component (or *ternary*) system. At high temperatures, on the other hand, the chemical reaction: $2H_2(g) + O_2(g) = 2H_2O(g)$ provides a relation between the concentrations of the three molecular species. This restraint effectively reduces the number of components from three to two.

1.12.4 Proof of the Phase Rule

Figures 1.19 and 1.20 suggest the existence of a fundamental relationship between the number of components in a system, the number of coexisting phases at equilibrium and the number of variables (temperature, pressure, composition) that can be independently fixed without altering the number of phases present. The number of independently controllable variables is called the *degrees of freedom* of the system, and is designated as **F**.

To be perfectly general, let the system contain **C** components, labeled A, B, C,... that are distributed among **P** phases designated α, β, γ,... Because the sum of the mole fractions is unity in each phase, there are **C** – 1 independent composition variables in each phase. Since the system contains **P** phases, the total number of independently adjustable compositions is **P**(**C**– 1).

Other independent variables are the temperature and pressure in each phase. For the moment, we allow these properties to be different in each phase. These quantities add **P** variable temperatures and the same number of variable pressures. The total number of potentially independent variables describing the system is thus:

$$\text{Number of variables} = 2\mathbf{P} + \mathbf{P}(\mathbf{C} - 1) = \mathbf{P}(\mathbf{C} + 1)$$

However, the system is required to be in a state of internal equilibrium, which provides restraints on the above property variations. The most obvious are the requirements of temperature and pressure equality in all phases. Each of these conditions provides **P** – 1 restraints (i.e., if the system is two-phase $\alpha + \beta$, the single temperature restraint is $T_\alpha = T_\beta$). Together, temperature and pressure equality in all phases provides 2(**P** – 1) restraints.

The final internal equilibrium requirement that the chemical potentials of each component must be the same in all phases provides **P** – 1 restraints for each component. For a system containing **C** components, this particular equilibrium condition contributes **C**(**P** – 1) restraints on the compositions in the phases of the system. Summing all of the restraints yields:

$$\text{Number of restraints} = 2(\mathbf{P} - 1) + \mathbf{C}(\mathbf{P} - 1) = (\mathbf{P} - 1)(\mathbf{C} + 2)$$

The difference between the number of potentially variable properties and the number of restraints is equal to the number of degrees of freedom actually allowed to the system. Subtracting the above two equations gives:

$$\mathbf{F} = \mathbf{P}(\mathbf{C} + 1) - (\mathbf{P} - 1)(\mathbf{C} + 2)$$

or:

$$\mathbf{F} = \mathbf{C} - \mathbf{P} + 2 \tag{1.21}$$

This equation is the famous *Gibbs phase rule*. Applying it to a pure substance (**C** = 1) gives **F** = 3 – **P**. If only one phase is present (**P** = 1), then **F** = 2. This result corresponds to the areas labeled solid, liquid, and vapor in Figure 1.19. In these regions, both p and T can be independently set. However, if two phases of a pure substance are to coexist (**P** = 2), the phase rule permits only one degree of freedom (**F** = 1). Consequently, if T is specified, p follows (or vice versa). This restriction applies along the sublimation, vaporization and melting curves in Figure 1.19. Finally, at the triple point, **P** = 3, so **F** = 0, which means that the conditions for simultaneous equilibrium of the three states of matter are unique.

Figure 1.20 represents a two-component system, so Equation (1.21) reduces to **F** = 4 – **P**. However, in phase diagrams involving only condensed phases, the total pressure is fixed (and has little influence in any case), thereby removing one degree of freedom. The phase rule then reduces to **F** = 3 – **P**. In the single-phase solid and liquid zones of Figure 1.20, **P** = 1 and **F** = 2. This means that both temperature and composition can be independently varied without altering the single-phase character of the system. In the α + L two-phase region of Figure 1.20, there is only one degree of freedom; if the temperature is specified and if the overall composition (including both phases) places the point inside the lens-shaped α + L region, the compositions of the coexisting solid and liquid phases are fixed. The simple binary phase diagram of Figure 1.20 has no invariant point analogous to the triple point in Figure 1.19. However, more complex binary phase diagrams exhibit points where three phases coexist at a unique temperature.

The phase rule is modified if species in the system engage in an equilibrium chemical reaction, which in effect removes one degree of freedom. However, the number of components is equal to the number of molecular species, not the number of elements. For **N** multiple simultaneous equilibria, the phase rule reads:

$$\mathbf{F} = \mathbf{C} - \mathbf{P} + 2 - \mathbf{N} \tag{1.21a}$$

Example: The gas-phase reaction CO + ½ O_2 = CO_2 contains a single phase and two components: either the elements C and O or the species CO, O_2 and CO_2 less one reaction connecting them. In either interpretation, **F** = 2 – 1 + 2 = 3. The degrees of freedom are temperature, total pressure and one composition.

The phase rule can be applied more generally than it has been to the examples discussed above. It is a powerful tool for making order from complex multicomponent, multiphase thermodynamic systems.

PROBLEMS

1.1 Determine the degradation of high-quality input work, W, or energy, E, into heat Q_{irr} (as the ratio Q_{irr}/W or Q_{irr}/E) for the following irreversible processes:

(a) Pulling at constant speed a solid block on a horizontal surface.
(b) Expanding a gas in a cylinder with a piston. Because of friction between cylinder and piston, movement of the piston requires that the pressure of the gas (p) be greater than the pressure of the surroundings (p_{surr}).
(c) Supplying a current i from a battery to a circuit containing a resistance R.

(d) Loading a mass m horizontally on the top of a vertical spring, which has a force constant k. After the initial oscillations are damped out, the spring reaches an equilibrium compression distance x.

1.2 The process in problem 1.1(d) can be made reversible in the following manner. A number n of smaller masses Δm that are arranged on vertically stacked shelves next to the spring. The small weights are loaded on to the spring sequentially, each one compressing the spring a small amount. Loading of each small mass is done horizontally as the top of the spring reaches the elevation of the shelf holding the small mass.

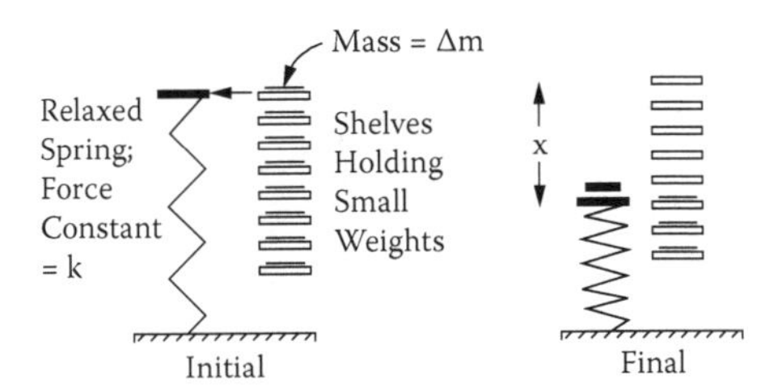

Prove that the energy stored in the spring is equal to the loss of potential energy of the distributed masses, which means that the process is reversible.

Hint: Construct a table showing the cumulative potential energy loss of the first few small masses added to the spring. From this table, deduce the equation for ΔE_P for n added masses.

1.3 A thermometer manufacturing company is experiencing difficulty filling the capillary tubes with mercury (see diagram below). While under vacuum, a supply reservoir attached to the top of the tube is filled with mercury (left-hand diagram). However, the mercury is prevented from entering the mouth of the tube by the surface tension force of the curved liquid surface. Company engineers propose a simple solution: admit gas to the reservoir until the gas pressure adds sufficiently to the liquid head to overcome the surface tension force at the capillary mouth and drive the liquid into the capillary (right-hand diagram).

(a) Write the pressure balance at the mercury surface at the top of the empty tube. The surface tension creates a pressure difference across a curved liquid interface equal to $2\gamma/S$, where γ is the surface tension and S is the radius of curvature of the surface, which must be equal to or larger than the radius of the capillary tube. As the gas pressure above the reservoir is increased, the radius of curvature of the liquid surface decreases. When it reaches a minimum value equal to the tube radius, the tube immediately fills with liquid. Develop the equation for the critical value of the external gas pressure, p_{crit}. Determine the critical pressure for the following parameter values: ρ = 7000 kg/m^3; γ = 0.5 N/m; r = 10^{-4} m; L = 0.1 m; H = 0.02 m; R = 0.01 m.

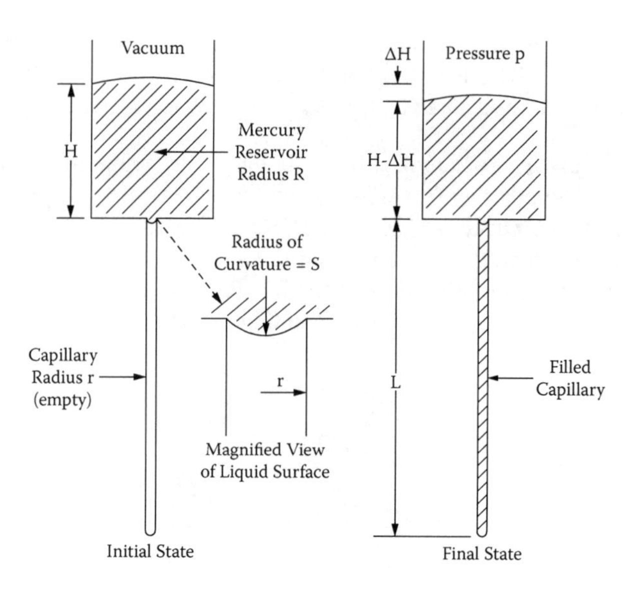

(b) In transforming from the initial state to the final state, the surface area of the liquid increases. Because mercury does not wet the container walls, the energy needed to form the additional surface is the product of the increment in surface area and the surface tension (which can be viewed as a surface energy per unit area). This energy is supplied from two sources. The first is the change in the potential energy as mercury drops into the tube. The second is the work done by the external gas pressure on the liquid surface as it moved down during filling of the capillary. Calculate the numerical values of these work/energy components. Do they conserve energy? If not, why? What is the form of the "missing" energy?

1.4 An ideal gas is contained in an adiabatic cylinder fitted with a frictionless piston and a stop for the piston. The piston is attached to a rigid structure by a spring. Initially, the piston rests on stops; the spring is at its equilibrium length, so there is no force on the piston.

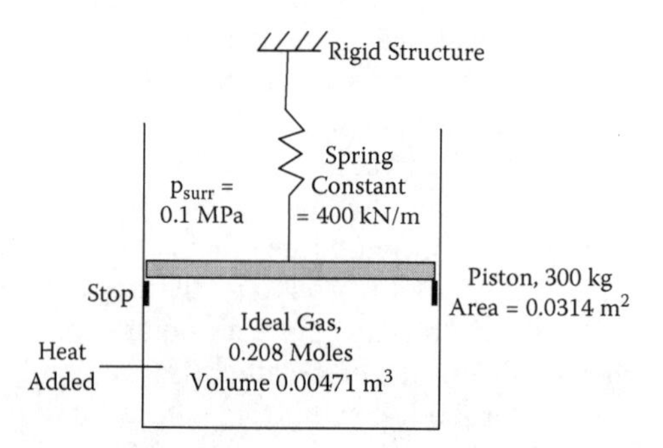

(a) Heat is added to the gas in the cylinder until the piston just leaves the stops. What is the gas temperature at this point?
(b) Heat is added reversibly to the gas until the final volume is V_2 = 0.00628 m^3. How much pV work is done by the system? Is the process isothermal?
(c) Show that the process is reversible by calculating the work done on the surroundings.

1.5 As shown in the p–T diagram below, a gas is expanded adiabatically from initial state 1 to a final pressure p_2. If the process is reversible, the final state is 2. If the process is irreversible, the final state is 3. For each of the four possible explanations for this result, answer *yes* or *no* whether you think the explanation to be correct. Give your reason for each choice.

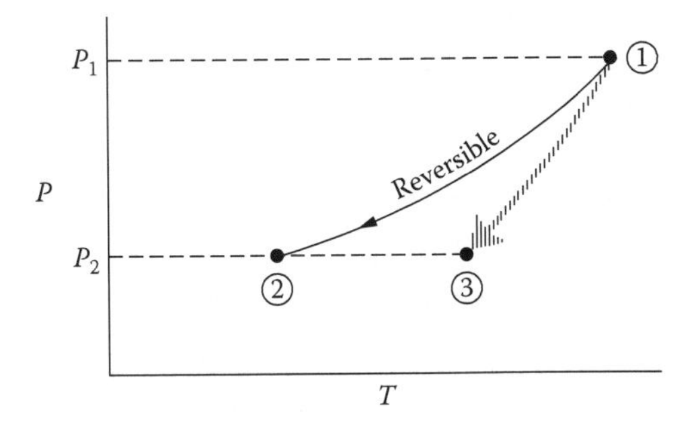

(a) $Q \neq 0$ for the irreversible process
(b) $W_{irr} > W_{rev}$
(c) The irreversible process cannot perform work
(d) Because $W_{irr} < W_{rev}$, $\Delta U_{irr} < \Delta U_{rev}$

1.6 Describe a version of each of the four processes in Problem 1.1 that is closer to reversible than the original process.

1.7 One mole of air is changed from state 1 (580 K, 17 atm) to state 2 (380 K, 8 atm) by two paths, both consisting of two parts.

Path A: proceeds from state 1 isothermally to state 3, then isobarically to state 2.
Path B: adiabatic expansion to state 4 followed by constant-volume cooling to state 2.

During the adiabatic expansion from state 1 to state 4, the temperature and pressure are related by:

$$(T/T_1) = (p/p_1)^{0.286}$$

(a) Determine T_4 and p_4 and plot the two paths on a p–T diagram.
(b) What is the molar volume at state 2?
(c) Determine the changes in internal energy between state 1 and states 2, 3 and 4. Air is an ideal gas, and its thermal equation of state is:

$$\Delta u = 2.5R\Delta T$$

where R is the gas constant.

(d) Calculate the heat and work for each stage and complete the second, third, and fourth columns of the table. Express Δu, Q, and W in Joules.

Path	Δu	Q	W	Δs
1→3				
3→2				
A				
1→4				
4→2				
B				

(e) Calculate the entropy (in J/K) changes during each stage and complete the last column of the table.

1.8 Joule is said to have determined the mechanical equivalent of heat by measuring the temperature difference between the water flowing over the top of a waterfall and the water in the pool at the bottom.

If the height of the waterfall is 100 m,

(a) What is the velocity at which the water enters the pool?
(b) What is the temperature rise between the two locations?

1.9 In Section 1.7.1.2, the work performed by the surroundings on the ideal gas in a cylinder by placing the entire mass M on the piston at once was calculated. In the present process, $\frac{1}{2} M$ is placed first, then the second $\frac{1}{2} M$ at the level where the first mass comes to equilibrium.

All three states in the diagram are at the same temperature and the external pressure is p_S. The mass M and the pressure of the surroundings are related by $p_S = \frac{1}{2} Mg/A$, where g is the acceleration of gravity and A is the area of the piston.

Calculate the ratio of the work W done by the surroundings for the above process to the work W_{irr} done by the surroundings when the entire mass is placed on the piston at once for the volume ratio $V_o/V_f = 3$.

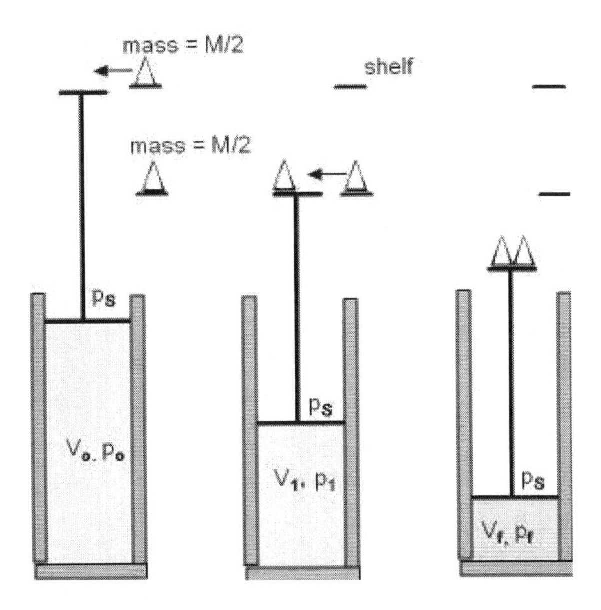

1.10 An elemental solid has a heat capacity of 1 J/mole-K. What is the atomic weight of the element?

1.11 What are the restraints for which the minimum Helmholz free energy is the criterion of equilibrium?

REFERENCES

Abbott, M. and H. C. Van Ness. 1989. *Theory and Problems in Engineering Thermodynamics*. New York: McGraw-Hill.

Potter, M. and C. Somerton. 1993. *Theory and Problems in Engineering Thermodynamics*. New York: McGraw-Hill.

Van Ness, H. C. 1983. *Understanding Thermodynamics*. Dover.

Von Baeyer, H. C. 1998. *Maxwell's Demon Why Warmth Disperses and Time Passes*. New York: Random House.

Van Wylen, G. R. Sonntag and C. Borgnakke. 1994. *Fundamentals of Classical thermodynamics*. 4th ed. New York: Wiley or later editions.

2 Equations of State

2.1 WHAT IS AN EQUATION OF STATE?

There are two types of *equations of state*, or EOS. In general, they are thermodynamic relationships between three properties of a pure (one-component) substance. In discussing the fundamental meaning of equations of state, it is useful to eliminate the quantity of the substance in order to deal only with intensive properties (see Section 1.6.2). Extensive properties can be converted to intensive properties by dividing the former by the quantity of the substance. (e.g., on a per-mole basis $v = V/n$, $u = U/n$).

The volumetric EOS refers to the relationship of the p-v-T properties of a gas. Solids and liquids are also described by p-v-T equations of state, although the quantitative forms are very different from those applicable to gases.

In functional form, the volumetric EOS can be written as $v = f(p,T)$, or $v(p,T)$ for short. This form indicates that the specific volume (or molar volume) is expressed as a function of pressure and temperature. However, the EOS can be equally well written as $p(v,T)$ or $T(p,v)$. Because pressure and temperature are usually specified in an experiment or in a process, the form $v(p,T)$ is most commonly employed.

The EOS relating v, p, and T provides no information about the other thermodynamic properties, in particular about the internal energy u and the entropy s. However, according to the phase rule (Section 1.12), specifying any two properties of a pure substance fixes all properties. Nonetheless, the EOS in the form $p(T,V)$, for example, gives no hint about the functions $u(T,v)$, $s(T,v)$, or of the remaining auxilliary properties h, f, and g (see Section 1.6.1). Knowledge of (T,v) requires information about the substance beyond that contained in its p-v-T relationship. Functional relationships such as $u(T,v)$ are sometimes called *thermal equations of state*. Two gases, helium and nitrogen for example, may follow the ideal gas law with reasonable accuracy, but show significantly different $u(T,v)$ behavior.

2.2 THE IDEAL GAS LAW

The ideal gas law,

$$pv = RT \tag{2.1}$$

is frequently used as an aid in explaining the fundamental concepts of thermodynamics. It is a volumetric equation of state that reasonably well describes most real gases at sufficiently low pressure.

R in this equation is the gas constant. Its numerical value is obtained from the results of the experiment in the constant-volume gas thermometer described in Section 1.1.1. Specifically, the slope of the line in Figure 1.1 yields the numerical value of R. This quantity turns out to be the product of Avogadro's number (6.023×10^{23} molecules/mole) and Boltzmann's constant (1.38×10^{-23} J/molecule-K), and so is a universal constant. Its value, in various units, is:

$$R = 8.314 \text{ J/mole-K or } 1.986 \text{ calories/mole-K or } 82.06 \text{ cm}^3\text{-atm/mole-K}$$

The first value is the SI* or metric value. The second value given above is convenient in many calculations where thermochemical properties from tables or plots are in the now-obsolete calorie or kilocalorie (kcal) units. The third form above is convenient in direct application of Equation (2.1) in which p and v are expressed in commonly used units (neither cubic centimeters nor atmospheres are SI units).

Although the gas constant R obtained from the slope of the line in Figure 1.1 is empirical, the fact that it is the product of two universal constants of physics suggests a deeper meaning. Indeed it does. The ideal gas law, together with the value of R, can be derived in statistical thermodynamics from the translational motion (i.e., kinetic energy) of a collection of particles subject to two conditions:

1. The actual volume occupied by the particles in a container must be small compared to the volume of the container. It is somewhat misleading to characterize a molecule as having a volume as if it had distinct boundaries like a soccer ball. In reality, a repulsive potential emanates from a molecule (Figure 2.1) and interacts with the repulsive fields of colliding molecules to cause the two to scatter from each other. This interaction of repulsive force fields gives each particle an effective volume. The requirement that the effective volume of the molecule be small compared to the container volume permits ideal gas behavior to be approached by all gases at sufficiently low pressure (which is equivalent to low density).
2. The above condition means that particles are essentially geometric points that collide with each other as well as impinge on container walls. In between collisions, the particles do not experience an attractive force from the presence of other particles. That long-range attractive forces (Figure 2.1) must exist is evident because all substances eventually condense into liquid or solid states at sufficiently high pressure and low temperature. However, for the gas to be ideal, these attractive interparticle forces must be negligible. This means that, on average, the particles must be far apart, which again implies low gas density.

If either of conditions 1 or 2 is not satisfied, the EOS does not follow the form given by Equation (2.1), and the gas is termed a *real*, or *nonideal*, gas.

---

* SI stands for (in French), "Système International."

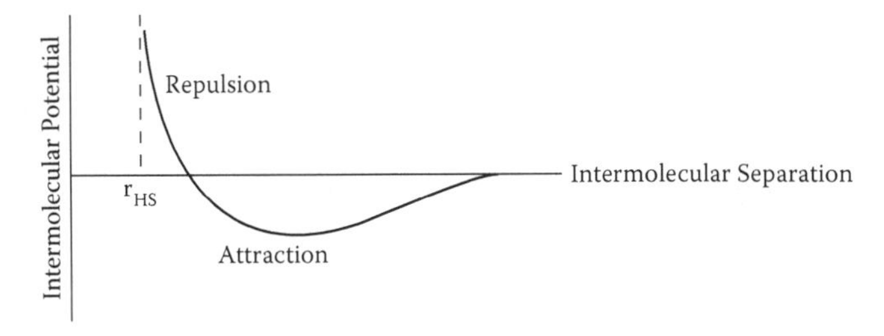

FIGURE 2.1 Potential energy of interaction between gas molecules. r_{HS} is the radius of the atom or molecule represented as a hard sphere.

The ideal gas has been referred to above as a collection of "particles" rather than specifically as "atoms" or "molecules." The reason is that particles other than atoms or molecules can exist in an ideal-gas state. Well-thermalized neutrons in the moderator of a nuclear reactor, for example, possess all of the properties of an ideal gas even though they are not analyzed using Equation (2.1). Similarly, a sufficiently-dilute collection of electrons exhibits classical ideal-gas behavior.

The microscopic thermodynamic view of the ideal gas (now considering ordinary gases of atoms or molecules) sheds light on the distinction between the volumetric equation of state, $v(T,p)$, and the thermal equation of state, say $u(T,v)$. In the ideal gas, the only molecular motion that affects the p-v-T behavior is translation in the space of the container. The thermal equation of state, on the other hand, depends on the kinetic energy of translation motion as well as on forms of energy internal to the molecule. These include rotation of the molecule around its symmetry axes, vibration of the atoms in the molecule relative to each other and, at very high temperatures, electronic excitation. All of these forms of internal energy influence the thermal EOS but not the p-v-T behavior of the gas.

Problems 2.5, and 2.11 provide exercises utilizing the ideal gas law.

2.3 NONIDEAL (REAL) GASES

As described above, deviations from ideal gas behavior are, on a microscopic level, due to the intermolecular forces or, equivalently, the potential energy between molecules in the gas. Figure 2.1 shows a generic intermolecular potential function with both a short-range repulsive interaction (positive potential energy) and a longer range attractive interaction (negative potential energy). At large separation distances, the potential energy between the two species approaches zero, which is the condition for ideal-gas behavior.

Characterization of nonideality can be accomplished with the simple cylinder-piston apparatus shown in Figure 2.2. The pressure of a fixed quantity of gas held at constant temperature is varied and the volume measured. The ratio pv/RT is plotted against pressure. For an ideal gas, this quantity should be equal to unity for all pressures. The graph in Figure 2.2 shows the behavior of a gas at two temperatures. At T_1, deviation from ideality is positive, or $pv/RT > 1$. At the lower temperature T_2, the deviation is negative, and $pv/RT < 1$. In all cases, however, the data extrapolate to the ideal gas value $pv/RT = 1$ as $p \to 0$.

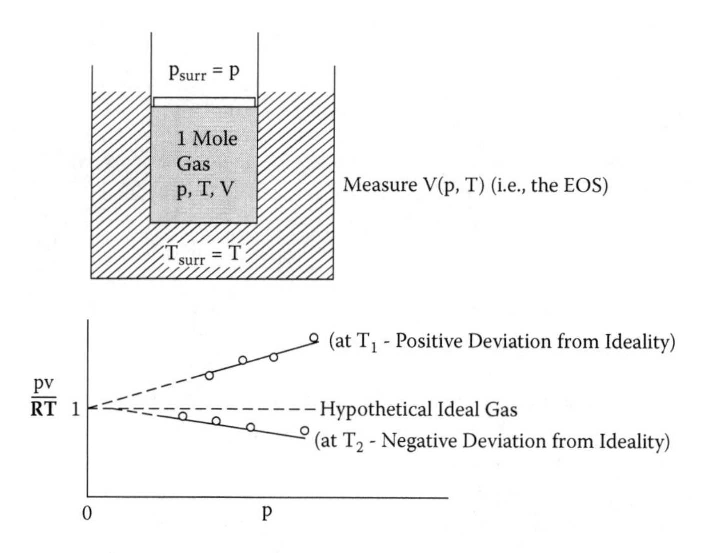

FIGURE 2.2 Deviations from ideality of a gas at two temperatures.

2.3.1 The Compressibility Factor

All gases, even He, deviate from ideality, the more so as the pressure increases and the temperature decreases. Deviation from ideality is commonly expressed by the *compressibility factor**:

$$Z = \frac{pv}{RT} = \frac{v}{v_{id}} \tag{2.2}$$

where $v_{id} = RT/p$ is the molar volume that the gas would have were it ideal. By the definition of Z, the compressibility of an ideal gas is unity. Nonideal gases, even the same one, can exhibit values of Z greater or smaller than unity, depending on the pressure and temperature conditions. As an example, Figure 2.3 shows the compressibility of nitrogen as a function of pressure for a number of temperatures. Figure 2.2 is a simplified version of Figure 2.3. The complexity of the curves in Figure 2.3 is a result of the interplay of the repulsive and attractive intermolecular forces. Three main points emerge from Figure 2.3.

1. Ideal behavior is favored by high temperature and low pressure. Ideality is reached at a higher gas pressure at 300 K (~7 MPa) than at the lower temperatures (~1 MPa at 200 K).
2. Point A at 2 MPa and 130 K represents a state with negative deviations from ideality ($Z < 1$). The reason for this behavior is the attractive forces between N_2 molecules, which act to contract the gas to a volume smaller than what it would have were it ideal.

* Not to be confused with the coefficient of compressibility defined in Equation (1.2).

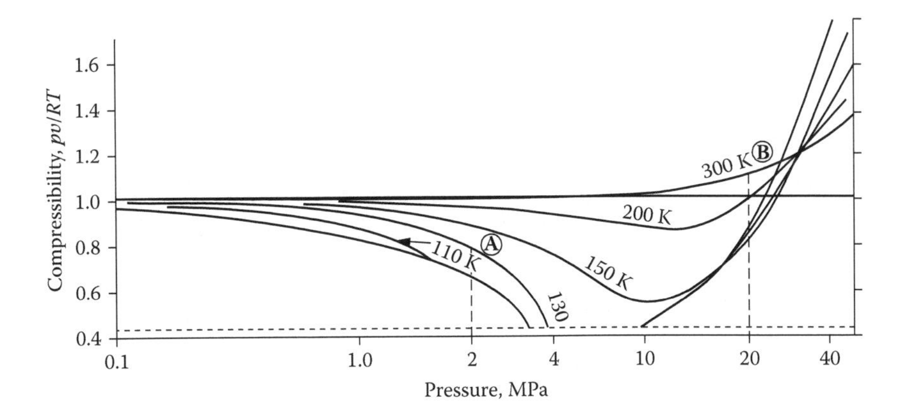

FIGURE 2.3 Compressibility of nitrogen.

3. Point B at 20 MPa and 300 K exhibits positive deviation from ideal behavior. This is due to the small specific volume of the gas, which causes the intermolecular repulsive forces to become significant.

Abbott and Van Ness (1989) present an extensive discussion of equations of state that reproduce plots such as that shown in Figure 2.3.

2.3.2 The Van der Waals Equation of State

We will consider mainly the simplest and oldest of these EOS, namely the one due to the Dutch physicist Van der Waals. This equation of state was developed in 1910, but is still in use for many applications.

The Van der Waals equation is most commonly written in the form of the ideal gas law of Equation (2.1) with corrections to the p and v terms:

$$(p + a/v^2)(v - b) = RT \tag{2.3}$$

where a and b are constants for the particular gas.

The radial extent of the repulsive portion of the intermolecular potential function shown in Figure 2.1 is the basis for assigning a hard-sphere radius to the molecule. This quantity determines the constant b, which is four times the "volume" of the molecule.*

The term $v - b$ in Equation (2.3) corresponds to the free volume available for unimpeded molecular motion after the effective volume occupied by the molecules has been deducted.

a in Equation (2.3) reflects the attractive portion of the intermolecular potential curve in Figure 2.1. If the parameter b is neglected for the moment, the physical meaning of the constant a can be seen by solving Equation (2.3) for the pressure:

---

* If r_{HS} is the hard-sphere radius, two molecules collide if their center-to-center distance is $2r_{HS}$ (Fig. 2.4). The volume around a molecule from which other molecules are excluded is $b = \frac{4}{3}\pi(2r_{HS})^3 = 4 \times \frac{4}{3}\pi r_{HS}^3$.

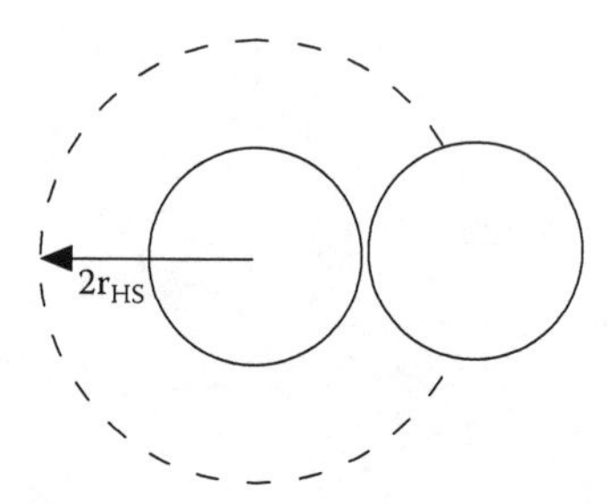

FIGURE 2.4 The excluded volume around hard-sphere molecules.

$$p = \frac{RT}{v} - \frac{a}{v^2} = p_{id} - \frac{a}{v^2}$$

where p_{id} is the pressure that would be exerted by the gas if it were ideal. Subtraction of the a/v^2 term suggests that the actual pressure is less than the ideal-gas pressure. This may be viewed as the result of the attractive intermolecular forces "holding together" the assembly of gas molecules and thereby reducing the intensity of the molecular impacts on the container walls, which is the manifestation of pressure.

Without neglecting the constant b, Equation (2.3) can be written in the form of the compressibility:

$$Z = \frac{pv}{RT} = \frac{1}{1 - b/v} - \frac{a/RT}{v} \tag{2.4}$$

Equation (2.3) is the $T(p,v)$ form of the Van der Waals EOS and Equation (2.4) can be converted to the $p(v,T)$ form by multiplying both sides by RT/V. However, being cubic in v, neither of these equations can be converted analytically to the $v(T,p)$ form, which is equivalent to the $Z(T,p)$ function that is plotted in Figure 2.3.

The explicit $v(T,p)$ function is useful in many practical problems involving nonideal gases. Fortunately, an accurate approximation provides a satisfactory solution. First, it is assumed that $b/v << 1$, so that the first term on the right in Equation (2.4) can be approximated by a one-term Taylor series expansion as $1 + b/v$. The b/v term so obtained is combined with the second term in Equation (2.4) by factoring the common term $1/v$, which is then further approximated by its ideal-gas value p/RT. The result of these mathematical manipulations is the explicit $v(p,T)$ form of the Van der Waals EOS:

$$v \cong \frac{RT}{p} + b - \frac{a}{RT} = v_{ideal} + \left(b - \frac{a}{RT}\right) \tag{2.5}$$

or, in terms of the compressibility:

$$Z = \frac{pv}{RT} \cong 1 + \frac{p}{RT}\left(b - \frac{a}{RT}\right) \tag{2.5a}$$

Despite its simplicity, Equation (2.5a) provides a reasonably good approximation to the compressibility curves for nitrogen shown in Figure 2.3 (see Problem 2.6b).

The Van der Waals constants for N_2 can be determined by fitting the curves in Figure 2.3 to Equation (2.5a). An exercise in this fitting is given by Problem 2.6a. The results of a detailed analysis yield (for N_2):

$$a = 0.14\ \frac{\text{J-m}^3}{\text{mole}^2} \qquad b = 3.9\times 10^{-5}\ \frac{\text{m}^3}{\text{mole}} \tag{2.6}$$

With these constants, Equation (2.5a) predicts that deviations from ideality should change sign at a temperature given by setting the parenthetical term equal to zero, which yields:

$$T(Z<1 \Leftrightarrow Z>1) = \frac{a}{bR} = \frac{0.14}{(3.9\times 10^{-5})(8.314)} = 432\ \text{K}$$

Although the actual switch-over temperature from Figure 2.3 appears to be closer to 300 K, the simplified form of the Van der Waals equation provides a result that is at least not unreasonable.

The Van der Waals EOS can be used to predict the critical temperature and pressure of a nonideal gas (Problem 2.4). Problem 2.10 illustrates how the Van der Waals equation can be used to account for deviations from ideality for steam. In this case, the experimental database is the steam tables.

2.3.3 The Virial Equation of State

In this EOS, the compressibility is expressed as a series in $1/v^n$:

$$Z = 1 + \frac{B}{v} + \frac{C}{v^2} + \ldots.. \tag{2.7}$$

B and C are the second and third virial coefficients, respectively. The chief advantage of this description is that the virial coefficients can be determined by the methods of molecular thermodynamics. The second virial coefficient, for example, is given by:

$$B = 2\pi N_{Av} \int_0^\infty \left[1 - e^{-\phi(r)/kT}\right] r^2 dr \tag{2.8}$$

where N_{Av} and k are Avogadro's number and the Boltzmann constant, respectively. ϕ is the intermolecular potential function shown in Figure 2.1 as a function of r, the separation of a pair of molecules. Values of B are shown in Figure 2.2 of Prausnitz et al. (1986) for a number of gases.

The virial EOS has the additional feature of accommodating binary gas mixtures, which the Van der Waals EOS cannot do. For an A-B mixture, the second virial coefficient is obtained from:

$$B = x_A^2 B_{AA} + x_A x_B B_{AB} + x_B^2 B_{BB} \tag{2.9}$$

B_{AA} and B_{BB} are the pure-component second virial coefficients while B_{AB} accounts for interactions between A and B molecules.

2.4 THERMAL EQUATIONS OF STATE: THE SPECIFIC HEAT OF GASES

We noted at the beginning of this chapter that having the *p-v-T* equation of a gas (or any pure substance, for that matter) does not provide information on the thermal equation of state, specifically, a function such as $u(T,v)$. The fundamental thermodynamic property that determines the thermal EOS is the specific heat at constant volume, whose definition is:

$$C_V = \left(\frac{\partial u}{\partial T}\right)_V \tag{2.10}$$

from which the thermal equation of state $u(T,v)$ can be obtained by integration.*

Because most practical applications of thermodynamics involve processes that occur at constant pressure rather than at constant volume, the specific heat at constant pressure is also needed. This property is defined by:

$$C_P = \left(\frac{\partial h}{\partial T}\right)_p \tag{2.11}$$

2.4.1 Specific Heats of Ideal Gases

For an ideal gas, the connection between C_P and C_V is easily determined from the definition of the enthalpy, $h = u + pv$. Because $pv = RT$, $h = u + RT$. Taking the temperature derivative of this relation yields:

$$C_P = C_V + R \tag{2.12}$$

Neglecting the effects of volume or pressure on the specific heats, the partial derivatives of Equations (2.10) and (2.11) can be replaced by ordinary derivatives, and the specific heats become:

---

* Strictly speaking, determination of $u\ (T,v)$ requires knowledge of $(\partial u/\partial v)_T$ in addition to C_V. However, for most substances, the volume dependence of u is small compared to the dominant effect of temperature. However, $(\partial u/\partial v)_T$ is evaluated as a function of v and T in Chapter 6.

$$C_V = du/dT \qquad \text{and} \qquad C_P = dh/dT \tag{2.13}$$

If the temperature dependence of C_V (or C_P) is known, the internal energy (or the enthalpy) follows by integration:

$$u(T) - u_{ref} = \int_{T_{ref}}^{T} C_V(T')dT'$$

$$h(T) - h_{ref} = \int_{T_{ref}}^{T} C_P(T')dT' \tag{2.14}$$

where T_{ref} is an arbitrary reference temperature where the internal energy or the enthalpy is set equal to u_{ref} or to h_{ref}. Either of these could be chosen as zero, but the two must be related (for the ideal gas) by $h_{ref} = u_{ref} + RT_{ref}$.

2.4.2 Temperature Dependence of the Specific Heats

The microscopic origins of the specific heat were in Section 1.6.2. The temperature dependence of C_V (or C_P) of ideal gases is due to the various ways that the particles store energy. Classical physics accords $^1/_2\, kT$ to each mode of energy storage. These energy-storage modes start to contribute at different temperatures, resulting in specific heats that either increase monatonically with temperature or exhibit plateaus if the onset temperatures of the internal modes are widely separated. Figure 2.5 shows these features in various gases.

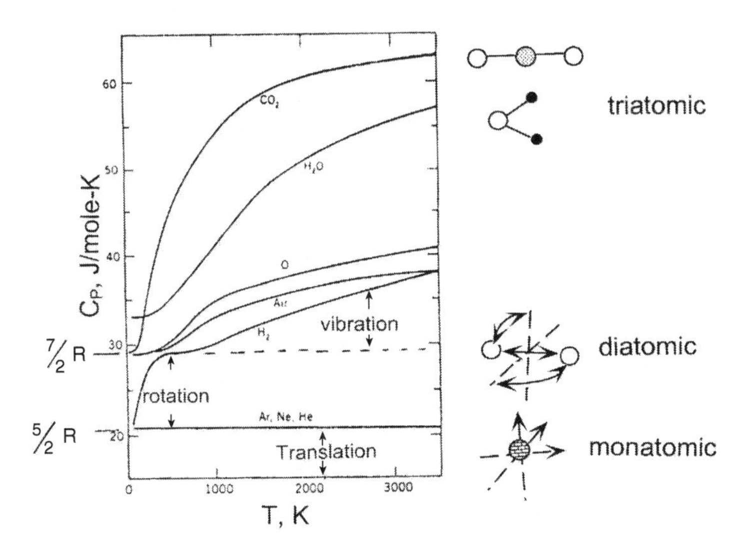

FIGURE 2.5 The specific heats of various gases (condensation to the liquid and solid states are not included in the plots).

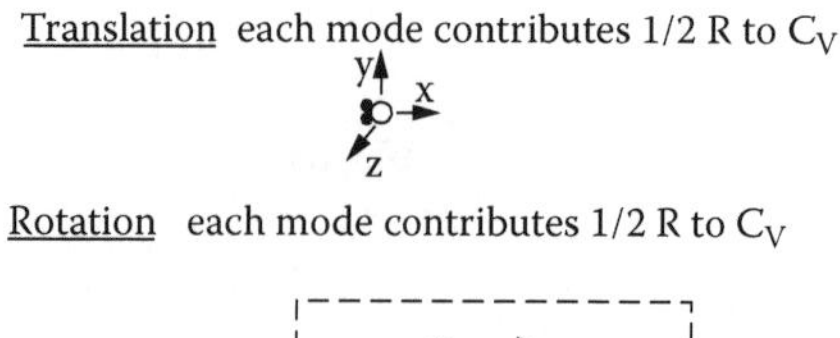

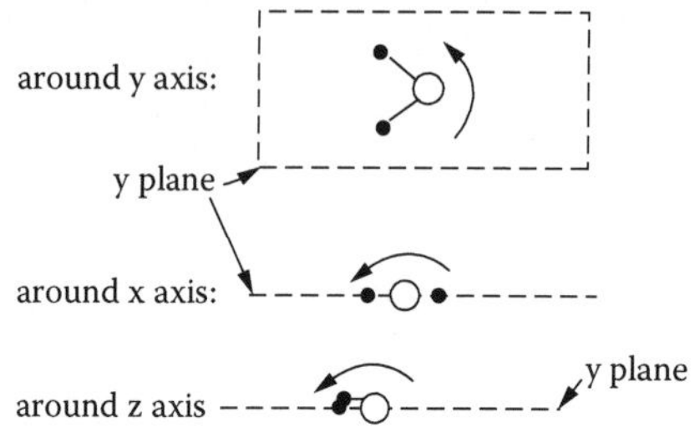

Vibration each mode contributes R to C_V

Quantized Vibrational States:

$$(C_V)_{vib} = \sum_{i=1}^{3} \left[\frac{w_i^2 e^{w_i}}{\left(e^{w_i} - 1\right)^2} \right] R \quad w_i = \frac{h\nu_{vib,i}}{kT}$$

ν_{vib} = Vibration frequency
h = Planck's constant
k = Boltzmann's constant
as $T \to \infty$ ($w \to 0$),
$(C_V)_{vib} = R$ (classical limit)
- as $T \to 0$ ($w \to \infty$),
$(C_V)_{vib} = 0$ (mode inactive)
$C_V = (C_V)_{trans} + (C_V)_{rot} + (C_V)_{vib}$
$= 3/2\ R + 3/2\ R + 0\ R = 3R$
$C_p = C_V + R = 4R = 33$ J/mole-K

FIGURE 2.6 Contributions to the specific heat of water vapor.

The atomic gases (Ar, Ne, He) possess only translational kinetic energy but no internal energy modes. Consequently, an atom moving in three-dimensional space possesses $3/2kT$ of translational energy, where k is Boltzmann's constant. Per mole of gas, $C_V = 3/2R$, and, by Equation (2.12), $C_P = 5/2R$. C_P for the atomic gases is independent of temperature because translation is the only mode of energy storage.

The diatomic gases O_2 and N_2 (or air in Figure 2.5) exhibit an apparent low-temperature limit of $C_P = 7/2R$. This is because rotation of these dumbbell-shaped molecules provides an additional energy-storage mode that adds one unit of R to C_V, and hence to C_P.

The curve for H_2 in Figure 2.5 shows that the rotational mode first appears at low temperature but does not fully contribute until ~400 K. The continued increase in C_P of the diatomic gases at higher temperatures is due to the contributions of the various vibrational modes of the molecules.

The triatomics CO_2 and H_2O have higher values of C_P because their more complex structures give these molecules a greater number of rotational and vibrational modes than the diatomic gases.

Example: Figure 2.6 illustrates the contributions of the different modes of energy storage to the specific heat of water vapor. In addition to three degrees of rotational freedom, the water vapor curve in Figure 2.5 includes three degrees of rotation, shown schematically in Figure 2.6. All three modes persist to near 0 K because transformation to liquid or solid states is not considered in the plot. At low temperatures, the vibrational modes are not active, so C_P includes $3/2R$ for translation, $3/2R$ for rotation and a final R to convert C_V to C_P. The total is $4R$, or 33 J/mole-K. At $T >$ ~250 K, energy storage by vibration begins. Each of the three modes of vibration is quantized. Depending on the frequency of the mode, each grows in differently with temperature.

In the high-temperature (classical) limit the quantum sum in Figure 2.6 approaches $3R$, at which point C_P should be $7R$, or 58 J/mole-K. This appears to be the high-temperature limit of the water vapor curve in Figure 2.5, but the slope is not zero. This means that other modes of energy storage are growing in.

The high-temperature contributions to C_P include ionization, illustrated for the diatomic molecule A_2:

$$A_2 = A_2^+ + e^-$$

The fraction of the A_2 molecules that are ionized absorbs energy (enthalpy) according to:

$$\Delta H_i = f_i x I$$

where I = ionization energy and f_i = fraction ionized, obtained from the equilibrium of the above reaction (see Chapter 9). The specific heat contribution from this source is:

$$\left(C_P\right)_{ioniz} = \frac{d\Delta H_i}{dT} = \frac{df_i}{dT} \times I \tag{2.15}$$

The other high-temperature energy-storage mode is dissociation:

$$A_2 = 2A$$

where E_{diss} = bond energy of A_2. The heat capacity addition from dissociation is:

$$\left(C_P\right)_{diss} = \frac{d\Delta H_{diss}}{dT} = \frac{df_{diss}}{dT} \times E_{diss} \tag{2.16}$$

where f_{diss} is the fraction of element A in atomic form.
The total heat capacity is the sum of all modes:

$$C_V = (C_V)_{trans} + (C_V)_{rot} + (C_V)_{vib} + (C_P)_{ioniz} + (C_P)_{diss} \tag{2.17}$$

2.5 THERMODYNAMIC PROPERTIES OF SOLIDS AND LIQUIDS

2.5.1 *p-v-T* Equation of State

The volumetric equations of state of condensed phases (solids and liquids) obey the same rules as gases but the equations are quite different in form. The differential of $v(T,p)$ is given by:

$$dv = \left(\frac{\partial v}{\partial T}\right)_p dT + \left(\frac{\partial v}{\partial p}\right)_T dp \tag{2.18}$$

TABLE 2.1
Equations-of-State Properties of Gases and Condensed Phases at Room Temperature and Atmospheric Pressure

Substance	v, m³/mole	α, K⁻¹	β, MPa⁻¹
Ideal gas	2.5×10^{-2}	3×10^{-3}	10
Water (liquid)	$\sim 10^{-5}$	$\sim 10^{-4}$	$\sim 10^{-4}$
Solid (typical)	$\sim 10^{-5}$	$\sim 10^{-5}$	$\sim 10^{-5}$

The partial derivatives in the above equation are related to the coefficient of thermal expansion and the coefficient of compressibility*

$$\alpha = \frac{1}{v}\left(\frac{\partial v}{\partial T}\right)_p \qquad \beta = -\frac{1}{v}\left(\frac{\partial v}{\partial p}\right)_T \tag{1.2}$$

Typical values of α and β are listed in Table 2.1. For ideal gases, substitution of Equation (2.1) into Equations (1.2) shows that $\alpha_{gas} = 1/T$ and $\beta_{gas} = 1/p$. Also given in Table 2.1 are specific volumes of water and typical solids.

With the above definitions of α and β, Equation (2.18) becomes:

$$dv = v(\alpha dT - \beta dp) \tag{2.18a}$$

Because α and β are approximately independent of temperature and pressure, the above equation can be integrated while holding these properties constant. In addition, α and β are small, so that volume changes due to changes in p and T are also small. Therefore, the term v on the right-hand side of Equation (2.18a) can also be taken as a constant and integration from initial state o to final state 1 yields:

$$\Delta v = v_o[\alpha(T_1 - T_o) - \beta(p_1 - p_o)] \tag{2.18b}$$

For greater accuracy in determining volume changes of condensed phases, allowance must be made for the variation of α and β with temperature and pressure. In this case, integration of Equation (2.18a) between initial state zero (0) and final state 1, still assuming small changes in v, yields:

$$\frac{\Delta v}{v_0} = \int_{T_0}^{T_1} \alpha(T, p)dT - \int_{p_0}^{p_1} \beta(T, p)dp \tag{2.18c}$$

Even if α and β are known functions of T and p, the integrals of Equation (2.18c) require that a path between states 0 and 1 be specified. That is, the first

* Not to be confused with the nonideality measure of the same name in Equation (2.2).

integral on the right requires that the variation of p with T be known, as does the second integral. Even if the integrals on the right-hand side of Equation (2.18c) are path-dependent, their difference is not because Δv represents a change in a thermodynamic property v. Thus, the relatively unimportant problem of calculating volume changes of a condensed phase when the properties involved are not constants serves as an illustration of a much more fundamental and general feature of thermodynamics: the independence of property changes on the path taken from the initial to the final state. This is explored in more detail in the following example.

Example: Liquid water at 20°C and atmospheric pressure (0.1 MPa) is heated to 30°C and 100 MPa. The detailed dependences of α and β on T and p are not known, but average values of these properties over ranges of one variable with the other held constant are given. With this sort of information available, natural choices of the path between the initial and final states are shown on the drawing below.

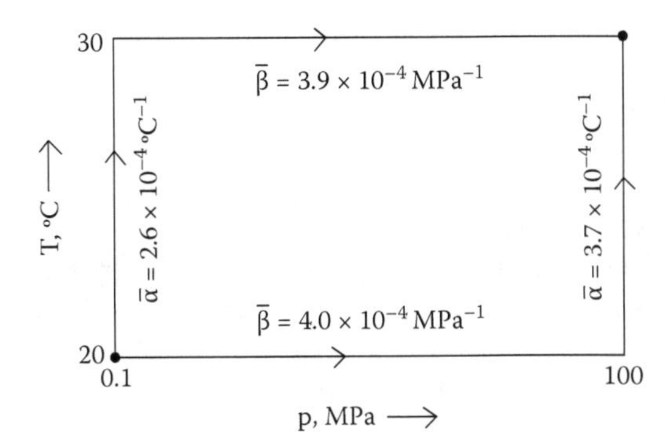

Upper path: isobaric heating at 0.1 MPa from 20°C to 30°C; isothermal compression to 100 MPa.

Lower path: isothermal compression at 20°C from 0.1 to 100 MPa; isobaric heating to 30°C.

The temperature-averaged thermal expansion coefficients ($\bar{\alpha}$) between 20°C and 30°C are shown on the drawing for the initial and final pressures; they differ by about 40%. The pressure-averaged compressibilities ($\bar{\beta}$) at the two temperatures differ by only 2.5%. For either path, the fractional volume change is given by:

$$\frac{\Delta v}{v_o} = \bar{\alpha}(T_1 - T_o) - \bar{\beta}(p_1 - p_0)$$

For the upper path, the above equation yields:

$$\left(\frac{\Delta v}{v_o}\right)_{upper} = 2.6 \times 10^{-4}(30 - 20) - 3.9 \times 10^{-4}(100 - 0.1) = -0.0364$$

For the lower path, the fractional volume change is:

$$\left(\frac{\Delta v}{v_o}\right)_{lower} = 3.7\times10^{-4}(30-20) - 4.0\times10^{-4}(100-0.1) = -0.0363$$

As required, the 3.6% volume decrease is calculated from either path. Integration along *any* p–T curve connecting the end states would give the same result. The path independence of $\Delta v/v_o$ implies that the functions $\alpha(T,p)$ and $\beta(T,p)$ cannot be totally independent of each other. Indeed they are not—see Problem 2.16.

2.5.2 Thermal Equations of State

The thermal equations of state of condensed phases are dependent chiefly on temperature, with pressure and volume changes having rather small effects on u or h. Consequently, Equations (2.13) apply reasonably well to condensed phases as well as to gases. Thus, the thermal properties of solids and liquids are determined by their specific heats. For solids the C_V–T relationship is typified by the lower curve in Figure 2.7, which is for copper.

As T approaches zero, so does C_V, which is a consequence of quantization of the vibrations of the atoms in a solid. At the other extreme of high temperatures, C_V remains at a plateau value of $3R$. This is the classical physics result (called the law of DuLong and Petit) for a collection of vibrating particles. At very high temperatures, C_V begins to increase above the value of $3R$ because of a combination of electronic excitation and creation of point defects in the crystal structure. Both of these processes absorb energy and so appear in the specific heat.

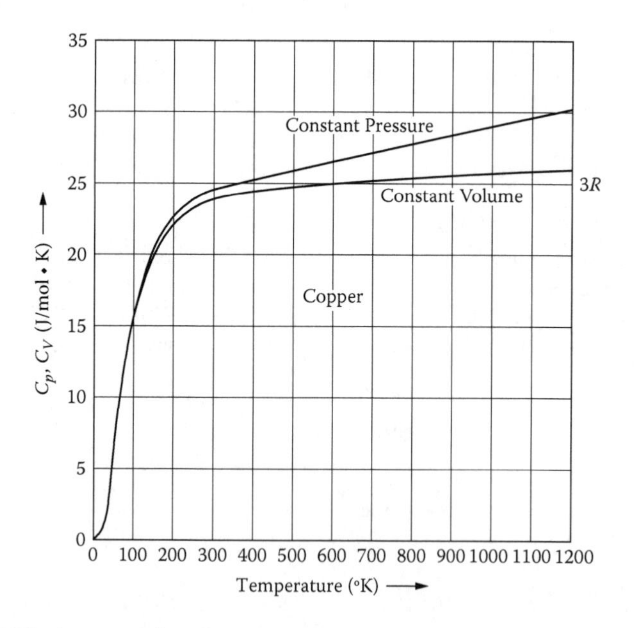

FIGURE 2.7 The heat capacity of copper.

Below a few hundred degrees Celsius, C_P and C_V are practically equal. Using the methods to be discussed in Chapter 6, the general relation between the heat capacities is:

$$C_P = C_V + \alpha^2 Tv/\beta \tag{2.19}$$

The small values of α and v for condensed phases account for the proximity of C_P and C_V. At high temperatures, however, the two heat capacities diverge significantly, as shown in Figure 2.7.

An *Incompressible solid* (or liquid) is one for which $\beta = \infty$, which, according to Equation (2.19), renders $C_P = C_V$. This is a common approximation for most condensed plates.

A first-law application utilizing the heat capacity of a solid is given in Problem 2.3.

2.6 GRAPHICAL REPRESENTATIONS OF THE EQUATION OF STATE OF WATER

Water is the prime example of a condensable substance, meaning that in many practical situations it can exist as a vapor or a liquid (and less frequently as a solid). Water is a goldmine of detailed thermodynamic data and applications of thermodynamic theory. Van Ness (1983) and Potter and Somerton (1993) provide very readable expositions of this topic with many illustrative problems.

The equations of state of water, both volumetric and thermal, exhibit features of the gas equations of state described in Sections 2.2 to 2.4 the condensed substances EOS of Section 2.5. However, because water vapor is often used at high pressures in many processes (e.g., a steam power plant), and because the water molecule is inherently fairly nonideal, the ideal gas law is usually a poor approximation of the EOS of steam. Similarly, liquid water is quite compressible compared to most solids (see Table 2.1), so the simple condensed-phase EOS of Equation (2.18b) is of marginal use when applied to water.

2.6.1 THERMODYNAMIC SURFACES

The EOS of water, like those of many other pure substances, expresses a property, pressure, for example, as a function of two other properties, temperature and volume. Thus, the EOS is given by the function $p(T,v)$, which forms a surface in three-dimensional space. This type of graphical representation of the p-v-T properties of water is shown in Figure 2.8.

Represented on this thermodynamic surface are regions where water exists as a single phase, as two coexisting phases, or all three phases simultaneously. The light surfaces in Figure 2.8 depict the single-phase gas, liquid, and solid states. The shaded surfaces represent two-phase regions, the most important being the liquid-vapor portion. The lines labeled with letters are isotherms, or cuts through the surface at constant temperatures.

The "triple line" in Figure 2.8 represents the unique three-phase state of water. The designation of a triple line may appear to be a contradiction of the usual

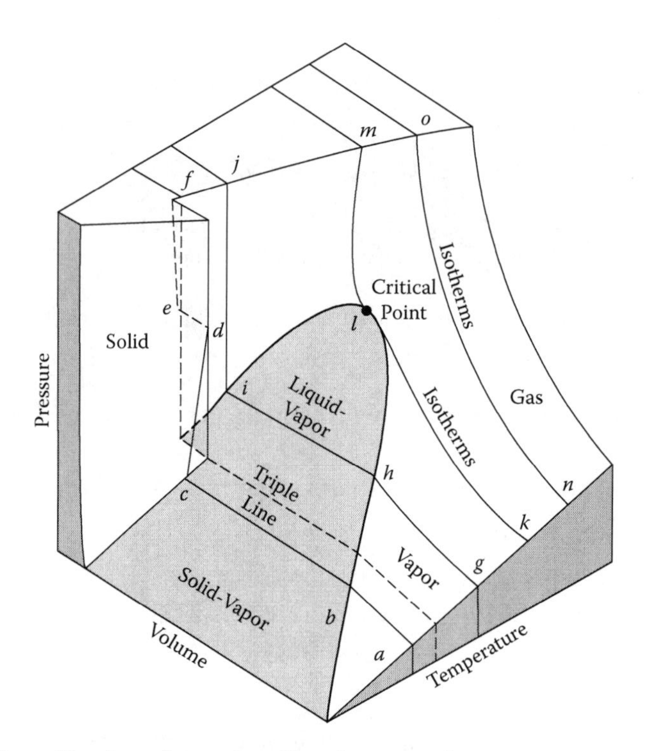

FIGURE 2.8 *p-v-T* surface of a condensable substance such as water.

terminology of a triple point introduced in Section 1.12. The explanation lies in the meaning of the volume coordinate axis in Figure 2.8. While there is no ambiguity about the pressure and temperature axes, the volume axis represents the mass-fraction-weighted average of the specific volumes of the phases present. Only in the single-phase regions does the volume denote that of a well-defined phase. In the two-phase liquid-vapor region, for example, the points along the volume axis are defined by:

$$v = xv_g + (1 - x)v_f \tag{2.20}$$

where

x = mass fraction of vapor in the mixture (also called the *quality*)
v_g = specific volume of the steam (gas) phase
v_f = specific volume of the liquid (fluid) phase

2.6.2 Two-Dimensional Projections of the EOS Surface

Better insight into the EOS of water is obtained by the projections on the three coordinate planes than by the three-dimensional thermodynamic surface. Figure 2.9 shows the *p-v*, the *T-v*, and the *p-T* projections of Figure 2.8.

The most commonly-used projection, the heavy bell-shaped curve in the *T-v* diagram (middle graph), is the envelope of the two-phase vapor-liquid region. To

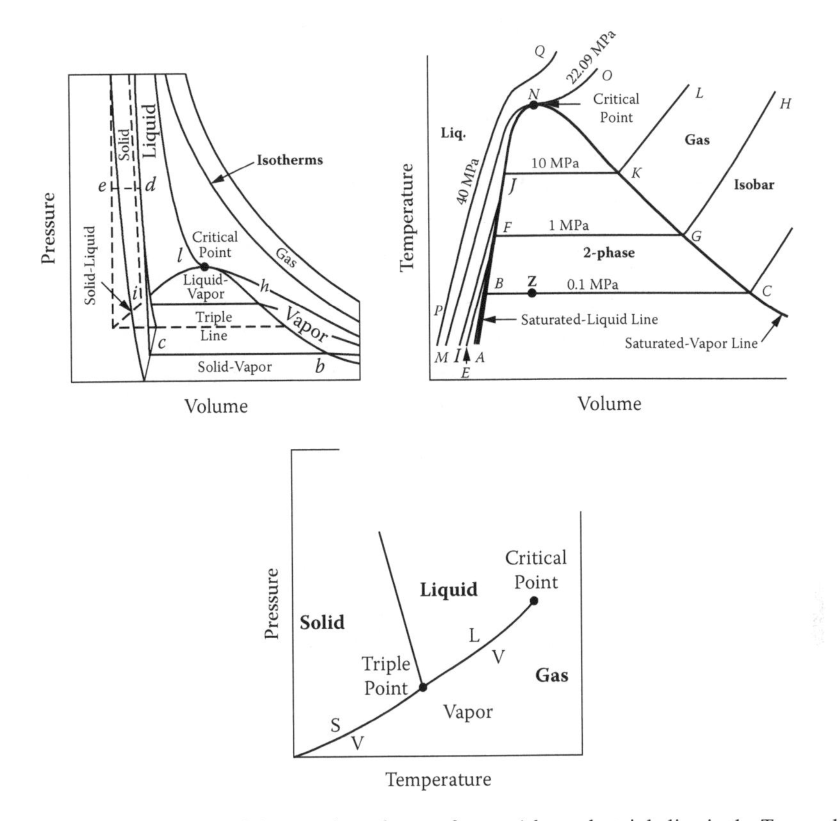

FIGURE 2.9 Projections of the equation of state of water (above the triple line in the T-v graph).

the left of this zone is the compressed-liquid single-phase region, and to the right of the envelope is the superheated-vapor region (labeled "gas" in the diagram). The left-hand and right-hand boundaries of the two-phase envelope are the locus of states of the saturated liquid and the saturated vapor, respectively. The maximum of the envelope is the critical point of water. At temperatures greater than that at this point, there is no distinction between liquid and vapor. Problem 2.9 is an example of analyzing a process using the T-v diagram.

The other lines in the T-v projection of Figure 2.9 are isobars for pressures ranging from 0.1 to 40 MPa. The slopes of the isobars in the compressed liquid region are all approximately the same and equal to the coefficient of thermal expansion of the liquid. In this range, Equation (2.18b) is a fair approximation of the EOS. In analogous fashion, isobars such as G-H are reasonably well described by the ideal gas law in the vicinity of H. However, near the saturation curve at G, a nonideal gas EOS such as the Van der Waals equation (Equation [2.3]) is required for an accurate analytic description of the p-v-T behavior of water vapor.

The isobars in the two-phase region of the T-v diagram of Figure 2.9 are horizontal. This means that the temperature is a function of pressure only, or equivalently, pressure is a function of temperature only. This p-T relationship in the two-phase region describes the vapor pressure of water. The p-T relationship is independent of the average volume of the mixture, which simply reflects the relative quantities of liquid and vapor, but does not affect the vapor pressure. For a one-component system with two phases present, the phase rule, Equation (1.21), permits only one degree of freedom. That is, only one thermodynamic property need be specified in order to fix all of the others.

Following the 0.1 MPa isobar in the *T-v* projection in Figure 2.9 illustrates the evolution of water as its average volume increases due to vaporization. Along the segment A-B, water is a single-phase liquid.* At point B, the liquid is just saturated with respect to the vapor at point C. Along the horizontal segment B-C, both temperature (100°C) and pressure (0.1 MPa) are constant. Moving from B to C represents vaporization of the liquid, terminating with the saturated vapor at C. At an intermediate point such as Z, the mass fraction of vapor in the mixture (i.e., the quality of the mixture) can be calculated from the definition of the average volume given by Equation (2.17). Here v_f and v_g are, respectively, the specific volumes of the saturated liquid at B and the saturated vapor at C. The average mixture volume v corresponds to the abcissa value at point Z. The ratio of the line segments B-Z and B-C gives the quality at point Z:

$$\frac{BZ}{BC} = \frac{v - v_f}{v_g - v_f} = \frac{xv_g + (1-x)v_f - v_f}{v_g - v_f} = x \tag{2.21}$$

Equation (2.21) is an example of the *lever rule* applied to phase diagrams. The name lever rule arises from the determination of the fraction of a phase in a two-phase mixture (in this case the quality x) as the ratio of the lengths of horizontal line segments on the diagram.

From point C in the *T-v* projection, the 0.1 MPa isobar continues into the superheated (gas) single-phase zone. The *p-v-T* surface here is the same as that described above for the G-H isobar.

The *p-T* projection in Figure 2.9 (right-hand panel) appears simpler than the other two views because isochores (constant-volume lines) have not been superimposed on it. In addition, the two-phase envelopes, prominent features of the other projections, are viewed on edge, and so are collapsed in the *p-T* view to curved lines. Similarly, the triple line in the *T-v* projection regains its proper status as the triple point in the *p-T* projection. The simpler appearance of the *p-T* projection is due to the elimination of the volume as a represented variable in the diagram.

The line labeled L/V in the *p-T* projection gives the temperature dependence of the vapor pressure of liquid water, also called the *vaporization curve*. The S/V line is the *sublimation curve*. It represents the equilibrium pressure of water vapor over ice. The S/L line is called the *melting line* and gives the combinations of pressure and temperature for which solid and liquid water coexist at equilibrium. Water is unusual among pure substances because its melting temperature is lowered as the pressure is increased. Most other substances behave in the opposite way; their melting

* The line AB in the *T-v* plot of Figure 2.9 is hidden in the saturated-liquid curve.

curves tilt to the right from the triple point rather than to the left. The unusual behavior of water is due to the higher density of the liquid compared to that of the solid.

Detailed analyses of the vaporization curve and the melting line are presented in Chapter 5.

2.7 THE STEAM TABLES

The graphical representations of Figures 2.8 and 2.9 are valuable for understanding the general features of the *p-v-T* properties of water but are of little use for quantitative analysis. For this purpose, extensive tables of the thermodynamic properties of water have been compiled. Such tabular information is available for many condensable substances in addition to water. For the latter, the property listings are called the *steam tables*, although they contain data for liquid water and ice as well as for the vapor phase. The steam considered in the tables is pure, undiluted by other gases such as air. Mixtures of water vapor and noncondensible gases are considered in Chapter 5.

In addition to listing the numerical values of the *p-v-T* properties illustrated in Figures 2.8 and 2.9, the steam tables present a complete listing of the thermal properties u, h and s. Tables 2.2–2.5 are condensed versions of these tabulations. Complete tables are in the appendices of References 4 and 5.

2.7.1 Two-Phase Saturated Liquid–Saturated Vapor

Table 2.2 contains properties of saturated water, both gaseous (denoted by subscript g) and liquid (subscript f). These are the numerical values along the heavy-line, two-phase envelope in the *T-v* projection of Figure 2.9. The first two columns of Table 2.2 gives the coordinates of the L/V line in the *p-T* projection of Figure 2.9. That is, these data represent the vapor pressure of liquid water. Like the line in right-hand graph of Figure 2.9, the *p-T* properties in the saturated-water table start at the triple point and end at the critical point.

The intervals of temperature in Table 2.2 are designed so that intermediate values can be obtained with reasonable accuracy by linear interpolation. For example, the vapor pressure at 73°C is determined by solving the following equation for p_{sat} (73°C):

$$\frac{p_{sat}(73°\text{C}) - p_{sat}(70°\text{C})}{p_{sat}(75°\text{C}) - p_{sat}(70°\text{C})} = \frac{p_{sat}(73°\text{C}) - 31.2}{38.6 - 31.2} = \frac{73 - 70}{75 - 70}$$

The result is p_{sat} (73°C) = 35.6 kPa.

The variable absent from Table 2.2 is the volume v. The reason is that in a two-phase region, Equation (2.20) shows that the system volume depends on the quality (steam mass fraction) in addition to the specific volumes v_f and v_g of the saturated liquid and vapor phases. The latter are listed in the third and fourth columns of the table. The quality x is not a thermodynamic property; it can vary between 0 and 1. Internal energy (u), enthalpy (h) and entropy (s) are handled in the same manner as described above for the specific volume (v).

TABLE 2.2
Saturated Two-Phase Water (Abbreviated)

°C	kPa	m^3/kg		kJ/kg		kJ/kg		kJ/kg-K	
T,	p	$v_f \times 10^3$	v_g	u_f	u_g	h_f	h_g	s_f	s_g
0.01	0.61	1.00	206	0	2375	0.0	2501	0	9.16
5	0.87	1.00	147	21.0	2381	21.0	2510	0.076	9.02
10	1.23	1.00	106	42.0	2389	42.0	2519	0.151	8.90
15	1.71	1.00	77.9	63.0	2396	63.0	2528	0.225	8.78
20	2.34	1.00	57.8	83.9	2402	83.9	2537	0.297	8.66
25	3.17	1.00	43.3	105	2409	105	2547	0.367	8.56
30	4.25	1.00	32.9	126	2416	126	2556	0.437	8.45
35	5.63	1.01	25.2	147	2423	147	2565	0.505	8.35
40	7.39	1.01	19.5	168	2429	168	2574	0.572	8.26
45	9.60	1.01	15.3	188	2436	188	2582	0.639	8.16
50	12.35	1.01	12.0	209	2442	209	2591	0.704	8.07
55	15.76	1.02	9.56	230	2449	230	2600	0.768	7.99
60	19.95	1.02	7.67	251	2456	251	2609	0.831	7.91
65	25.04	1.02	6.19	272	2462	272	2618	0.894	7.83
70	31.20	1.02	5.04	293	2469	293	2626	0.955	7.75
75	38.60	1.03	4.12	314	2475	314	2635	1.016	7.68
80	47.42	1.03	3.41	335	2482	335	2643	1.076	7.61
85	57.87	1.03	2.83	356	2489	356	2651	1.135	7.54
90	70.18	1.04	2.54	377	2494	377	2660	1.193	7.48
95	84.61	1.04	1.98	398	2500	398	2668	1.250	7.42
100	101.4	1.04	1.67	419	2506	419	2676	1.307	7.35
110	143.4	1.05	1.21	461	2518	461	2691	1.419	7.24
120	198.7	1.06	0.891	504	2529	504	2706	1.528	7.13
130	270.3	1.07	0.668	546	2540	546	2720	1.635	7.03
140	361.5	1.08	0.509	589	2550	589	2734	1.739	6.93
150	476.2	1.09	0.392	632	2559	632	2746	1.842	6.84
160	618.2	1.10	0.307	675	2568	675	2758	1.943	6.75
170	792.2	1.11	0.243	718	2576	719	2768	2.042	6.67
.	.	.	.	.	.	.	.	.	.
.	.	.	.	.	.	.	.	.	.
.	.	.	.	.	.	.	.	.	.
374.1	22054	3.11	0.0031	2016	2016	2084	2084	4.41	4.41

Notes:

The table starts at the triple point and ends at the critical point
$u_f = 0$ at the triple point is the reference internal energy but $h_f = 0.001$ kJ/kg here
$s_f = 0$ at the triple point is the reference for the entropy
u_g and h_g achieve maximum values at ~235°C

Example: Often the quality is implicit in the specification of a problem. For example, suppose we want to calculate the quality of 2 kg of water contained in a vessel of 3 m^3 volume at 75°C. The average specific volume of the mixture is $v = 3/2 = 1.5$ m^3/kg, and at 75°C, the specific volumes of the individual phases are (from Table 2.2) $v_f = 0.001$ m^3/kg and $v_g = 4.12$ m^3/kg. Solving Equation (2.20) for x yields:

$$x = \frac{v - v_f}{v_g - v_f} = \frac{1.5 - 0.001}{4.12 - 0.001} = 0.36$$

Overall conditions such as those in the above example must be approached with caution; the implicit assumption that two phases are present and that Table 2.2 is appropriate may not be valid. For instance if the mass of steam in the previous example were 0.5 kg instead of 2 kg, the mixture specific volume would have been 6 m^3/kg. This value is greater than that of the saturated vapor, and application of Equation (2.20) would give $x > 1$, which is impossible. In this case, the conclusion is that the system is a single-phase superheated vapor, and Table 2.3 rather than Table 2.2 must be used. The use of Table 2.3 will be taken up in due course.

The steam tables provide complete thermal property data in the form of u, h, and s. Enthalpy data are presented even though this property can be computed from the pressure, volume and internal energy using the definition $h = u + pv$. The difference $h_{fg} = h_g - h_f$ is called the *enthalpy of vaporization* and is an important thermodynamic property in its own right. In many problems only h_{fg}, and not h_f and h_g individually, is needed. In addition, h_{fg} modestly simplifies the calculation of average values in two-phase mixtures. Thus, the enthalpy of a liquid-vapor mixture of specified quality x is, by analogy to Equation (2.20),

$$h = xh_g + (1 - x)h_f = h_f + xh_{fg} \tag{2.22}$$

The later form is somewhat easier to use.

$$h = h_1 + xh_{fg} \tag{2.22a}$$

An interesting feature of the thermal-property data in Table 2.2 is the choice of reference states. In the discussion of u and h in Section 1.6, it was noted that either of these properties can be set equal to an arbitrary value at an arbitrary temperature and pressure, but both u and h cannot be so specified. Table 2.2 appears to violate this dictum because both u_f and h_f are equal to zero at the triple point ($T = 0.01$°C). If u_f is taken as the reference value of zero, then h_f at this condition should be 0 + (611.3 Pa)(0.001 m^3/kg) = 0.61 J/kg, or 0.00061 kJ/kg. Since the tables provide u_f and h_f in kJ/kg units only to three significant figures, listing h_f at the triple point as 0.0 is not a contradiction.

The value of zero assigned to the entropy of liquid water in Table 2.2 appears to violate the third law of thermodynamics. This law states that the zero of entropy is attained only for crystalline solids at the absolute zero temperature. On this basis, the entropy of liquid water at 0.01°C is clearly greater than zero. However, all calculations of thermodynamic processes involve differences in s (as well as of u and h), so the reference value chosen at the triple point cancels out. Again, the contradiction is not of practical importance.

Example: The average specific volume and enthalpy are given as $v = 0.12\ m^3/kg$ and $h = 1500$ kJ/kg. To be determined is whether the system is in the two-phase region and if so, to calculate the quality and the temperature.

To solve this problem, a two-phase system is assumed. At a series of temperatures in Table 2.2, the quality is calculated by two methods.

$$x(\text{from } v) = \frac{v - v_f}{v_g - v_f} \approx \frac{v}{v_g} \qquad x(\text{from } h) = \frac{h - h_f}{h_g - h_f}$$

At each temperature, v_g, v_f, h_f, and h_{fg} are read from Table 2.2. The following table shows the details of the calculation.

T, °C	V_g	*x* (from *v*)	h_f	h_g	*x* (from *h*)
150	0.392	0.332	632	2746	0.411
160	0.307	0.388	675	2758	0.396
170	0.243	0.494	719	2768	0.381

x(from h) is greater than x(from v) at 150 and 160°C, so the solution cannot lie between these two temperatures. However, at 170°C, x(from h) is smaller than x(from v), so the solution must lie between this temperature and 160°C. Linear interpolation provides the final solution.

$$\frac{x(\text{from } v) - 0.388}{0.494 - 0.388} = \frac{x(\text{from } h) - 0.396}{0.381 - 0.396}$$

$$\frac{T - 160}{170 - 160} = \frac{0.395 - 0.388}{0.494 - 0.388}$$

The result is T = 160.6°C.

This problem illustrates the potential benefits of a computer program containing the thermodynamic data for water in the steam tables and algorithms for calculating the state of the system given any two properties (or one property and the quality). Fortunately, a number of sites on the Web provide numerical values of water properties just by entering the conditions (e.g., saturated steam at 150°C).

2.7.2 Single-Phase Superheated Vapor—Compressed Liquid

Tables 2.3 and 2.4 list the properties of water in its single-phase *superheated vapor* and *compressed liquid* states. The term "superheated vapor" means that the vapor temperature is greater than the saturation temperature at a specified pressure. The isobars KL and GH in the *T-v* diagram of Figure 2.9 arise from the entries and equations in Table 2.3. The term "compressed liquid" applies to liquid water at temperatures below the saturation temperature at a specified pressure. Alternatively, if the temperature is specified, a compressed liquid is one that is overpressurized compared to the saturated liquid. The numbers in Table 2.4 correspond to the isobars

TABLE 2.3
Superheated Water Vapor (Abbreviated)

p = 10 kPa

T, K	v, m³/kg	h, kJ/kg	s, kJ/kg-K
318.8	**14.67**	**2584**	**8.15**
323	**14.87**	**2592**	**8.17**
373	**17.20**	**2688**	**8.85**
423	**19.51**	**2783**	**8.69**

For T > 423 K:
use ideal-gas law for v
$h = 2783 + 1.62 \times (T - 423) + 3.33 \times 10^{-4} (T^2 - 423^2)$
$s = 8.69 + 1.62 \times \ln(T/423) + 6.65 \times 10^{-4} (T - 423)$

p = 50 kPa

T, K	v, m³/kg	h, kJ/kg	s, kJ/kg-K
354.3	3.24	2645	7.59
373	3.42	2682	7.70
423	3.89	2780	7.94
473	4.36	2878	8.16

For T > 473 K:
use ideal-gas law for v
$h = 2878 + 1.58 \times (T - 473) + 3.59 \times 10^{-4} (T^2 - 473^2)$
$s = 8.16 + 1.58 \times \ln(T/473) + 7.17 \times 10^{-4} (T - 473)$

p = 200 kPa

T, K	v, m³/kg	h, kJ/kg	s, kJ/kg-K
393.2	0.87	2706	7.13
423	0.96	2769	7.28
473	1.08	2871	7.51
523	1.20	2971	7.71

For T > 523 K:
use ideal gas law for v
$h = 2971 + 1.67 \times (T - 523) + 3.14 \times 10^{-4}(T^2 - 523^2)$
$s = 7.71 + 1.67 \times \ln(T/523) + 6.28 \times 10^{-4} (T - 523)$

p = 500 kPa

T, K	v, m³/kg	h, kJ/kg	s, kJ/kg-K
424.8	**0.37**	**2748**	**6.82**
473	**0.43**	**2856**	**7.06**
523	**0.47**	**2961**	**7.27**
573	**0.52**	**3064**	**7.46**

For T > 573 K:
use ideal gas law for v
$h = 3065 + 1.71 \times (T - 573) + 2.94 \times 10^{-4} (T^2 - 573^2)$
$s = 7.71 + 1.71 \times \ln(T/573) + 5.87 \times 10^{-4} (T - 573)$

p = 1000 kPa

T, K	v, m³/kg	h, kJ/kg	s, kJ/kg-K
452.9	0.194	2777	6.59
473	0.206	2828	6.70
523	0.233	2943	6.93
573	0.258	3052	7.12

For T > 573 K:
use ideal gas law for v
$h = 3052 + 1.75 \times (T - 573) + 2.82 \times 10^{-4} (T^2 - 573^2)$
$s = 7.12 + 1.75 \times \ln(T/573) + 5.63 \times 10^{-4} (T - 573)$

p = 3000 kPa

T, K	v, m³/kg	h, kJ/kg	s, kJ/kg-K
506.9	0.067	2803	6.19
523	0.071	2857	6.29
573	0.081	2994	6.54
623	0.091	3116	6.75

For T > 623 K:
use ideal gas law for v
$h = 3116 + 1.82 \times (T - 623) + 2.63 \times 10^{-4} (T^2 - 73^2)62$
$s = 6.75 + 1.82 \times \ln(T/623) + 5.25 \times 10^{-4} (T - 623)$

Note: The first row in each table gives saturation values.

PQ, MN, etc. in the *T-v* diagram of Figure 2.9. In Table 2.3, the first row in each subtable corresponds to saturation conditions. In Table 2.4, the last row of the subtables correspond to saturation.

Formulas replace numerical entries for condition beyond the range of the latter. For example, in the $p = 10$ kPa subtable in Table 2.3, the ideal-gas law very accurately yields the specific volume v for temperatures above 423 K. In this same range, the heat capacity of water vapor is linearly-dependent on temperature. This linear function gives h and s from Equations (2.14) and (3.22):

$$h = h_{ref} + \int_{T_{ref}}^{T} C_P(T')dT' \qquad s = s_{ref} + \int_{T_{ref}}^{T} \frac{C_P(T')}{T'}dT'$$

For the purpose of analytically extrapolating h and s, the reference temperature, along with the property value at this temperature, can be arbitrarily chosen. T_{ref} in these tables has been selected as the temperature that separates the tabular values from the analytical representations of the two properties.

The formulas for h and s in Table 2.4 result from a quadratic variation of C_P with temperature.

In using the single-phase tables, the complications associated with the quality of a two-phase mixture are absent. Given any two of the five thermodynamic properties treated in these tables, the others can be found. By the way that the tables are constructed, determination of the state is most easily done if pressure and temperature are given, although even in this case, double interpolation in the table is generally required. If the specified pair of variables is v and h, for example, determining the state of the system requires considerable trial-and-error hunting in either Table 2.3 or Table 2.4. For problems involving single-phase water, the computer-based steam table calculation is a vast improvement over the hand calculation. Problems 2.7, 2.8, and 2.12 to 2.15 involve manipulations of the steam tables.

The tabular mode of thermal-property data representation in the single-phase tables for water is quite different from the usual method of presenting EOS information for gases or single condensed phases (solids or liquids). For gases, p-v-T behavior more commonly expressed by an equation of state such as the Van der Waals formula. Because such formulas have a limited number of parameters, their accuracy for steam is not as high as the values listed in Table 2.3. Problem 2.10 shows how to use data from Table 2.3 to determine the Van der Waals coefficients for steam.

For nonvolatile liquids and solids, the p-v-T equation is contained in the coefficient of thermal expansion and the coefficient of compressibility. Isobars in the liquid region of the T-v plot of Figure 2.9 all have about the same slope, which means that the coefficient of thermal expansion of compressed water is approximately constant. Examination of Table 2.4 shows that α varies by about a factor of two over the p and T ranges covered. Similarly, the coefficient of compressibility of water also varies about twofold in the p-T range covered by Table 2.4.

Tables 2.3 and 2.4 do not provide α and β nor the specific heats C_V and C_{VP}. However, these properties can be estimated from the steam tables by numerical differentiation, as shown in the following examples from Table 2.4.

Example: The coefficient of thermal expansion at 5 MPa between 40 and 60°C is obtained by using the formula for v below the 5 MPa subtable:

$$\alpha = \frac{1}{v \times 10^3}\frac{\Delta(v \times 10^3)}{\Delta T} = \frac{1}{1.0}\frac{1.018 - 1.009}{60 - 40} = 4.5 \times 10^{-4}\ °\mathrm{C}^{-1}$$

TABLE 2.4
Compressed Liquid Water (Abbreviated)

p = 5 MPa

T, K	v × 10³, m³/kg	h, kJ/kg	s, kJ/kg-K
493	1.19	944	2.51
513	1.23	1038	2.70
533	1.28	1134	2.88
536.9	1.29	1148	2.92

For $T < 493$ K:

$v \times 10^3 = 1.23 - 1.78 \times 10^{-3}T + 3.43 \times 10^{-6}T^2$

$h = 89 + 5.48(T - 293) - 4.1 \times 10^{-3}(T^2 - 293^2) + 4.2 \times 10^{-6}(T^3 - 293^3)$

$s = 0.30 + 5.48\ln(T/293) - 8.2 \times 10^{-3}(T - 293) + 6.3 \times 10^{-6}(T^2 - 293^2)$

p = 10 MPa

T, K	v × 10³, m³/kg	h, kJ/kg	s, kJ/kg-K
513	1.22	1083	2.69
533	1.27	1134	2.87
553	1.32	1235	3.06
573	1.40	1343	3.24
584.0	1.45	1408	3.36

For $T < 513$ K:

$v \times 10^3 = 1.22 - 1.77 \times 10^{-3}T + 3.40 \times 10^{-6}T^2$

$h = 93 + 5.62(T - 293) - 4.5 \times 10^{-3}(T^2 - 293^2) + 4.6 \times 10^{-6}(T^3 - 293^3)$

$s = 0.29 + 5.62\ln(T/293) - 9.0 \times 10^{-3}(T - 293) + 7.0 \times 10^{-6}(T^2 - 293^2)$

p = 20 MPa

T, K	v × 10³, m³/kg	h, kJ/kg	s, kJ/kg-K
533	1.25	1134	2.85
553	1.30	1232	3.03
573	1.36	1334	3.21
593	1.45	1446	3.40
613	1.57	1572	3.61
633	1.82	1740	3.89
638.8	2.04	1827	4.01

For $T < 533$ K:

$v \times 10^3 = 1.27 - 2.02 \times 10^{-3}T + 3.72 \times 10^{-6}T^2$

$h = 103 + 5.92(T - 293) - 5.35 \times 10^{-3}(T^2 - 293^2) + 5.3 \times 10^{-6}(T^3 - 293^3)$

$s = 0.29 + 5.92\ln(T/293) - 1.07 \times 10^{-2}(T - 293) + 8.0 \times 10^{-6}(T^2 - 293^2)$

p = 50 MPa

T, K	v × 10³, m³/kg	h, kJ/kg	s, kJ/kg-K
553	1.24	1230	2.95
573	1.29	1324	3.12
593	1.34	1421	3.29
613	1.40	1523	3.46
633	1.48	1631	3.63
653	1.59	1747	3.81
*	—	—	—

For $T < 553$ K:

$v \times 10^3 = 1.28 - 2.0 \times 10^{-3}T + 3.42 \times 10^{-6}T^2$

$h = 130 + 4.99(T - 293) - 2.95 \times 10^{-3}(T^2 - 293^2) + 3.1 \times 10^{-6}(T^3 - 293^3)$

$s = 0.28 + 4.99(T/293) - 5.9 \times 10^{-3}(T - 293) + 4.7 \times 10^{-6}(T^2 - 293^2)$

Note: The last row in each table gives saturation values.

Example: The coefficient of compressibility at 240°C (513 K) between 5 and 10 MPa is determined from the numerical entries in the 5 MPa and 10 MPa subtables:

$$\beta = -\frac{1}{v \times 10^3}\frac{\Delta(v \times 10^3)}{\Delta p} = -\frac{1}{1.2}\frac{1.22 - 1.23}{10 - 5} = 1.7 \times 10^{-3}\ MPa^{-1}$$

Because $v \times 10^3$ is given to only three significant figures, the above value of β is not very accurate. Using values from the tabulation in Reference 5, $\beta = 1.3 \times 10^{-3}$.

Example: The heat capacity at constant pressure of liquid water at 5 MPa between 220°C (493 K) and 100°C (513 K) is obtained by the approximating the derivative dh/dT by:

$$C_P = \frac{\Delta h}{\Delta T} = \frac{1038 - 944}{240 - 220} = 4.20 \frac{\text{kJ}}{\text{kg-°C}}$$

This value of C_P is equivalent to 1 cal/g-°C. The old calorie unit is the quantity of heat needed to raise the temperature of 1 gram of water by 1 degree Celsius.

The single-phase tables often require double interpolation, an example of which is given below.

Example: The enthalpy of steam at 230°C (503 K) and 240 kPa is determined in two steps:

Step 1: Interpolate on 503 K at bounding pressures:

p = 200 kPa		p = 500 kPa	
T,K	h	T,K	h
523	2971	523	2961
503	h_{200}	503	h_{300}
473	2870	473	2856
h = 2931		h = 2919	

Step 2: Interpolate on pressure at 503 K:

p	h
500	2919
240	h
200	2931
h = 2929	

2.7.3 Ice–Vapor

The last of the steam tables, Table 2.5, is an abbreviated summary of the thermodynamic properties of ice and vapor along the S/V line in the p-T projection of Figure 2.9. The subscript i denotes ice. This table is utilized in Problem 2.2.

TABLE 2.5
Saturated Two-Phase Ice-Water Vapor (Abbreviated)

°C	kPa	m³/kg		kJ/kg		kJ/kg-K	
T	p	$v_i \times 10^3$	v_g	h_i	h_g	s_i	s_g
0.01	0.612	1.091	206	−333	2501	−1.22	9.16
−4	0.437	1.090	284	−342	2493	−1.25	9.28
−8	0.310	1.090	395	−350	2486	−1.28	9.41
−12	0.217	1.089	554	−358	2478	−1.31	9.55
−16	0.151	1.088	788	−366	2471	−1.34	9.69
−20	0.103	1.087	1131	−374	2464	−1.37	9.84
−24	0.085	1.087	1645	−381	2456	−1.41	9.99
−28	0.047	1.086	2421	−389	2449	−1.44	10.14
−32	0.031	1.086	3611	−397	2441	−1.47	10.30
−36	0.020	1.085	5460	−404	2434	−1.50	10.47
−40	0.013	1.084	8377	−412	2427	−1.53	10.64

Note: The reference states for the enthalpy and entropy are the same as in Table A.2.

PROBLEMS

2.1 One kilogram of steam, initially at 200°C and 50 kPa (state 1), is cycled through the following processes:

1→2. Heat to 500°C at constant pressure.
2→3. Cool to 400°C with simultaneous pressure increase to 500 kPa along a straight-line path in p-v coordinates.
3→4. Return to 200°C at 200 kPa via two subprocesses: decrease pressure at constant temperature, then decrease temperature at constant pressure. (Hint: you will need to perform a numerical integration for the first subprocess of step 3→4.)

What are the volume changes and heat requirements for each step? Draw the cycle in p-T coordinates.

2.2 The specific volume of a three-phase mixture of ice, liquid water, and water vapor at the triple point is v m³/kg and the specific internal energy is u kJ/kg.

(a) Using Table 2.5, prepare a table giving v and u for the three phases.
(b) For given u and v what are the equations that determine the mass fractions of water in the three phases (x, y, z for vapor, liquid and solid, respectively)? Assume that the specific volumes of the condensed phases are negligible compared to that of the vapor.
(c) Plot an existence diagram that shows the region of permissible values of u and v.

2.3 A 10-kg specimen of copper is quenched at constant pressure from 700°C in a 100-liter, adiabatic water bath initially at 20°C. Neglecting the heat capacity of the container and vaporization of water, calculate the final temperature of the system using heat capacities from Figure 2.6:

(a) C_V
(b) C_p

2.4 The isotherm passing through the critical point in the *p-v* diagram of Figure 2.9 has the following mathematical properties:

$$(\partial p / \partial v)_{T_C} = (\partial^2 p / \partial v^2)_{T_C} = 0.$$

(a) Use the values of the Van der Waals constants a and b given in the text to determine the critical pressure and temperature of N_2. (Hint: to solve the equations, use the dimensionless critical volume $X = V_c/b$ and dimensionless critical temperature $Y = T_c/(a/Rb)$.)
(b) Compare the values of p_c and T_c so calculated with the values reported in the *Handbook of Physics and Chemistry* (CRC Press, 2006) of $p_c = 3.35$ MPa, $T_c = 126$ K.

2.5 An initially deflated balloon is connected by a valve to a tank containing helium at 1 MPa and 20°C. When the valve is opened and the gas equilibrates between the tank and the balloon, the diameter of the balloon is 4 m, the pressure is 4 kPa and the temperature is unchanged. What is the volume of the tank?

2.6 Nitrogen behaves as a Van der Waals nonideal gas. The experimentally-determined compressibility of N_2 is shown in Figure 2.3.

(a) Using Equation (2.4), determine the coefficients a and b from points A and B in Figure 2.3.
(b) Using the parameters determined in part (a), compare the accuracy of the approximate form, Equation (2.5a), with the full form, Equation (2.3), at 150 K and 1 MPa.

2.7 A pump delivers 50 kg/s of water at 300°C and 20 MPa.

(a) What is the volume flow rate in m³/s?
(b) If the properties of saturated liquid water at 300°C were used in the calculation, what is the error in the calculation of (a)?
(c) If the properties of saturated liquid water at 20 MPa were used, what would the error be?

2.8 Determine whether water in the following states is a compressed liquid, a superheated vapor, or a two-phase mixture of liquid and vapor:

Temp . °C	Pressure, kPa	Specific vol., m³/kg
150	120	—
350	—	0.4
—	160	0.4
200	110	—

2.9 (a) A container contains half liquid water and half vapor (by volume) at 1 atm pressure. Estimate the quality of this mixture.

(b) A steam/water mixture of 50% quality is contained in a rigid vessel. If heat is added to this system, does the quality increase or decrease?

(c) A 15-liter rigid vessel contains 10 kg of water (liquid + vapor). If the vessel is heated, will the liquid level rise or fall? At what temperature does one of the two phases disappear?

2.10 For the p-v-T values in table 2.2 at 100°C and 150°C, determine:

(a) The percentage deviations of v_g from the value obtained from the ideal gas law.

(b) The constants a and b in the Van der Waals equation.

2.11 An ideal gas is contained in a cylinder by a piston. The cylinder cross-sectional area is 0.2 m^2 and is equipped with a set of stops 1 m from the bottom. The gas initially in the cylinder is at 200 kPa and 500°C and holds the piston at a height of 2 m.

(a) The gas is cooled until the piston reaches the stops. What is the temperature at this stage?

(b) The gas is further cooled to 20°C. What is the pressure in this state?

2.12 The piston in the cylinder in the sketch is attached to a rigid wall by a spring, which has a force constant k = 10^4 N/m. The piston has an area of 0.5 m^2. The system contains 1 kg of steam of 90% quality at 110°C. Under these conditions, the spring is at its equilibrium length. As heat is added to the cylinder, the piston rises and compresses the spring.

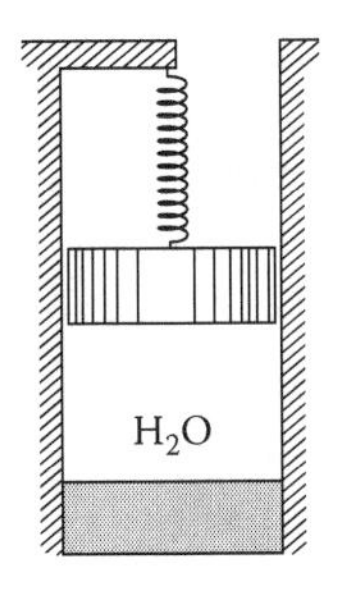

(a) Assuming the final state is superheated vapor, calculate the pressure in the cylinder when the temperature reaches 200°C.

(b) What would the final pressure be if steam were assumed to be ideal?

2.13 A vessel is filled with liquid water at 250°C. At the top of the vessel is a small capillary tube in which the liquid forms a concave (upward-curving) surface with a 1 mm radius of curvature. The space in the capillary above the liquid surface is occupied by steam. The distance from the bottom of the vessel to the vapor–liquid interface is 1 m. What is the absolute pressure at the bottom of the tank? The surface tension of water is 0.1 N/m.

2.14 A steam of water mixture is contained in a rigid container at a pressure of 2 MPa. The initial quality x is chosen so that the water-vapor mixture reaches the critical point when heated.

(a) On *T-v* axes, sketch the following:
1. The saturation lines for the liquid and the vapor, and the two-phase region (i.e., the two-phase vapor dome). Label the critical point and then initial and final temperatures. Sketch several isobars.
2. The process path connecting the initial and final (i.e., critical) states.

(b) What is the initial quality of x?

2.15 Water undergoes the following cyclic process:
State 1 (sat'd liquid at 300°C) → (const.pressure)→ state 2 (sat'd vapor) → (const. volume) → State 3 (2-phase at 0.075 MPa) → (const.pressure)→ state 4 (sat'd liquid) → (along sat'd liquid line) → State 1

(a) Sketch this cycle on a *T-v* diagram along with the two-phase dome of the water equation of state. Label the four states and give their *T,v* coordinates on the sketch.
(b) Determine the quality and internal energy of the mixture at state 3.
(c) What is the mass density of the liquid phase at state 3?
(d) For the cycle, what is the change in the enthalpy of water. Explain your reasoning.

2.16 By taking the appropriated mixed derivatives of α and β defined in Equation (1.2), show that the values of $\bar{\alpha}$ and $\bar{\beta}$ given in the figure on p. 61 are internally consistent.

2.17 Derive the equation for the coefficient of thermal expansion (α) of a gas obeying the Van der Waals EOS. Assume that in the definition of α only the v term can be approximated by the ideal gas law.

2.18 Determine the following:

(a) The specific volume of a steam-water mixture of 50% quality at 0.1 MPa.
(b) The temperature at which the liquid phase disappears when heat is added to this system in a rigid vessel.

2.19 In the diagram below, water in the upper vessel (1) is at 20°C. The height difference between the surface of the water in (1) and the inlet to the large rigid vessel (2) is h. The temperature of the two-phase mixture in vessel (2) is 100°C. The valve between the two vessels is opened briefly and then closed. The contents of (2) are maintained at 100°C during this process.

(a) Does water leave or enter (2)?
(b) Does the pressure in (2) increase, decrease, or remain the same?

Explain your choice for each part.

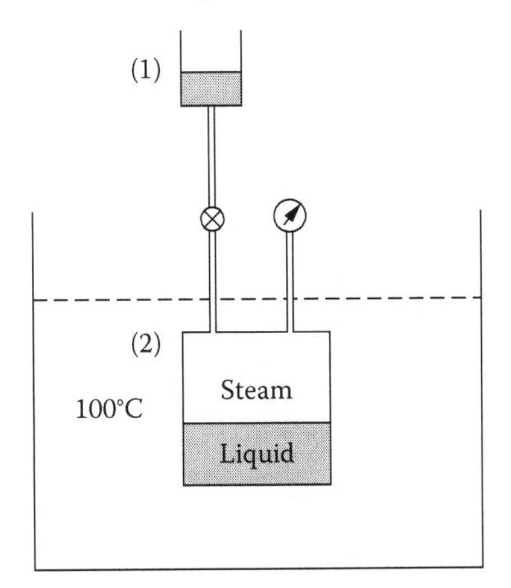

2.20 A 1 m^3 rigid vessel contains 4 kg of water. Heat is added until the temperature is 150°C. Using the *p-v* diagram below, estimate:

(a) the pressure
(b) the quality

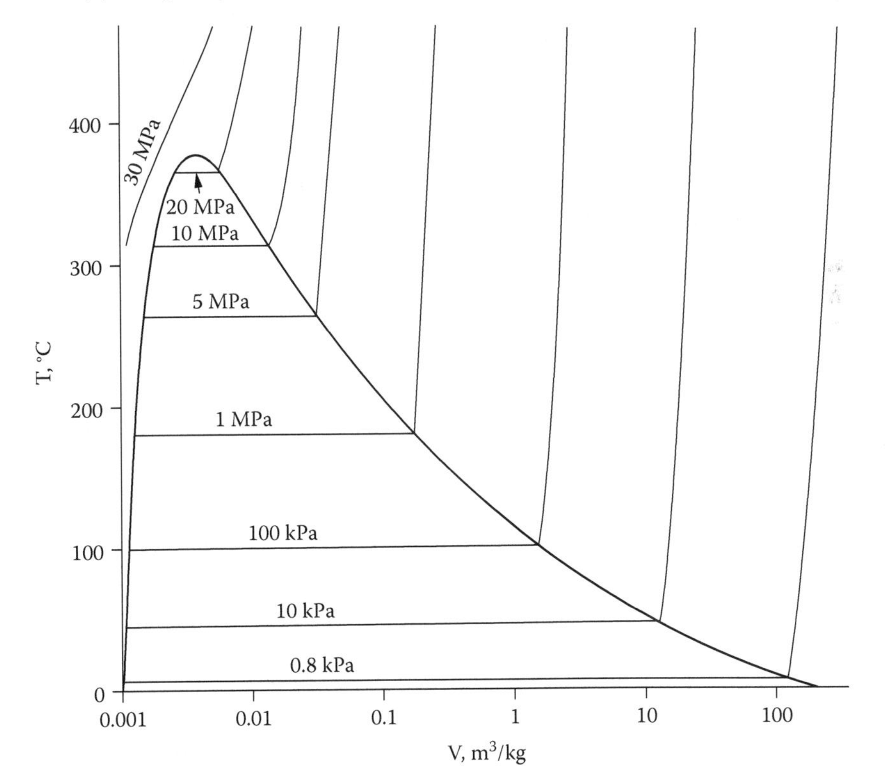

2.21 A 5-liter, insulated vessel contains 3 kg of water. A valve and a pressure gauge are connected to the vessel. With the valve closed, the pressure gauge reads 1.5 MPa.

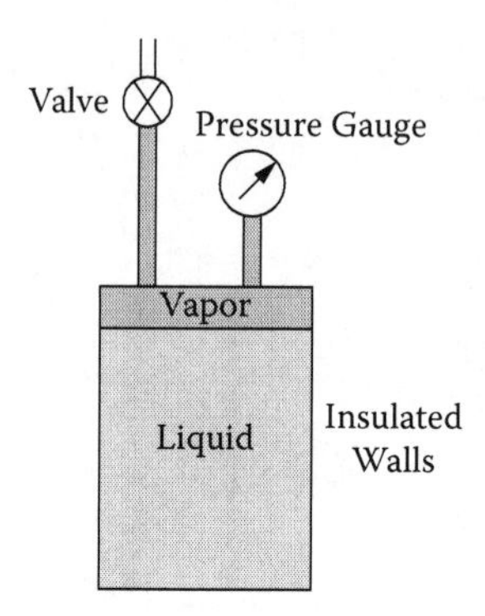

The valve is slightly opened and steam slowly (reversibly) escapes until the pressure has dropped to 0.75 MPa. Complete the following table.

State	*p*, MPa	*T*. °C	*m*, kg	x	s	V_f, m³
Initial	1.5		3.0			
Final	0.75					

2.22 The graph shows the heat capacity at constant pressure of a common gas.

(a) Identify the modes of energy storage represented by the numbered zones.
(b) State reasons for rejecting or accepting each of the following gases as the one in the graph: Ar; O_2; H_2O; CO_2.

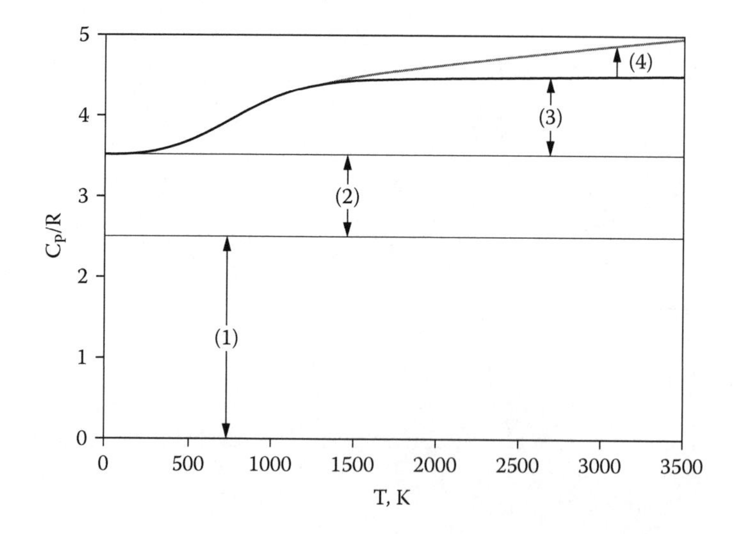

2.23 The following data for liquid water are taken from table 2.4.

T, °C	Specific Volume, m³/kg × 10³		
	v (at 20 MPa)	v (at 30 MPa)	v (at 40 MPa)
40	0.999	0.995	0.987
50	1.008	1.004	0.996
60	1.020	1.016	1.007

Find the coefficient of:

(a) Thermal expansion at 50°C and 25 MPa.
(b) Compressibility at 50°C and 40 MPa.

2.24 An adiabatic vessel contains 1 mole of subcooled liquid tin at 495 K. Scratching the vessel wall causes the tin to revert to its equilibrium melting temperature of 505 K, accompanied by solidification of a fraction f of the liquid. Determine f using the following physical data for tin:

Heat of fusion = 7.1 kJ/mole
$C_{PL} = 34.7 - 9.2 \times 10^{-3}\ T$
$C_{PS} = 18.5 + 2.6 \times 10^{-2}\ T$ J/mole-K

Hint: Use a two-step path for the process: one temperature-increase step at constant fraction solid and one isothermal step during which solidification occurs. It does not matter in which order these steps are taken. An efficient method of treating these paths is to first derive the formulas for $h_S(T)$ and $h_L(T)$.

2.25 An elemental solid has a heat capacity of 1 J/g-K. What is the atomic weight of the element?

2.26 Consider the tin-crystallization process of Problem 2.24.

(a) Is the entropy change positive or negative? Explain your reasoning.
(b) Write the equations for the temperature dependences of the entropies of the liquid and solid.
(c) Calculate the entropy change for both paths.

REFERENCES

Abbott, M. and H. C. Van Ness. 1989. *Theory and Problems in Engineering Thermodynamics.* New York: McGraw-Hill.
Cengel, Y. & M. Boles, *Thermodynamics*, 5th Ed., McGraw-Hill (2006).
Potter, M. and C. Somerton. 1993. *Theory and Problems in Engineering Thermodynamics.* New York: McGraw-Hill.

Prausnitz, J. et al. 1986. *Molecular Thermodynamics of Fluid-Phase Equilibria*, 2nd ed. Englewood Cliffs: Prentice-Hall.

Van Ness, H. C. 1983. *Understanding Thermodynamics*. Mineola, NY: Dover.

Van Wylen, G. R. Sonntag & C. Borgnakke, *Fundamentals of classical thermodynamics*, 4th Ed, Wiley (1994) or later editions.

3 Application of the First and Second Laws to Processes in Closed Systems

3.1 SCOPE

In this chapter, the two basic laws of thermodynamics are applied to changes in closed-system properties and to the work and heat requirements for several common processes. These include the four discussed in Section 1.4, namely, isothermal, isobaric, isochoric, and isentropic processes. These processes have practical applications in their own right, and they also constitute component steps in thermodynamic cycles considered in Chapter 4.

In the analyses, only reversible pV work is considered, so that the expression for the work done by the system is:

$$w = \int_{v_1}^{v_2} pdv \tag{3.1}$$

where the subscripts 1 and 2 denote the initial and final states of the system. These states are specified, as is the *restraint* that fixes the path between them. If the paths are sketched on p-v coordinates, the area under the curve is the work performed by the system. Because the work is considered to be performed reversibly, the pressure in Equation (3.1) is the common pressure of the system and surroundings. When the system is a fluid confined in a piston/cylinder apparatus, the pressure of the surroundings may be augmented by weights or springs (see Problems 3.3 and 3.14).

For the sake of simplicity, calculations are based on one mole of the substance that comprises the system. Thus, v is the specific volume and w is the work performed per mole. In a similar manner, heat absorbed from the surroundings is written on a per-mole basis and denoted by q.

In applications involving only the first law, the starting point is Equation (1.4) written on a per-mole basis:

$$\Delta u = u_2 - u_1 = q - w \tag{3.2}$$

with w given by Equation (3.1).

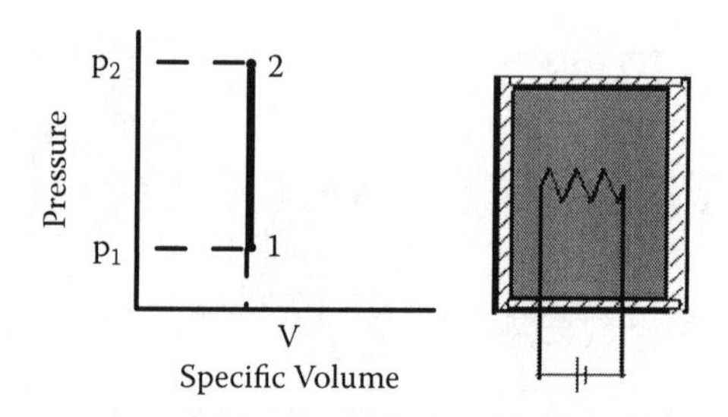

FIGURE 3.1 An isochoric process. The contents of the rigid vessel are heated respectively.

Numerical examples of each of the four "iso" processes are presented. The examples will be based on the properties of ideal gases, water, and simple solids. An example of a process that does not follow any of the four "iso" restraints is also analyzed.

3.2 THE ISOCHORIC PROCESS

An isochoric process in p-v coordinates is shown in Figure 3.1. Because the system boundary is rigid, no work is performed ($w = 0$). Passage of the system from initial state 1 to final state 2 is induced by adding a quantity q of heat per mole of substance. The initial state is specified by a property pair such as p_1 and v, where v is the constant specific volume maintained during the process.

The first law for the isochoric process is:

$$u_2 - u_1 = q \tag{3.3}$$

The final state is determined by v and u_2. To translate this information to T_2 and p_2, both p-v-T and thermal equations of state for the substance involved are needed.

The diagram in Figure 3.1 shows a rigid vessel with an internal electrical resistor inside to heat the contents. The first law (or energy conservation) for this system is:

$$E_{el} = I^2R\ t = \Delta u = q$$

where I is the electric current, R is the resistance and t is time.

Example 1. 1 mole of air (an ideal gas) initially at p_1 = 5 MPa and T_1 = 60°C heated in a rigid vessel until p_2 = 15 MPa. How much heat is required?

The vessel volume is obtained from the ideal-gas law: $v = RT_1/p_1$:

$$v = \frac{8.314\ Pa - \text{m}^3/\text{mole-K} \times 333\ \text{K}}{5 \times 10^6\ Pa} = 5.5 \times 10^{-4}\ \text{m}^3$$

The final temperature is $T_2 = p_2v/R = 15 \times 10^6 \times 5.5 \times 10^{-4}/8.314 = 999$ K = 726°C.

For an ideal gas, u is independent of pressure. Assuming a temperature-independent heat capacity, the increase in internal energy is related to the increase in temperature by:

$$q = u_2 - u_1 = C_V(T_2 - T_1) = \tfrac{5}{2} \times 8.314\,(726 - 333) = 8170 \text{ J/mole} \tag{3.4}$$

Problem 3.1a explores a slightly modified version of this type of process.

Example 2. Compressed liquid water—the same process as in Example 1.

For p_1 = 5 MPa, T_1 = 60°C, steam table A.4 gives v = 0.001015 m³/kg and u_1 = 250.2 kJ/kg. For the same specific volume and a final-state pressure of 15 MPa, interpolation in Table 2.4 gives T_2 = 67.7°C and u_2 = 280.3 kJ/kg. Finally, substitution of u_1 and u_2 into Equation (3.3) gives:

$$q = 280.3 - 250.2 = 30.1 \text{ kJ/kg or } 542 \text{ J/mole}$$

Note that compressing liquid water requires a much smaller heat addition than the same process with an ideal gas.

Example 3. Simple solid—the same process as in Example 1; properties from Table 2.1.

A simple solid means one (a) for which the effect of pressure on internal energy is negligible; and (b) that obeys the equation of state given by Equation (2.18b).

Because $v_2 = v_1$ for an isochoric process, the EOS gives:

$$T_2 - T_1 = \frac{\beta}{\alpha}(p_2 - p_1) = \frac{10^{-5}}{10^{-5}}(15 - 5) = 10 \text{ K}$$

This solid obeys Equation (3.3) (with $C_V = 3R$), so that

$$q = u_2 - u_1 = C_V(T_2 - T_1) = 3 \times 8.314 \times 10 = 250 \text{ J/mole}$$

A comparison of these examples is shown in Table 3.1.

TABLE 3.1
Property Changes Due to Constant-Volume Compression of 1 Mole from 5 to 15 MPa

Substance	v, m³/mole*	ΔT, °C	Δu, J/mole
Air	5.5×10^{-4}	666	8170
Water	1.8×10^{-5}	7.7	542
Simple solid	$\sim 10^{-5}$	10	250

* at 60°C and 5 MPa

The order of the heat inputs (or Δu) is the inverse of the coefficients of compressibilities (β) of the three substances. The gas is highly compressible and so requires significant heat addition to be pressurized over the 10 MPa range. The solid, on the other hand, is quite incompressible, so achieving the same pressure increase requires a much smaller heat addition.

The temperature increases in the table are inversely proportional to the specific heats of the substances. For gas:water:solid, the relative values of C_V are 1:6:2. Water has a high heat capacity, which is why its ΔT for the isochoric pressurization process is the smallest of the three substances. For the same reason water is an effective coolant in innumerable industrial processes.

Other examples of heating of a system in a rigid container are given in Problems 3.7, 3.8, and 3.11.

3.3 THE ISOTHERMAL PROCESS

Figure 3.2 shows the path of a process during which the temperature is maintained constant. The detailed shape of the curve depends on the equation of state of the substance. The area under the curve between v_1 and v_2 is the work performed by the system (Equation [3.1]).

While work is done on the system by compressing the piston with a spring, heat is removed by the thermal reservoir, which maintains the system at constant temperature. For most substances, constant temperature means constant internal energy so the first law reduces to:

$$w = q$$

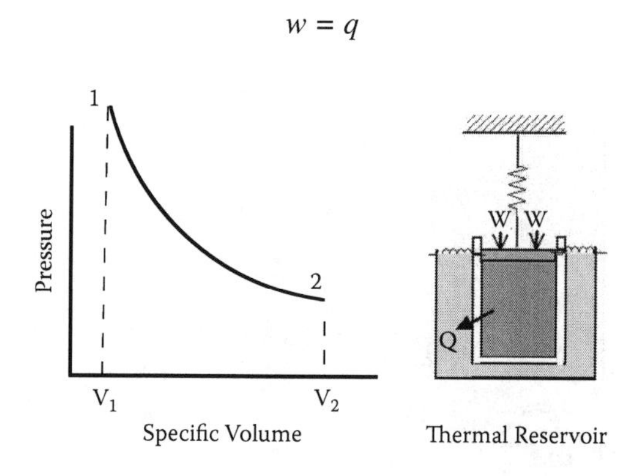

FIGURE 3.2 An isothermal process.

Example 1. 1 mole of an ideal gas at a constant temperature of 800°C compressed from 0.2 MPa to 0.6 MPa.

For this case, the curve in Figure 3.2 is hyperbolic, or $p \propto 1/v$. The work per mole is obtained by using $p = RT/v$ in Equation (3.1) and integrating from v_1 to v_2:

$$q = w = RT\int_{v_1}^{v_2} \frac{dv}{v} = RT \ln\left(\frac{v_2}{v_1}\right) \tag{3.5}$$

For the specified initial state, the ideal gas law gives $v_1 = 0.0149$ m³/mole. For the final state, the gas law gives $v_2 = 0.0446$ m³/mole.

Equation (3.5) gives $w = q = 9800$ J/mole.

Example 2. Superheated water vapor.

For this case, the temperature and the initial and final pressures are chosen to be the same as those in Example 1. The specific volumes and internal energies of these states are obtained from steam table A.3:

$$v_1 = 0.825 \text{ m}^3/\text{kg } (0.0148 \text{ m}^3/\text{mole}); \qquad u_1 = 3661.8 \text{ kJ/kg}$$

$$v_2 = 2.475 \text{ m}^3/\text{kg } (0.0446 \text{ m}^3/\text{mole}); \qquad u_2 = 3663.1 \text{ kJ/kg}$$

The above numbers show that in this particular process, the p-v-T behavior of steam very closely follows the ideal gas law. However, the slight change in the internal energy is a sign of residual nonideality. In calculating the heat absorbed using the first law, the ideal gas law can be used to compute the work, but Δu cannot be neglected in Equation (3.2). As in Example 1, $w = 9800$ J/mole and $u_2 - u_1 = (3663.1 - 3661.8) \times 18 =$ 23 J/mole. From Equation (3.2), the heat absorbed is:

$$q = 23 + 9800 = 9823 \text{ J/mole}$$

Example 3. Solid. See Problem 3.10.

3.4 THE ISOBARIC PROCESS

Figure 3.3 shows a generic constant-pressure, or isobaric, process. The means of energy input is the same as in the isochoric examples. The rigid container is replaced by a cylinder with a frictionless, massless piston, which assures equality of the system pressure and the external pressure.

In p-v coordinates, an isobaric process appears as a horizontal line as in Fig. 3.3. The work done is the rectangular area between v_1 and v_2. Because p is constant, Equation (3.1) reduces to $w = p(v_2 - v_1)$ and Equation (3.2), yields:

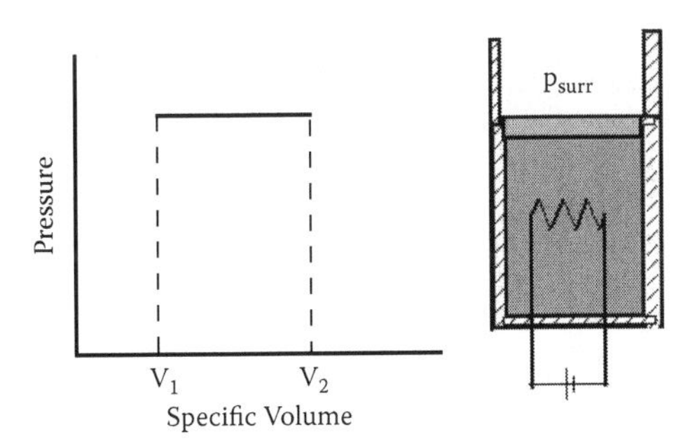

FIGURE 3.3 An isobaric process.

$$q = \Delta u + w = u_2 - u_1 + p(v_2 - v_1) = u_2 - u_1 + p_2v_2 - p_1v_1$$

Because the enthalpy is defined by $h = u + pv$, combining the first and third terms and the second and fourth terms simplifies the above formula to:

$$q = h_2 - h_1 \qquad (3.6)$$

which provides the following useful generalization:

Heat transferred in an isobaric process in which only reversible expansion work is performed is equal to the change in the system's enthalpy.

The simplification afforded by combining u and pv for the constant-pressure case with a closed system does not in itself warrant naming the combination a new thermodynamic property h. However, as will be shown in Chapter 4, in applications to open, or flow, systems, the analog of Equation (3.6) takes on considerably greater importance.

Example 1. Vaporization of water.

When saturated liquid water is vaporized to saturated steam at constant temperature, the path followed is a horizontal line such as BC in the T-v projection of the EOS of water in Figure 2.9. The enthalpy difference in converting saturated liquid to saturated vapor is the heat of vaporization. In steam-table terminology, Equation (3.6) becomes:

$$q = h_g - h_f = h_{fg}$$

In many applications, the heat of vaporization is assumed to be a temperature-independent property of the substance. For water, however, h_{fg} varies sufficiently over the range of practical interest that its temperature dependence generally must be taken into account. At 100°C, h_{fg} = 2257 kJ/kg, but at 300°C h_{fg} is 1405 kJ/kg.

Example 2. Heating of superheated steam.

2 m^3 of steam initially at 200°C are contained in a cylinder by a frictionless piston. 3500 kJ of heat are added and the gas expands at a constant pressure of 400 kPa. Find the final temperature.

From Table 2.3, the pertinent properties of steam in state 1 are:

$$v_1 = 0.534 \text{ m}^3/\text{kg} h_1 = 2860.5 \text{ kJ/kg}$$

The mass of steam in the cylinder is 2/0.534 = 3.75 kg, so the heat added per unit mass is q = 3500/3.75 = 934.5 kJ/kg. Applying the first law in the form of Equation (3.6), the enthalpy in state 2 is:

$$h_2 = h_1 + q = 2860.5 + 934.5 = 3797 \text{ kJ/kg}$$

This value of h is located between 600°C and 700°C in the enthalpy column of the 400 kPa subtable of Table A.3. Linear interpolation yields T_2 = 641°C. A variant of isobaric processes involving steam is given in Problem 3.9.

Example 3. Heating of a solid. See Problem 3.12.

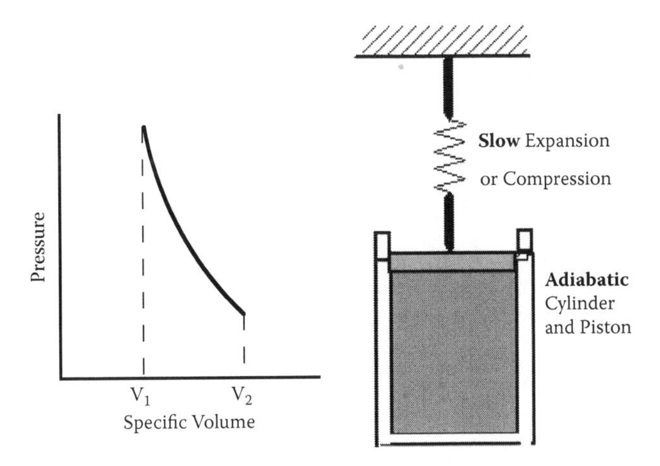

FIGURE 3.4 An isentropic process.

Many processes do not fit the simple restraints of constant volume, temperature, or pressure. These are considered in Problems 3.2 and 3.4.

3.5 THE ISENTROPIC PROCESS

The last of the important "iso" processes are those taking place at constant entropy. Figure 3.4 shows a simple version of this type of state change. A less evident example of an isentropic process is the blowdown of gas from a high-pressure cylinder, as illustrated in Figure 1.8. As emphasized on the diagram, such a process must be both adiabatic and reversible.

No real process can be perfectly reversible, so isentropic analyses of adiabatic processes are idealizations. However, nearly isentropic processes include the following: expansion or compression of gases (except through a valve or an orifice); liquid-vapor phase changes of water encountered in practical devices such as turbines of power plants; compression of liquids by efficient pumps. Entropy changes accompanying reversible and irreversible processes in gases and liquids are treated in Problems 3.1 and 3.8.

The starting point for thermodynamic calculations involving this particular restraint is the pair of formulas relating entropy changes in a process to changes in pressure, temperature and specific volume. These are rearranged forms of Equations (1.15a) and (1.16a):

$$ds = \frac{du}{T} + \frac{p}{T}dv \tag{3.7a}$$

and

$$ds = \frac{dh}{T} - \frac{v}{T}dp \tag{3.8a}$$

3.5.1 Ideal Gases

The functional dependence of s on p, v and T can be determined from the above equations by using the following ideal-gas properties:

$$du = C_V dT \qquad p/T = R/v$$

$$dh = C_p dT \qquad v/T = R/p$$

The resulting transformed equations are:

$$ds = C_V \frac{dT}{T} + R\frac{dv}{v} \tag{3.7b}$$

and

$$ds = C_p \frac{dT}{T} - R\frac{dp}{p} \tag{3.8b}$$

Assuming (for simplicity) that C_V and C_p are independent of temperature, these differentials can be integrated directly from state 1 to state 2:

$$s_2 - s_1 = C_V \ln\left(\frac{T_2}{T_1}\right) + R\ln\left(\frac{v_2}{v_1}\right) \tag{3.9}$$

$$s_2 - s_1 = C_p \ln\left(\frac{T_2}{T_1}\right) - R\ln\left(\frac{p_2}{p_1}\right) \tag{3.10}$$

These equations apply to both reversible and irreversible changes because they contain only state functions.

Isentropic changes in ideal gases depend on the ratio of the specific heats:

$$\gamma = C_p/C_V \tag{3.11}$$

This ratio is 1.67 for monatomic gases and 1.4 for diatomic gases. Since the difference in specific heats of an ideal gas is $C_p - C_V = R$, the following formulas express the specific heats in terms of their ratio and their difference:

$$C_V = R\frac{1}{\gamma - 1} \qquad C_p = R\frac{\gamma}{\gamma - 1} \tag{3.12}$$

For an isentropic process, $s_1 = s_2$ in Equations (3.9) and (3.10). Multiplying Equation (3.9) by γ and subtracting Equation (3.10) cancels the temperature term and produces the relation:

$$\frac{p_2}{p_1} = \left(\frac{v_1}{v_2}\right)^{\gamma} \tag{3.13}$$

The temperature change in an isentropic process is obtained by setting $s_1 = s_2$ in Equations (3.9) and (3.10):

$$\frac{T_2}{T_1} = \left(\frac{p_2}{p_1}\right)^{\frac{\gamma-1}{\gamma}} = \left(\frac{v_1}{v_2}\right)^{\gamma-1} \tag{3.14}$$

The path of an isentropic process on p-v coordinates can be inferred from Equation (3.13):

$$pv^{\gamma} = \text{constant} \tag{3.15}$$

The constant in this equation is determined by the properties of the initial state. The isentrope (Figure 3.4) is of the form $p \propto 1/v^{\gamma}$, while the isotherm (Figure 3.2) varies as $p \propto 1/v$. Since $\gamma > 1$ for all gases, the isentrope decreases more rapidly with v than does the isotherm.

The work done by isentropically expanding an ideal gas is obtained by substituting Equation (3.15) into Equation (3.1):

$$w = pv^{\gamma}\int_{v_1}^{v_2}\frac{dv}{v^{\gamma}} = \frac{pv^{\gamma}}{\gamma-1}\left(\frac{1}{v_1^{\gamma-1}} - \frac{1}{v_2^{\gamma-1}}\right) = \frac{p_1v_1 - p_2v_2}{\gamma-1} = \frac{R}{\gamma-1}\left(T_1 - T_2\right) \tag{3.16}$$

Problems 3.4 to 3.6 are examples of isentropic expansion calculations involving ideal gases.

Table 3.2 summarizes the work and heat effects in an ideal gas in the four iso processes discussed above.

TABLE 3.2
Iso Processes in an Ideal Gas

Process	pV Work	Heat
$\Delta v = 0$	0	$C_V\Delta T$
$\Delta p = 0$	$p\Delta v(v_2 - v_2)$	$C_P\Delta T$
$\Delta T = 0$	$RT\ln(v_2/v_2)$	$RT\ln(v_2/v_1)$
$\Delta s = 0$	$(p_1v_1 - p_2v_2)/(\gamma - 1)$	0

3.5.2 Water

Although Equations (3.9) and (3.10) can be employed to approximate entropy changes for processes in superheated steam, use of the table 2.3 is preferred.

Similarly, entropy changes of single-phase (compressed) liquid water are conveniently handled using Table 2.4. Many practical problems require calculation of entropy changes when water is partially vaporized. Such calculations utilize the entropies of the saturated liquid and saturated vapor provided in tables A.1 and A.2. The average entropy of a two-phase mixture of quality x is:

$$s = xs_g + (1 - x)s_f = s_f + xs_{fg} \tag{3.17}$$

The *entropy of vaporization* (s_{fg}), although not listed in the steam tables, can be obtained from h_{fg}. The vaporization process occurs at constant temperature and pressure. Under these restraints, the second term on the right of Equation (3.8a) is zero and the remainder of the equation can be directly integrated from the saturated liquid state to the saturated vapor state, yielding:

$$s_{fg} = h_{fg}/T \tag{3.18}$$

This equation applies to any pure substance, not just to water. The temperature in Equation (3.18) must be in Kelvins even though the steam tables use degrees Celsius.

Example 1. Verify Equation (3.18) from steam table entries at 150°C (423 K).

At 150°C, Table 2.2 gives h_{fg} = 2114 kJ/kg. Applying Equation (3.18), s_{fg} should be:

$$s_{fg} = 2114/423 = 5.00 \text{ kJ/kg-K}$$

Table 2.2 gives $s_{fg} = s_g - s_f = 6.84 - 1.84 = 5.00$ kJ/kg-K, verifying the value obtained by Equation (3.18).

Example 2. Superheated steam at 1 MPa and 300°C is isentropically expanded to 10 kPa. What are the temperature and quality of the final state?

From Table 2.3, the entropy of the initial state is 7.12 kJ/kg-K. This is also the entropy in the final state. Assuming that the final state is a two-phase mixture, interpolation in Table 2.2 gives a final temperature of 45.8°C for the specified final pressure of 10 kPa. Interpolating in Table 2.2 to obtain s_f and s_g at 10 kPa, the quality is determined by solving Equation (3.17) for x:

$$x = \frac{s - s_f}{s_g - s_f} = \frac{7.12 - 0.65}{8.15 - 0.65} = 0.86$$

3.6 MORE COMPLICATED PROCESSES

Many complex processes can be broken up into two or more "iso" steps. A simple example is changing the state of water from one (p,T) combination to another analyzed in Section 2.5.1. Here the prescribed change in state was achieved via an isothermal step and an isobaric step.

Not all processes are amenable to decomposition into iso steps. Even if the process could be so treated, the heat and work effects would be dependent upon the path chosen.

An example of a process restraint that must be treated in full is given below.

Example 1. A frictionless, massless piston is held in place inside an air-containing cylinder by a spring attached to a rigid wall. Heat is added until the volume is doubled. Find the changes in the other properties and the heat and work required to effect the change of state.

The process is diagrammed below:

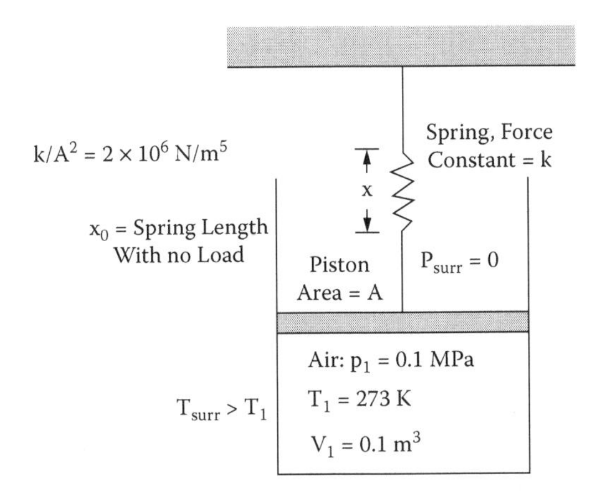

Heat is added to the air inside the cylinder until the volume is 0.2 m³. The changes in all other properties and the heat and work are to be determined.

From the conditions given in the diagram, the number of moles of air in the cylinder is:

$$n = \frac{p_1 V_1}{RT_1} = \frac{10^5\ Pa \times 0.1\ \text{m}^3}{8.314 \dfrac{Pa-\text{m}^3}{\text{mole-K}} \times 273\text{K}} = 4.41 \text{ moles}$$

The p-V path during the process is obtained from a force balance on the piston: $p_1A = k(x_o - x_1)$ initially and $pA = k(x_o - x)$ during expansion. Subtracting yields $p - p_1 = (k/A)(x_1 - x)$. Eliminating x using $V - V_1 = A(x_1 - x)$ yields the p-V relation:

$$p = p_1 + (k/A^2)(V - V_1)$$

Substituting into Equation (3.1):

$$W = \int_{V_1}^{V_2} pdV = p_1(V_2 - V_1) + \tfrac{1}{2}\left(\frac{k}{A^2}\right)(V_2 - V_1)^2 = 10^5 \times 0.1 + 0.5 \times 2 \times 10^6 (0.1)^2 = 2 \times 10^4 \text{ J}$$

From the force balance on the piston:

$$p_2 = 0.1 + 2(0.2 - 0.1) = 0.3 \text{ MPa.}$$

From the ideal gas law:

$$T_2 = T_1 \frac{p_2 V_2}{p_1 V_1} = 273 \frac{0.3 \times 0.2}{0.1 \times 0.1} = 1638 \text{ K}$$

The increase in internal energy is:

$$\Delta U = nC_V(T_2 - T_1) = 4.41 \times \tfrac{5}{2} \times 8.314(1683 - 273) = 1.3 \times 10^5 \text{ J}$$

From the first law:

$$Q = \Delta U + W = 1.5 \times 10^5 \text{ J}$$

From Equation (3.9):

$$\Delta S = 4.41 \times 8.314\left[(\tfrac{5}{2})\ln(1638/273) + \ln(0.2/0.1)\right] = 190 \text{ J/K}$$

Example 2. Show that thermal equilibration of two identical solids results in an entropy rise.

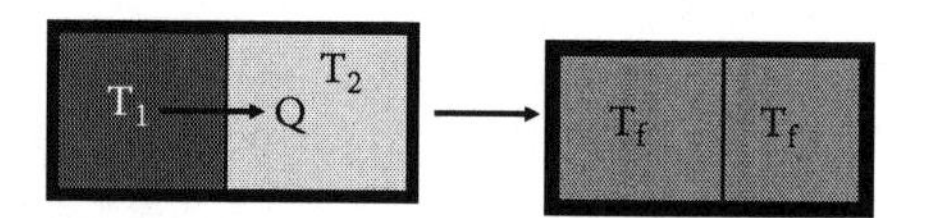

The two solids are identical, so energy conservation is $\Delta U_{tot} = \Delta U_1 + \Delta U_2 = 0$, or:

$$C_V(T_f - T_1) + C_V(T_f - T_2) = 0, \text{ which yields } T_f = \tfrac{1}{2}(T_1 + T_2)$$

The entropy change is:

$$\Delta S_{tot} = \Delta S_1 + \Delta S_2 = C_V \ln(T_f / T_1) + C_V \ln(T_f / T_2)$$

Combining:

$$\Delta S_{tot} = C_V \ln\left[\frac{(T_1 + T_2)^2}{4T_1 \times T_2}\right]$$

T_2/T_1	$\Delta S_{tot}/C_V$
0.9	0.0028
1.0	0
1.1	0.0023

The table shows that thermal equilibration (i.e., $T_1 \rightarrow T_2$ for $T_1 > T_2$ or $T_1 < T_2$) in an isolated system results in an increase in entropy, as indicated in Figure 1.18.

3.7 EFFECT OF PHASE CHANGES ON THE THERMAL PROPERTIES OF CONDENSED PHASES

This topic, which properly belongs in Chapter 2, has been postponed until this point in order to complete the account of the effect of temperature on the properties of solids and liquids.

3.7.1 Heat Capacities

The discussion in Section 2.5.2 alluded to the small effect of pressure on the thermal properties of solids and liquids. One consequence of this fact is the close proximity of C_P and C_V. Another feature of the heat capacities of condensed phases is their relative insensitivity to temperature as well (at least at room temperature and above, see Figure 2.7). For the present purposes, specific heat will be taken to mean C_P and for most applications will be assumed to be independent of both p and T.

However, the difference between the heat capacities of the liquid (C_{PL}) and that of the solid (C_{PS}) is significant; for all substances, C_{PL} is greater than C_{PS}. The right-hand plot in Figure 3.5 shows this behavior schematically. In accord with the approximations stated above, ΔC_P is taken to be independent of both temperature and pressure.

3.7.2 Enthalpy

The left-hand plot of Figure 3.5 shows the temperature effect on the enthalpies of the individual phases and on the enthalpy of fusion. Because of the underlying assumption of constant C_{PS} and C_{PL}, these enthalpy variations are linear in temperature. The dashed lines are extrapolations of h_L into the subcooled-liquid region ($T < T_M$) and of h_S into the superheated-solid region ($T > T_M$). Even though the dashed lines do not represent thermodynamically stable states, some phase equilibrium and chemical equilibrum calculations are based on estimates of the heat of fusion, Δh_M, at temperatures other than the melting point.

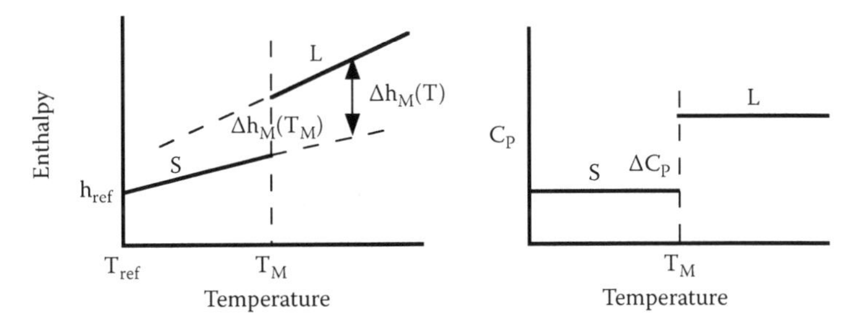

FIGURE 3.5 Effect of temperature and melting on enthalpy and specific heat.

As expressed by Equation (2.14), the enthalpy of a substance is the temperature integral of its heat capacity. To perform the integration, a reference state (temperature T_{ref} and enthalpy h_{ref}) must be selected. This choice is arbitrary, but usually T_{ref} is taken as 298 K (room temperature) and the enthalpy of the solid at T_{ref} is specified as $h_{ref,s}$, which is usually set equal to zero. The enthalpy of the solid phase at other temperatures (including $T > T_M$) is:

$$h_S(T) = h_{ref,S} + \int_{T_{ref}}^{T} C_{PS}(T')dT' \cong h_{ref,S} + C_{PS}(T - T_{ref}) \tag{3.19}$$

The enthalpy of the liquid must reflect the same reference state as that of the solid, so that h_L is given by:

$$h_L(T) = h_{ref,S} + C_{PS}(T_M - T_{ref}) + \Delta h_M(T_M) + C_{PL}(T - T_M) \tag{3.20}$$

$\Delta h_M(T_M)$ accounts for the enthalpy increase on melting at the normal melting point, T_M. The heat of fusion changes with temperature because the slopes of the h_L versus T and h_S versus T differ. Even though melting physically occurs only at T_M, later analyses of phase and chemical equilibria will require estimates of Δh_M at $T \neq T_M$. This is obtained by subtracting Equation (3.19) from Equation (3.20). In so doing, Equation (3.19) is written as: $h_s(T) = h_{ref,S} + C_{PS}(T_M - T_{ref}) + C_{PS}(T - T_M)$, whether T is larger or smaller than T_M. The result is:

$$\Delta h_M(T) = \Delta h_M(T_M) + \Delta C_P(T - T_M) \tag{3.21}$$

where $\Delta C_P = C_{PL} - C_{PS}$.

3.7.3 Entropy

In Section 2.5.2, it was shown that except at high temperatures, C_p and C_V are approximately equal for most solids and liquids. The chief reason for this is the small specific volumes of condensed phases. If the difference between the two heat capacities is ignored, du and dh in Equations (3.7a) and (3.8a) are about equal as well. In addition, the second terms on the right-hand sides of these equations can be neglected because v is small. Hence, both of these equations are effectively only the single equation $ds = dh/T = C_p dT/T$. Assuming temperature-independent heat capacities and integrating between a reference temperature T_{ref} and temperature T the analogs of Equations (3.19) to (3.21) for the entropies of solids and liquids are:

$$s_S(T) = s_{ref,S} + \int_{T_{ref}}^{T} \frac{C_{PS}(T')}{T'} dT' \cong s_{ref,S} + C_{PS} \ln\left(\frac{T}{T_{ref}}\right) \tag{3.22}$$

$$s_L(T) = s_{ref,S} + C_{PS} \ln(T_M / T_{ref}) + \Delta s_M(T_M) + C_{PL} \ln(T / T_M) \tag{3.23}$$

$$\Delta s_M(T) = \Delta s_M(T_M) + \Delta C_P \ln(T / T_M) \tag{3.24}$$

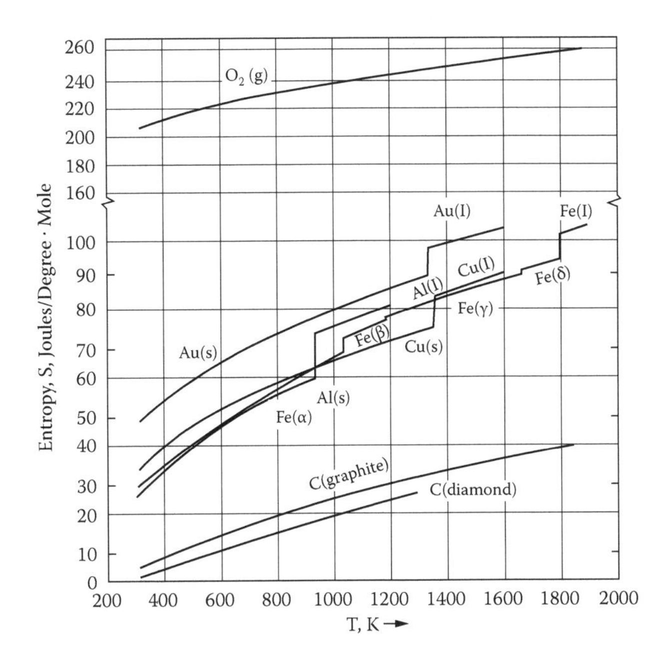

FIGURE 3.6 Molar entropies of selected elements at 1 atm pressure.

According to the third law of thermodynamics, if T_{ref} is chosen to be zero Kelvin, then the reference entropy is zero. The entropies of all solids exhibit this behavior. However, this assignment of the reference state is not necessary in order to use Equation (3.19) in thermodynamic calculations. For entropy calculations to be consistent with the enthalpy analysis, the same reference temperature must be chosen for each.

By analogy to Equation (3.18), the entropy of fusion at the normal melting point is equal to $\Delta h_M(T_M)/T_M$. Figure 3.6 shows the entropies of a number of common elements.

The entropy of the diamond form of carbon shown in Figure 3.6 is lower than that of graphite. This difference favors graphite as the stable form of carbon, but the relative stability of the two forms is reversed at ultrahigh pressures. The graphite-diamond conversion is analyzed in Section 5.8.

Discontinuities in the curves for Au, Al and Cu represent the entropies of fusion of these elements. The curve for iron contains four jumps, the last of which represents melting. The other three at lower temperatures reflect the thermal effects of crystal structure transitions. Each is accompanied by an enthalpy change and an entropy change. These solid–solid transition effects can be included in Equations (3.20) to (3.25) in the same way as the solid–liquid phase change is treated.

The curve for gaseous O_2 is also shown in Figure 3.6. The entropy of the gas is several times larger than those of the condensed phases, reflecting the more highly ordered states of the latter. Changing pressure from the 1 atm value to which Figure 3.6 applies would have little effect on the entropy curves for the condensed phase. However, such a change would significantly affect the entropy of O_2. Equation (3.10) shows that increasing the O_2 pressure from 1 atm to, say, 10 atm would lower the entropy by $R\ln 10 \sim 20$ J/mole-K.

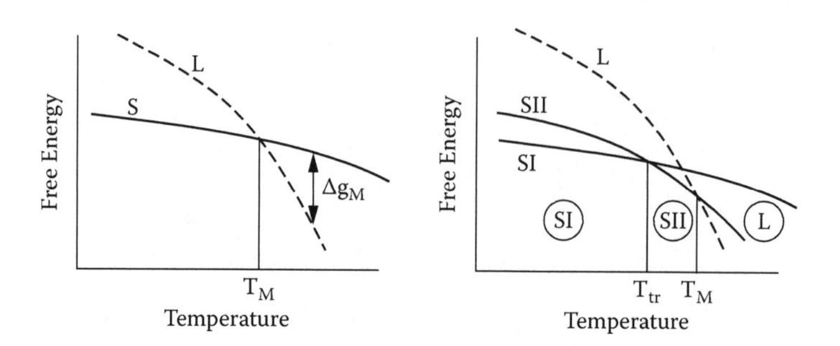

FIGURE 3.7 Free energy–temperature curves for a simple substance.

3.7.4 Free Energy of Phase Transitions

The temperature effect on the Gibbs free energy must be different for liquids and solids in order that there be a unique melting temperature at which both phases coexist at equilibrium. The source of this behavior lies in the different entropies of the two states, as can be demonstrated by the fundamental differential $dg = -sdT + vdp$ (Equation [1.18a]). At constant pressure, the second term vanishes, and dividing what remains by dT gives:

$$\left(\partial g / \partial T\right)_p = -s \tag{3.26}$$

This equation (in integrated form) is shown schematically in the left-hand graph of Figure 3.7 for a substance with no solid–solid phase transitions. Because entropy is a measure of order (or disorder) of a system, the entropy of the liquid is always greater than that of the solid at the same temperature. Since the entropy is always positive, Equation (3.26) requires that the slope of the plot of g_L versus T be steeper than the corresponding plot for the solid. At the melting point the equilibrium criterion of minimum free energy (Equation [1.20a]) is satisfied and the two phases coexist. For $T < T_M$, the free energy of the solid is lower than that of the liquid, so minimization of the system's free energy causes the liquid to disappear. Conversely, for $T > T_M$, the solid phase is unstable with respect to the liquid.

The analysis of binary (two-component) phase diagrams discussed in Chapter 8 requires as input the free energy difference $\Delta g_M = g_L - g_s$ for each species at temperatures other than their melting points.

Combining the enthalpy and entropy of melting given by Equations (3.21) and (3.24), with $\Delta s_M(T_M)$ in Equation (3.24) replaced by $\Delta h_M(T_M)/T_M$, the free energy change on melting is:

$$\Delta g_M(T) = \Delta h_M(T) - T\Delta s_M(T) = \Delta h_M(T_M)\left(1 - \frac{T}{T_M}\right) + \Delta C_P\left[T - T_M - T\ln\left(\frac{T}{T_M}\right)\right] \tag{3.27a}$$

This equation assumes a temperature-independent heat-capacity difference between the two phases. The first term on the right-hand side of Equation (3.27a) is the lowest-order approximation:

$$\Delta g_M(T) \cong \Delta h_M(T_M)\left(1-\frac{T}{T_M}\right) \tag{3.27b}$$

To assess the importance of the second term, the following properties of water are employed: Δh_M = 6008 J/mole at T_M = 273 K and $\Delta C_P = C_{PL} - C_{Ps}$ = 37 J/mole-K. For T = 263 K (–10°C), Equation (3.27b) gives Δg_M = 220 J/mole while Equation (3.27a) yields Δg_M = 213 J/mole. The approximately 3% effect of the second term is small enough to warrant use of the simpler form given by Equation (3.27b) for applications requiring evaluation of Δg_M at temperatures other than the melting point.

In the right-hand graph of Figure 3.7, the solid experiences a phase transition prior to melting. The solid–solid transition is a consequence of free-energy minimization just as is the melting process. Solid phase SI has a lower entropy than SII, and so is stable at low temperature. These two curves cross at the SI–SII phase-transition temperature, T_{tr}. The SII phase remains stable up to the melting temperature, where its free energy curve crosses the liquid curve.

PROBLEMS

3.1 A diatomic gas in a rigid container of 2 m^3 volume initially at 293 K and 200 kPa is heated until it reaches 500 K. Although the *p-v-t* behavior of the gas is ideal, its heat capacity is the ideal-gas value plus a term aT, where a = 0.03 J/mole-K^2. The temperature of the thermal reservoir that supplies heat to the system is 600 K.

(a) How much heat is absorbed during the process?
(b) What is the entropy change of the system? Why is it positive?
(c) What is the total entropy change (system + surrounding) during the process?
(d) Is the process reversible? If not, identify the source of the irreversibility.

3.2 One mole of a monatomic ideal gas at 273 K and 0.1 MPa is subjected to the following cycle, each step of which is conducted reversibly:

1. Isothermal increase in volume by a factor of 10.
2. Adiabatic pressurization by a factor of 100.
3. Return to the initial state along a straight-line path on the *p-v* diagram.

Calculate the work and heat in each step and verify that the sum of the work values is equal to the sum of the heat additions.

3.3 Consider a frictionless, massless piston of area A in a cylinder containing a fixed quantity of air. The top of the piston is connected to a rigid fitting above the assembly by a spring with a force constant k (see figure). The system is moved reversibly from state 1 to state 2 by adding heat.

State	p (kPa)	V (m³)	T (K)
1	100	0.1	273
2	300	0.2	—

(a) What is the variation of p with V during this process? This relation

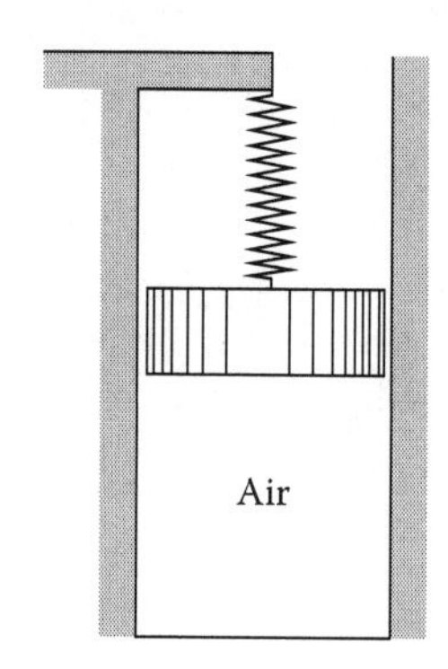

may be expressed in terms of p_1 and V_1. (Hint: a force balance on the piston is needed. Neglect p_{surr}. The spring is initially compressed.)

(b) How much work is done by the air inside the cylinder in moving from state 1 to state 2?

(c) How much energy is stored in the spring during the process? Is the stored energy equal to the work done by the system? (Hint: find the cylinder volume without gas when the spring is at its equilibrium length then derive the equation relating the system volume to the spring displacement.)

(d) What is the final temperature? How much heat has been added to the gas?

3.4 Air in an insulated cylinder is reversibly compressed from a volume of 0.17 m³ to a volume of 0.034 m³. The initial temperature and pressure are 10°C and 0.2 MPa, respectively. For air, C_P and C_V can be taken to be $7R/2$ and $5R/2$, respectively.

(a) Calculate the work needed to accomplish this process.

(b) Calculate the final temperature of the air.

(c) Show that the result of part (a) satisfies the first law.

3.5 Two adiabatic vessels each containing a diatomic ideal gas are connected by a valve. The initial state is shown in the diagram, with $p_{10} > p_{20}$. The valve is opened and the gas in vessel 1 expands reversibly (as in the blowdown of a gas in a cylinder). Vessel 2 contains a frictionless, adiabatic piston which is pushed to the right as the gas in it is compressed reversibly. The final state is in mechanical equilibrium ($p_{1f} = p_{2f} = p_i$) but not in thermal equilibrium ($T_{1f} \neq T_i \neq T_{2f}$).

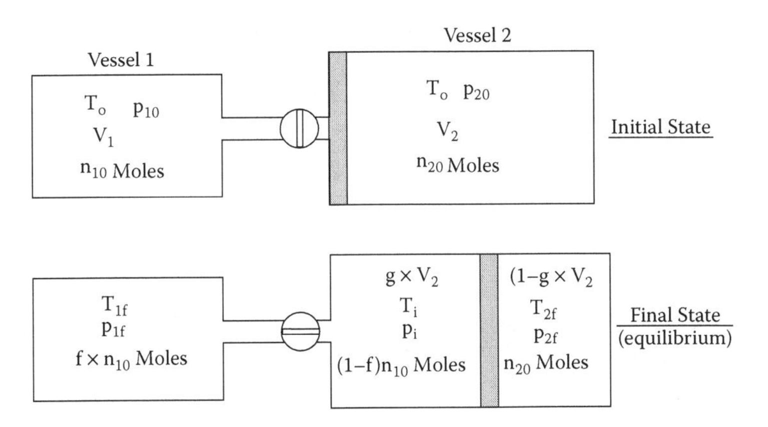

(a) Write all of the relevant equations that determine the parameters of the final state.
(b) Solve the equations with the initial conditions: $V_1 = 0.1$ m^3; $V_2 =$ 0.2 m^3; $T_o = 298$ K; $p_{10} = 10$ MPa; $p_{20} = 1$ MPa.

3.6 A monatomic ideal gas at temperature T_1 and pressure p_1 in an insulated container of volume V is connected by a valve to an insulated cylinder-piston that initially contains no gas. The valve is opened and the gas in the container flows into the cylinder. The massless piston rises reversibly against external pressure p_{surr} until the volume of gas in the cylinder is V_c. At this point the valve is closed. The process is adiabatic and the gas that remains in the cylinder at the end can be assumed to have expanded reversibly. What fraction of the gas remains in the container at the end of the expansion and what are the temperatures of this gas and of that in the cylinder? (Hint: use the results of the cylinder blowdown example.)

3.7 Helium gas in a 1 m^3 container is initially at 20°C. What quantity of heat transferred from the surroundings is required to raise the gas temperature by 1°C in the following three cases, in which the container is:

(a) Rigid and initially at 1 atm pressure.
(b) Rigid and initially at 3 atm pressure.
(c) Fitted with a frictionless piston and the pressure is maintained at 1 atm.

3.8 Two insulated tanks containing water are connected by a valve. The conditions in the tanks are:

Tank A: $V_A = 0.6$ m^3; $p_A = 200$ kPa; $T_A = 500$°C
Tank B: $V_B = 0.3$ m^3; $p_B = 500$ kPa; $x_B = 0.9$ (quality)
The valve is opened and the two tanks achieve a uniform state. Assume that the process is adiabatic.

(a) Is the process reversible or irreversible? Give reasons for your answer.
(b) What two thermodynamic properties of the system are conserved (i.e., remain unchanged) in this process?
(c) Is the final state a superheated vapor or a two-phase mixture?
(d) What is the final pressure?
(e) What is the entropy change?

3.9 2000 kJ of heat are removed from a container at constant pressure. The container initially was filled with 1 kg of steam at 0.6 MPa and 400°C. What is the final temperature and quality of the steam? To avoid a trial-and-error solution, use the p-h diagram for steam (below) and the steam tables.

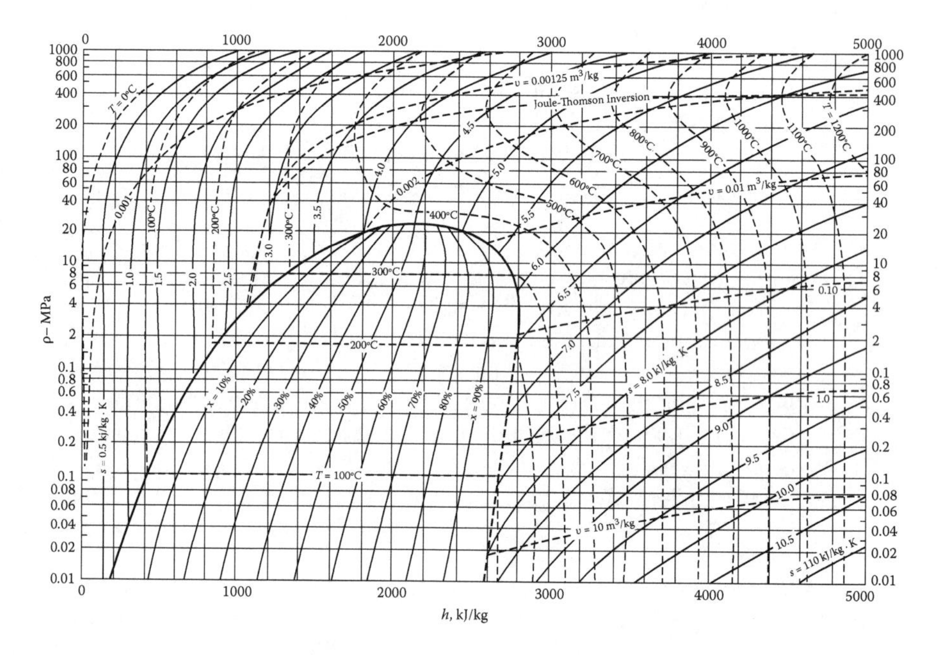

3.10 What is the work required to compress a solid isothermally from pressure p_1 where the specific volume is v_1 to pressure p_2 where the specific volume is v_2? The coefficient of compressibility of the solid is β and the equation of state is given in differential form by Equation (2.18a).

3.11 A 1.5 l rigid vessel contains 30 g of water and no air or other gas. The initial state is liquid-vapor equilibrium at 25°C. Heat is added to the vessel until the water is just in the saturated-vapor state. What is the final temperature and how much heat must be added to reach the final state?

3.12 How much heat is required to melt 1 mole of a metal starting at 25°C? The process is conducted at 1 atm pressure, and the relevant properties of the metal are:

$$C_V = 24 \text{ J/mole-K}; \; C_P = C_V + 0.01T; \; \Delta h_M = 10 \text{ kJ/mole}; \; T_M = 1000 \text{ K}.$$

3.13 A reversible process involving 1 mole of a diatomic ideal gas consists of two steps. In step 1, starting from an initial state at 200°C and 0.3 MPa, heat is added at constant pressure until the temperature reaches 400°C. In step 2, the gas is expanded adiabatically until the temperature returns to 200°C.

(a) How much heat is added in step 1?
(b) How much work is done in both steps combined?

3.14 One mole of an ideal gas is contained in a cylinder fitted with a massless, frictionless piston to which a spring is attached. The gas is initially at 0.1 Mpa pressure and 298 K.

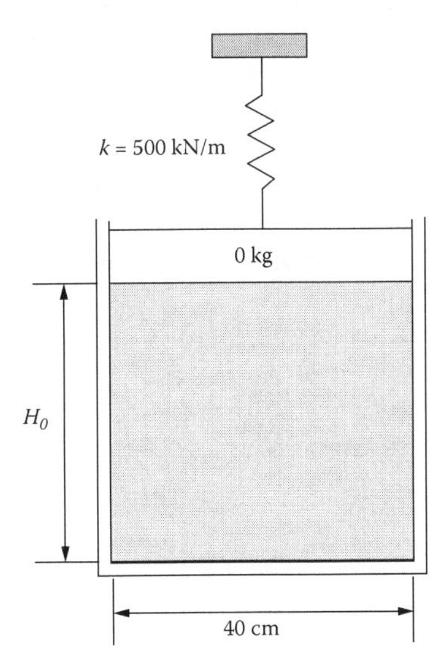

(a) Initially the spring is at its equilibrium length and so exerts no force on the piston. What is the height H_o of the gas in the cylinder?
(b) The cylinder is heated until the height of the piston increases from H_o to 0.25 m. What are the final pressure and temperature of the gas?

3.15 The heat capacity of a metal is given by: $C_P = 22.6 + 6.6 \times 10^{-3}\, T$ J/mole-K where T is the temperature in K. Two 1-mole pieces of the metal, one at $T_{A0} = 873$ K and the other at $T_{B0} = 573$K, are placed in an adiabatic container maintained at constant pressure.

(a) What process occurs within the container? Is it spontaneous and why?
(b) Derive the equation for T_B as a function of T_A during the process. Sketch this function. At equilibrium, what is the common final temperature of the two pieces?
(c) Derive the equation for the entropy change of the contents of the container as a function of T_A. Sketch this function. At what temperature is the entropy change a maximum? What is the entropy increase of the system at equilibrium?
(d) What is the entropy change of the surroundings during this process? How does the entropy of the universe increase as a result of this process?

3.16 A rigid container with a volume of 0.2 m³ is divided into two equal volumes by a partition. Both sides contain nitrogen, volume A at 2 MPa and 200°C and volume B at 200 kPa and 100°C. The partition is ruptured and the gas comes to equilibrium at 70°C, which is the temperature of the surroundings. Determine the work done and the entropy changes of the system and the surroundings during the process. Is the process reversible? Assume N_2 behaves ideally.

3.17 The following problem is analogous to the irreversible expansion of an ideal gas described in Section 1.9.2. The object is to demonstrate that attainment of equilibrium resulting from fluid expansion in an isolated system always occurs with an increase in entropy, no matter what substance is involved.

Consider 1 kg of liquid water initially at 120°C and 5 MPa pressure held in the small section of a rigid adiabatic container separated by a membrane from a section 100 times larger. The larger section is initially a vacuum. The membrane is ruptured and the water uniformly fills the entire container.

(a) What is the final state of the water?
(b) Show that the entropy of the system has increased in the process of changing from the initial state to the equilibrium state.

3.18 An ideal diatomic gas is contained in two tanks connected by a valve. Tank A is 600 liters in volume, insulated, and initially has gas at 1.4 MPa and 300°C. Tank B is not insulated, is 300 liters in volume, and initially contains gas at 200 kPa and 200°C.

The valve is opened and gas flows from tank A to tank B until the temperature in tank A reaches 250°C, at which time the valve is closed. During this process, heat transfer from tank B to the surroundings (at 25°C) maintains the gas in this tank at 200°C at all times. Assuming that the gas remaining in tank A has undergone a reversible adiabatic expansion, determine the following:

(a) The final pressure in tank A.
(b) The number of moles of gas transferred from tank A to tank B and the final pressure in tank B.
(c) The entropy change of the gas and that of the gas plus the surroundings.

3.19 This problem is an irreversible version of Problem 3.3. An ideal gas is contained in a cylinder by a piston of area A. A spring is connected at one end to the piston and at the other to a rigid fitting in the cylinder housing. The force constant of the spring is denoted by k. The spring and cylinder are encased in a rigid, adiabatic container, thus forming an isolated system. The heat capacities of the spring, the piston and the cylinder walls are negligible compared to that of the gas. Initially, the gas conditions are p_o, T_o, and V_o, and the spring is at its equilibrium length. The piston is held in place by blocks, which are removed to allow the gas to expand against the spring. In the final state, the gas has expanded to a volume V and its final pressure and temperature are p and T, respectively.

(a) Show that the final values of the ratios V/V_o and T/T_o are functions of the specific heat ratio of the gas $\gamma = C_P/C_V$ and a dimensionless parameter $\alpha = A^2p_o/kV_o$.
(b) For $\gamma = 5/3$ and $\alpha = 1$, what is the final state of the gas?
(c) Has the gas expanded isentropically in this process?

3.20 An ideal monatomic gas in a closed cylinder is separated into two parts by a frictionless piston that is held in place by stops.

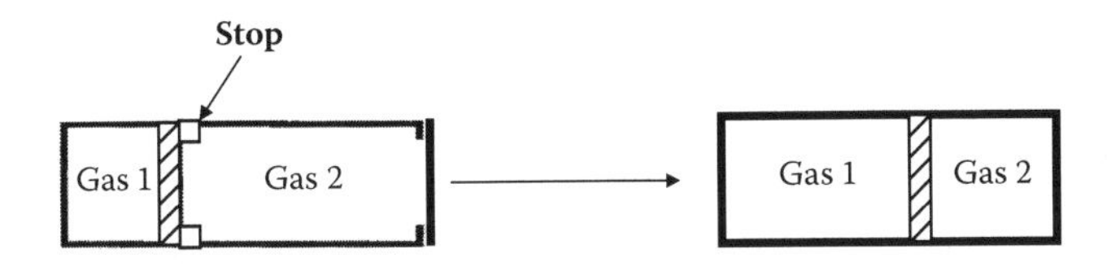

The initial conditions of the gas in the two sections are:

$p_{10} = 2$ atm, $V_{10} = 1$ cm^3, $T_{10} = 400$ K
$p_{20} = 1$ atm, $V_{20} = 3$ cm^3, $T_{20} = 300$ K

The cylinder walls and the piston are perfect insulators, so that no heat is transferred from one section of the gas to the other, or between the surroundings and the gas. By removing stops holding the piston in place, the system is allowed to achieve mechanical equilibrium at a final pressure p_f. Determine the state of the gas in the two parts of the cylinder after pressure equilibrium has been achieved. (Hint: derive equations for the total volume, the number of moles in each compartment, the first law, and the entropy increase. The principle of maximum entropy in an isolated system needs to be used. A trial-and-error solution is required.)

3.21 Pressure equilibration in the process described in Problem 3.20 can be done reversibly (but still adiabatically) by allowing the piston to do work on the surroundings via a shaft.

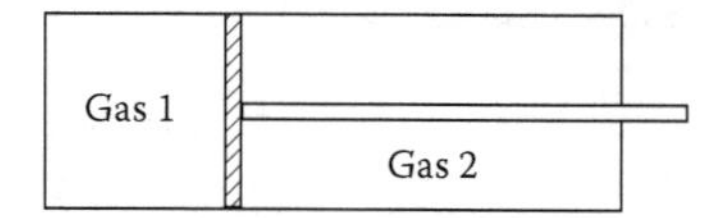

(a) What is the final common pressure of the gas in the two chambers?
(b) How much net (external) work is done by the gas?

3.22 The heat capacity of a solid is $C_{PS} = A + BT$. Its liquid phase heat capacity is a constant C_{PL}. The heat of fusion is Δh_M and the melting temperature is T_M. What is the change in entropy per mole when heated from T_1 as a solid to T_2 in the liquid region?

3.23 An ideal monatomic gas undergoes the cycle shown below.

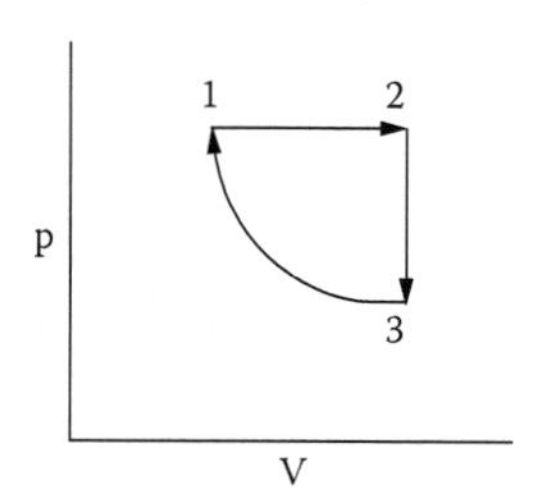

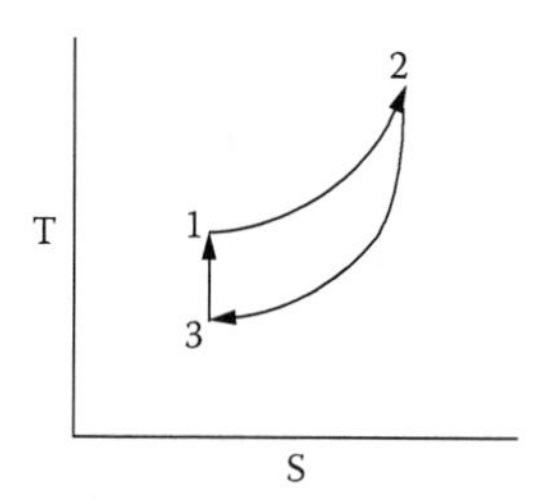

(a) What are the "iso" constraints on each process?
(b) What pressure p_3 assures that the $3 \rightarrow 1$ step returns the system to its initial state?
(c) What is the net work of the cycle?

3.24 A monatomic ideal gas undergoes the three-process cycle shown below.

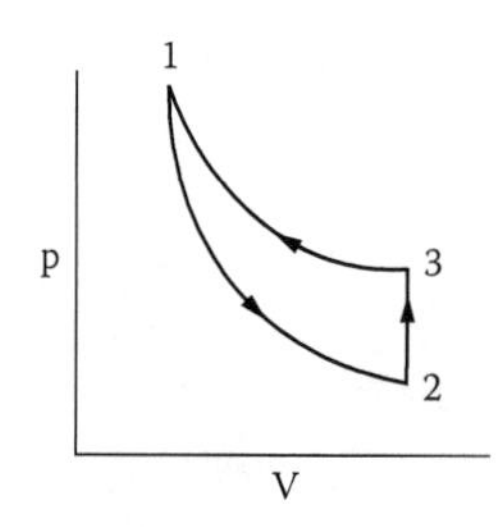

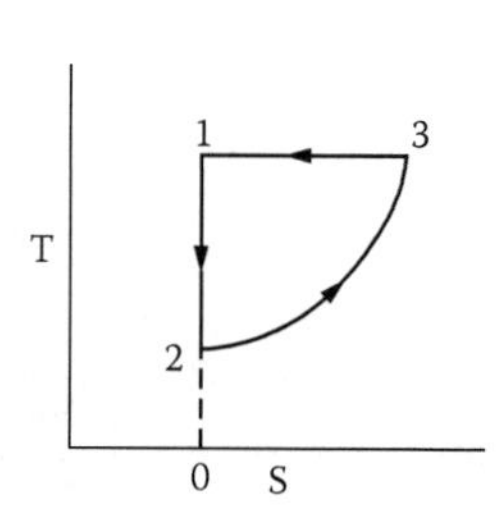

(a) Complete the following table:

State	$v \times 10^4$, m³/mole	p, MPa	T, K	s/R
1	5	10		0*
2	10			
3				

* Arbitrary reference value.

(b) What is the equation for the step 2 → 3 curve in the *T*-*s* diagram?

3.25 Consider the adiabatic, steady-flow device shown below:

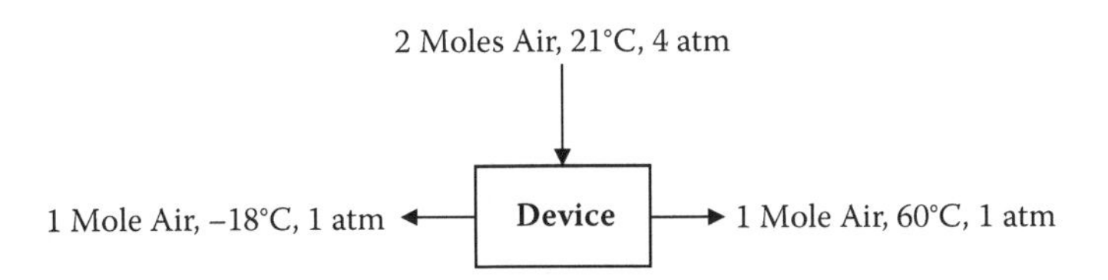

(a) Does it violate the first law of thermodynamics?
(b) Does it violate the second law of thermodynamics?

3.26 An insulated tank is attached to an insulated cylinder with a frictionless piston via tubing containing a valve.

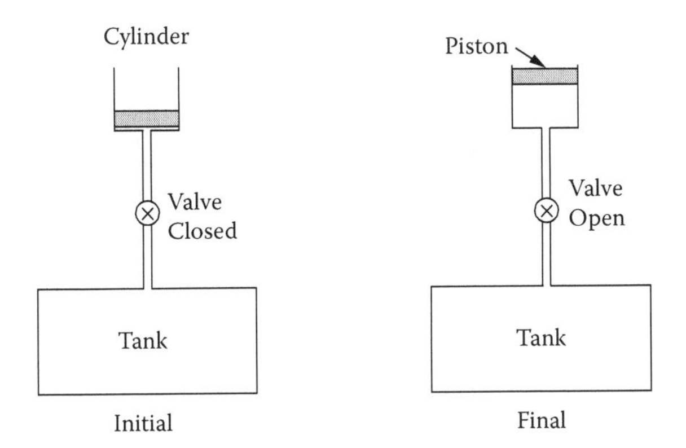

Initially, the tank contains superheated steam and the valve is closed. The valve is opened, steam escapes from the tank into the cylinder and the piston rises. An equilibrium is reached when the pressure of the surroundings is equal to that in the tank and cylinder. The valve is immediately closed. The following table gives known quantities and blanks for quantities to be calculated. Fill in the blanks.

Quantity	Tank Initial	Tank Final	Cylinder Final
Temperature (°C)	500	250	
Pressure (Mpa)	2	0.3	
Specific volume (m^3/kg)			
Steam mass (kg)			
Volume (m^3)	0.2	0.2	
Specific internal energy (kJ/kg)			

* At 60°C and 5 MPa.

4 Heat Engines, Power Cycles, and the Thermodynamics of Open Systems

4.1 HEAT ENGINES

In Section 1.9, it was noted that the first law regards heat and work as completely interchangeable; if a certain number of Joules of heat added to a system increases the internal energy of a body by, say, ΔU, the same number of Joules of work performed on the body would produce the same ΔU. In addition, work can be completely converted to heat, as everyday experience with friction attests. However, the reverse is not true; heat cannot be completely transformed into work. This limitation, which is a consequence of the second law, is best demonstrated by studying the properties of *heat engines*.

A heat engine is a system operating in a cycle that receives heat from a high-temperature source (called a thermal reservoir) and produces useful work. However, since the efficiency of conversion must be less than 100%, some of the input heat is rejected to a cold reservoir. Figure 4.1 shows a schematic of a heat engine/heat pump and their associated thermal reservoirs. The reservoirs supply or receive heat without alteration of their temperatures. Heat flows in the reservoirs are reversible whether or not the engine is.

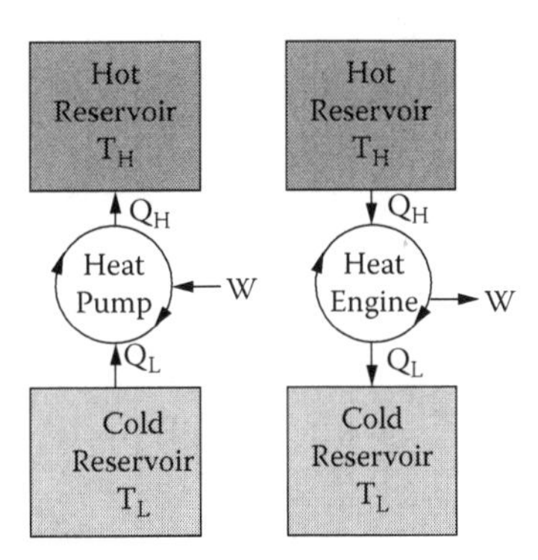

FIGURE 4.1 A schematic of a heat engine or heat pump. The heat pump is a heat engine running in reverse.

The circle with the arrows in Figure 4.1 is a shorthand representation of the heat engine. It is intended to signify that the working substance (a fluid such as an ideal gas or water) moves through many thermodynamic states in a never-ending cyclic process. The detailed structure of the heat engine can vary greatly, but the simplest version contains the following four steps:

1. One in which heat is absorbed isothermally from the high-temperature reservoir.
2. The next, in which work is produced adiabatically.
3. This is followed by isothermal rejection of heat to the low-temperature reservoir.
4. The last, in which work is done on the working substance to return it to the state at the start of step 1.

The heat engine can operate in either of two ways: (1) as a single device that moves sequentially through the four processes described above, or (2) with a fluid flowing through four distinct devices, each assigned to one of the four steps.

4.1.1 Single Device, Sequential States, Ideal Gas

The sequential type is illustrated in Figure 4.2 by a single piston-cylinder that performs each of the four steps. (See Van Ness, 1983, 36–40 for an explanation of this cycle.)

Proceeding clockwise in the diagram follows the device in time. If the gas in the cylinder were ideal *and* if the four steps were conducted reversibly, the cycle would appear in the *process diagrams* shown in Figure 4.3. The cycle consists of two isotherms (like the one in Figure 3.2) and a pair of isentropes (see Figure 3.4). Work is exchanged with the surroundings in each step. The net work produced by the cycle is $W = W_{1\text{-}2} + W_{2\text{-}3} - W_{3\text{-}4} - W_{4\text{-}1}$. Because each of the component work terms is the integral of pdV, W is the area inside the p-V graph in Figure 4.3. Because each step is assumed to be reversible, the work produced by an ideal gas in the piston/cylinder is:

$$W = RT_H \ln\left(\frac{v_2}{v_1}\right) + \frac{p_2v_2 - p_3v_3}{\gamma - 1} - RT_L \ln\left(\frac{v_3}{v_4}\right) - \frac{p_1v_1 - p_4v_4}{\gamma - 1} \tag{4.1}$$

The work terms originate from Equations (3.5) and (3.16). The net heat of the cycle is:

$$Q = T_H(S_2 - S_1) - T_L(S_3 - S_4) = (T_H - T_L)(S_2 - S_3) \tag{4.2}$$

Here T_H is the temperature of the hot reservoir that delivers heat during the 1–2 step and T_L is the temperature of the cold reservoir that receives heat from the engine during the 3–4 step.

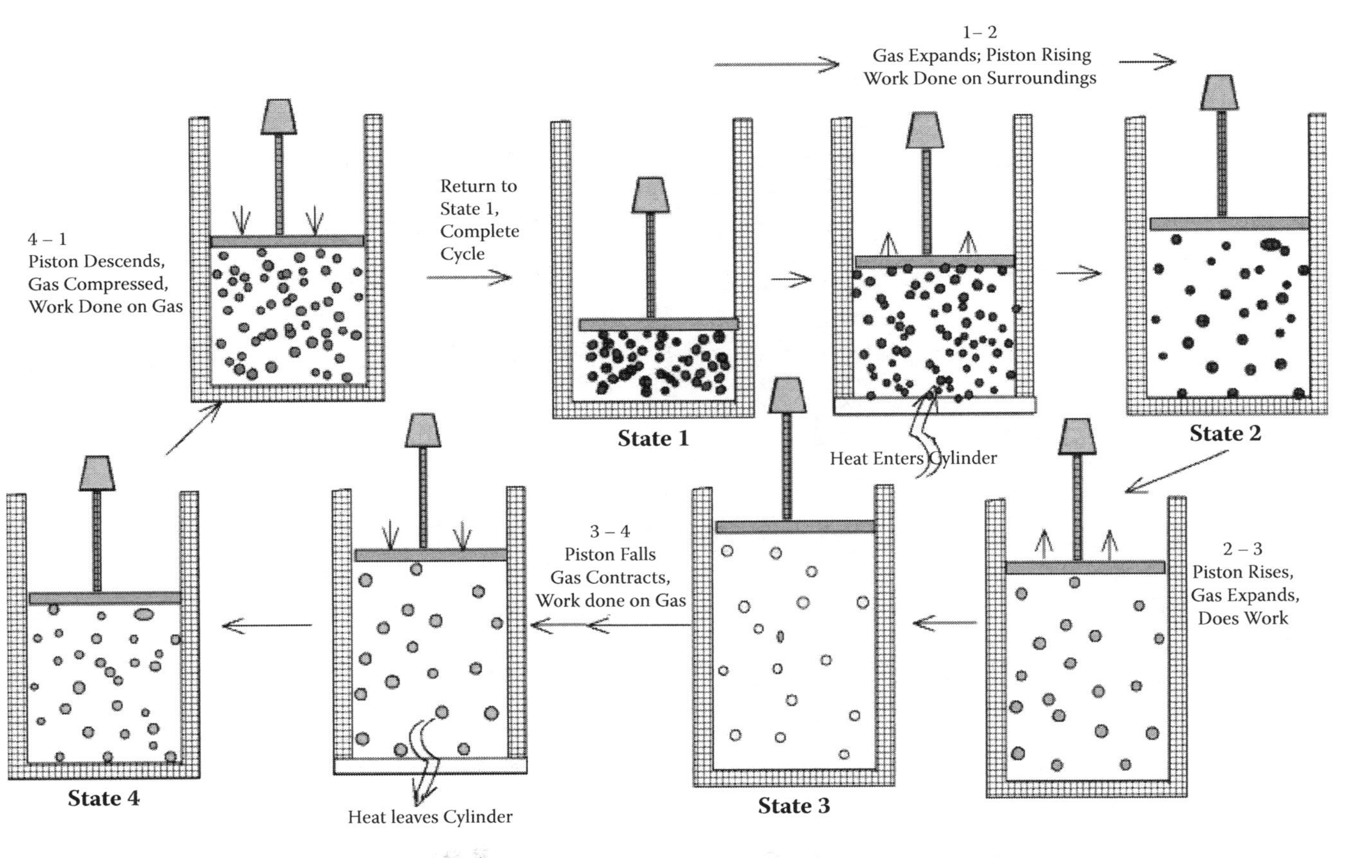

FIGURE 4.2 A heat engine consisting of a single piston/cylinder performing four operations.

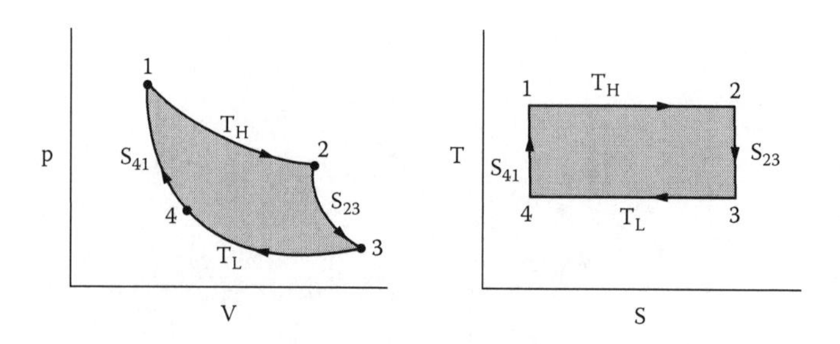

FIGURE 4.3 Process diagrams for the power cycle shown in Figure 4.2.

The same cyclical process is represented in T-S coordinates on the right in Figure 4.3. The net heat exchanged with the surroundings is $Q = Q_{1\text{-}2} - Q_{3\text{-}4}$. Because each of the component heat exchange terms is an integral of TdS, the shaded area inside the rectangle is Q.

In describing this power cycle, the convention for the signs of Q and W are different from that used for the first law (Section 1.8), in which heat added to the system is positive and work done by the system is positive. For analyzing power cycles, heat and work terms are positive in the direction in which they act (indicated by arrows). In this way, a welter of minus signs is avoided. Thus, $Q_{1\text{-}2}$ is written as Q_H because heat is delivered to the gas from a high-temperature reservoir, and for a similar reason, $Q_{3\text{-}4}$ is denoted as Q_L, heat rejected to a low-temperature reservoir.

4.1.2 Four Devices, Circulating Fluid, Water

The flow-cycle heat engine shown in Figure 4.4 contains the four processes in separate units that act on the circulating working fluid (see also Van Ness, 1983, Chapter 4). This heat engine is an idealized steam power plant, either nuclear or fossil. The boiler (or steam generator in a nuclear plant) receives heat from the primary source. The condenser rejects heat to a sink such as a river, lake, or cooling

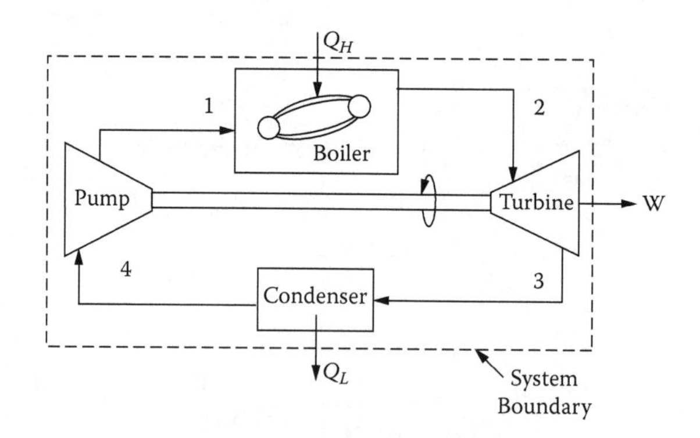

FIGURE 4.4 A heat engine based on a fluid circulating through four heat and work devices (step 1–2 converts saturated liquid to saturated vapor).

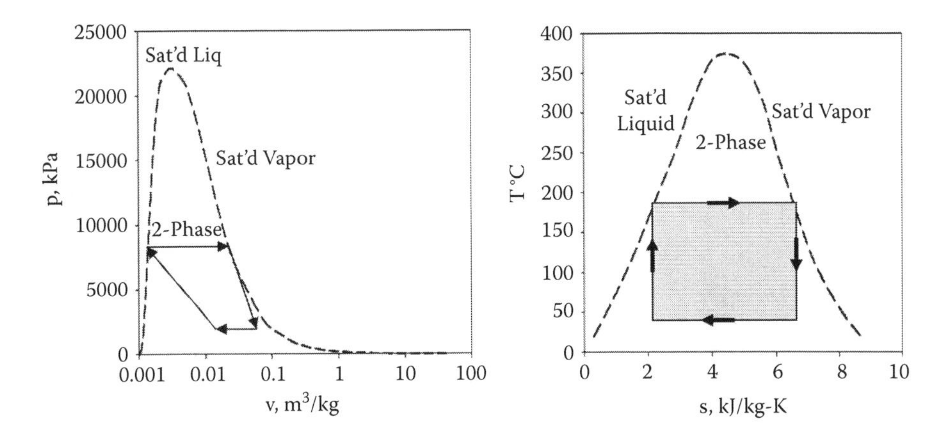

FIGURE 4.5 Process diagrams for the power cycle shown in Figure 4.4 on the equation of state of water.

tower. The turbine produces shaft (i.e., rotating) work, which is then converted to electric power with high efficiency. However, some of the turbine power is consumed internally by the pump needed to keep the fluid circulating in the loop. The net work of the cycle is $W = W_{2\text{-}3} - W_{4\text{-}1}$.

Process diagrams analogous to those in Figure 4.3 for the sequential piston/cylinder power cycle are shown in Figure 4.5 for water/steam as the working fluid. The areas of the four-sided process figures represent the net work W on the left and the net heat Q on the right.

4.1.3 The First Law for Heat Engines

In all four prior illustrations of heat engines, the first law is:

$$Q_H = Q_L + W \tag{4.3}$$

Because of the cyclic nature of the system, there is no change in internal energy or any other property of the working fluid in each cycle.

4.2 THE SECOND LAW APPLIED TO HEAT ENGINE CYCLES

The following qualitative constraints on the heat engine cycles discussed in Section 4.1 were discovered in the nineteenth century and eventually led to the concept of entropy. These constraints restrict the functioning of cycles more severely than does the first law. Both are equivalent statements of the second law, and are ultimately based on empirical evidence. By comparison to Figure 4.1, the disallowed cyclical engines are shown in Figure 4.6.

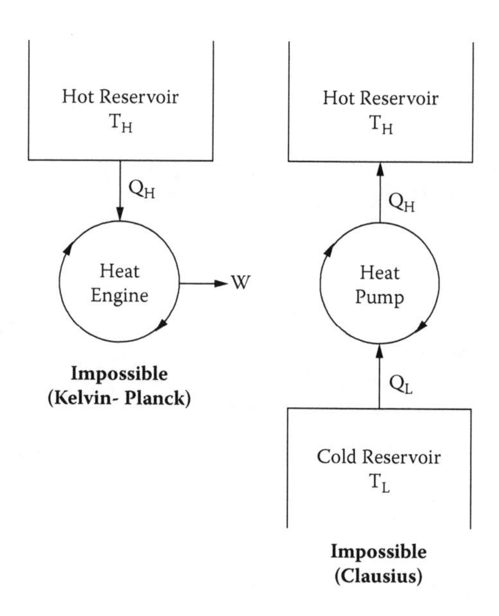

FIGURE 4.6 Power cycles inadmissible by the second law.

Kelvin-Planck: No cycle can produce net work with only a single thermal reservoir.

This is a formal phrasing of a fact that has been noted earlier: heat cannot be completely converted to work. The Kelvin-Planck statement says that Q_L in the power cycle Figure 4.1 cannot be zero.

Clausius: No cycle can only transfer heat from a cold reservoir to a hot reservoir.

This statement essentially prohibits heat from flowing from cold to hot bodies, something that was demonstrated (also using the second law) in Section 1.9. With reference to Figure 4.1, the Clausius statement (right-hand sketch in Figure 4.6) says that the direction of Q_H and Q_L cannot be reversed and at the same time W set equal to zero. This version of the second law does not, however, entirely prohibit transfer of heat from a cold body to a hot body; it simply requires that external work must be expended in order to do so.

The two statements of the second law appear to be quite different, but in fact they are equivalent. This equivalence can be shown with the aid of Figure 4.7, which contains heat engine A and heat pump B operating between the same hot and cold reservoirs.

The devices are arranged so that the heat withdrawn from the cold reservoir by engine B is the same magnitude as the heat rejected by engine A. Therefore, the two cancel and the cold reservoir experiences no net heat transfer. Cycle B violates the Clausius statement. It remains to show that the combination of cycle A and cycle B violates the Kelvin-Planck statement.

Cycle A receives a quantity of heat Q_H from the hot reservoir that is greater than delivered by cycle B. By the first law applied to the system within the dashed box, $Q_H - Q_L$ units of heat are completely converted to work without rejecting heat to the

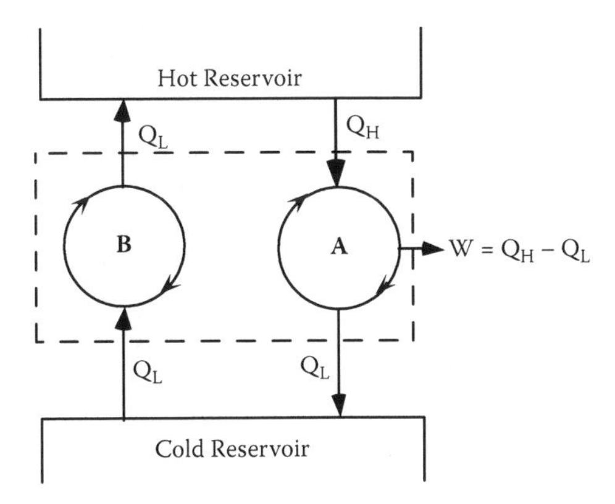

FIGURE 4.7 Demonstration of the equivalence of the two statements of the second law.

cold reservoir, in violation of the Kelvin-Planck statement. The device in Figure 4.7, which is known as a perpetual-motion machine of the second kind, thus fails both the Clausius and Kelvin-Planck versions of the second law.

4.3 THE CARNOT CYCLE

If the Kelvin-Planck form of the second law prohibits complete conversion of heat to work, what then is the maximum value of the efficiency of the cycle shown in Figure 4.1? The cycle efficiency is defined by:

$$\eta = \frac{W}{Q_H} = 1 - \frac{Q_L}{Q_H} \tag{4.4}$$

where the second equality arises from elimination of W using Equation (4.3).

The maximum-efficiency cycle is the one containing the four steps listed in Section 4.1 with the additional restriction that the heat engine be reversible; heat exchange with the reservoirs must take place over infinitesimally small temperature differences, all motions must be frictionless, and rapid compression of the working fluid is prohibited. This idealized heat engine was first investigated by Carnot in 1824, and the cycle and its efficiency bear his name. Carnot showed that the efficiency of this cycle is simply related to the temperature of the hot and cold reservoirs and, moreover, that no other engine operating between the same two temperatures can have a greater efficiency.

Determination of the efficiency of the Carnot cycle starts from the form of the second law given by Equation (1.13) for reversible operation:

$$\Delta S_{engine} + \Delta S_{surroundings} = 0$$

The thermodynamic system in this usage is the Carnot engine and the surroundings are the two reservoirs. The surroundings also contain a mechanism for receiving the work performed by the Carnot engine but this mechanism is not involved in exchange of entropy. Since the working fluid returns to its original state after each cycle, its entropy change is zero. Because all aspects of the heat engine are reversible, no entropy is generated from this device. Consequently, $\Delta S_{engine} = 0$, and by the above equation, $\Delta S_{surroundings} = 0$ as well.

The null entropy change of the surroundings consists of two canceling terms, each of the form given by Equation (1.9). The hot reservoir delivers entropy in the amount Q_H/T_H to the engine and the reject heat transfers entropy equal to Q_L/T_L to the cold reservoir. If the surroundings do not experience an entropy change as a result of operation of the Carnot engine, these two entropy flows must be equal, or:

$$\frac{Q_H}{T_H} = \frac{Q_L}{T_L} \tag{4.5}$$

Combining this result with Equation (4.4) gives the Carnot efficiency:

$$\eta_C = 1 - \frac{T_L}{T_H} \tag{4.6}$$

The above derivation did not utilize the details of the four-step cycle that comprises the Carnot engine. However, the net work and net heat of the cycle in Figure 4.2 given by Equations (4.1) and (4.2) result in the efficiency of Equation (4.6).

Demonstration that the Carnot engine has the maximum possible efficiency is based on the pair of engines shown in Figure 4.8. The method posits a higher efficiency for cycle A than for the Carnot Cycle C operating as a heat pump, then shows that this supposition violates the second law.

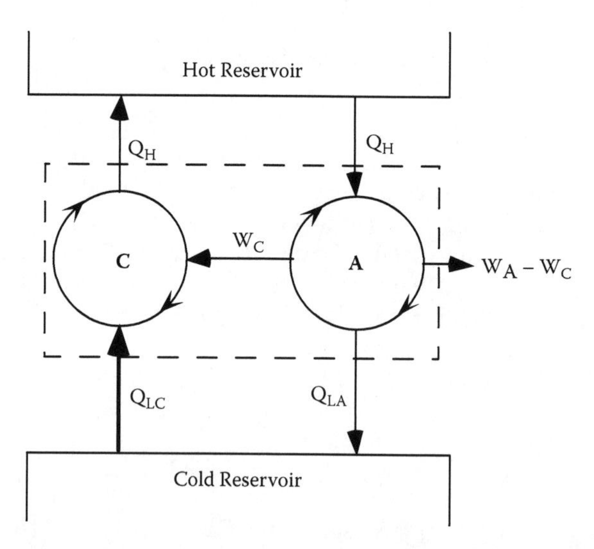

FIGURE 4.8 Diagram to demonstrate that the Carnot efficiency is the highest possible.

Cycles A and C are set to exchange equal quantities of heat with the hot reservoir, which therefore experiences zero net heat transfer. Because cycle A is presumed to be more efficient than Carnot cycle C, the work W_A is greater than the work W_C needed to operate cycle C. Overall, cycles C and A taken as a single system within the dashed box in the drawing receive an amount of heat net $Q_{LC} - Q_{LA}$ from the cold reservoir and produce an equal amount of net work, $W_A - W_C$. This is accomplished without rejecting heat to another reservoir, thus constituting a violation of the Kelvin-Planck statement of the second law. This result demonstrates that the initial premise was false; heat engine A cannot have an efficiency greater than the Carnot efficiency.

Example: Calculate the overall efficiency of a Carnot engine placed between two identical blocks, one initially at temperature T_{10} and the other at temperature T_{20}.

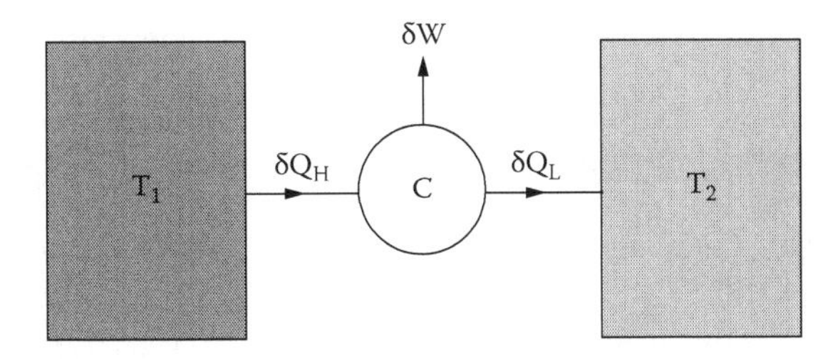

Because the blocks do no work, the first law for each is:

$$du_1 = \delta Q_H = -C_V dT_1 \qquad \text{and} \qquad du_2 = \delta Q_L = -C_V dT_2$$

The second law for the heat engine is:

$$\delta Q_H / T_1 = \delta Q_L / T_2,$$

or, with the first law:

$$dT_1 / T_1 = -dT_2 / T_2$$

Integrating from the initial temperatures to a common final temperature T_f gives:

$$\ln(T_f / T_{20}) = \ln(T_{10} / T_f) \qquad \text{or} \qquad T_f = \sqrt{T_{10} T_{20}}$$

The heat delivered to the engine from the hot source and from the engine to the cold sink is:

$$Q_H = C_V(T_{10} - T_f); \qquad Q_L = C_V(T_f - T_{20})$$

From which the overall engine efficiency follows:

$$\varsigma_c = 1 - \frac{Q_L}{Q_H} = 1 - \frac{T_f - T_{20}}{T_{10} - T_f} = 1 - \sqrt{\frac{T_{20}}{T_{10}}}$$

The table below shows several features of this process for $T_{20}/T_{10} = 0.67$. The middle column gives results for the heat exchange between the two blocks without the heat engine (Figure 1.16). The entropy changes for each block are computed using Equation (3.23).

Property	w/o engine	w/ engine
T_f/T_{10}	0.883	0.816
η_C	0	0.18
$\Delta S_{tot}/C_V$	0.04	0

The final temperature of the blocks with the engine running is less than in the absence of the engine because energy is removed from the overall system in the form of work. The entropy change of the combined system without the engine increases, but there is no change in entropy with the heat engine.

Other problems involving heat engines operating on the Carnot cycle include Problems 4.1 to 4.5, 4.11 and 4.12, and 4.15 and 4.16.

4.4 THERMODYNAMICS OF OPEN SYSTEMS

Before embarking on analyses of practical power cycles, understanding of open (or flow) systems is necessary. The laws of thermodynamics can be applied to fluids flowing through devices that change the properties of the fluid by exchanging heat and/or work with the surroundings. Examples of such devices include pumps, boilers, turbines, valves, nozzles (used for increasing fluid velocity) and orifices in pipes. Some of these devices are included in the simple steam cycle shown in Figure 4.4. In this cycle, the components are connected in series with the working fluid circulating continuously through them. Each device in the cycle is subjected to a first law analysis, and in some, application of the second law provides additional information on their performance.

4.4.1 THE FIRST LAW FOR OPEN (FLOW) SYSTEMS

All of the devices mentioned above can be represented by the schematic open system shown in Figure 4.9. This generalized device has a rigid casing forming a boundary that does no pV work. However, the unit may perform or accept shaft work, W_g, as do the pump and turbine in Figure 4.4. It may also exchange heat with the surroundings, which is the function of the boiler and condenser in the simple steam power cycle.

The principal difference between the system of Figure 4.9 and the closed systems that were analyzed in Chapter 3 is the presence of inlet and outlet flow ports in the

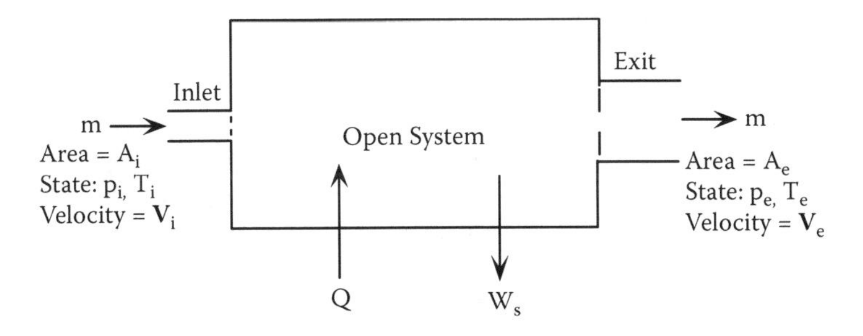

FIGURE 4.9 Schematic of an open system.

open system. The boundary of the open system in Figure 4.9 is composed of the inner surface of the physical device plus the imaginary meshes on the inlet and outlet ports through which fluid enters and leaves the unit. The mass flow rate through the device, expressed in kg/s, is assumed to be constant in time. The properties of the fluid at the entrance and exit are fixed by any two thermodynamic properties at these locations. Pressure and temperature are indicated in Figure 4.9, although any two properties suffice for a single-phase substance and any one property plus the quality fixes all others if the fluid is a vapor-liquid mixture. In addition, the velocities of the fluid entering and leaving are important because they appear in the first law applied to the open system.

The volumetric flow rate is the product of the area of the port, A, and the velocity of the fluid, $\mathbf{V}$:

$$\text{Volumetric flow rate} = A\mathbf{V}$$

The volumetric flow rate can be converted to the mass flow rate $\dot{m}$ by dividing by the fluid's specific volume, giving:

$$\dot{m} = \frac{A\mathbf{V}}{v} \tag{4.7}$$

In the following development, only steady-state operation is treated, so $\dot{m}$ is constant in time and the same at the inlet and exit ports. These are designated by subscripts i and e, respectively. However, the quantities on the right-hand side of Equation (4.7) at the inlet may be different from those at the exit port.

Application of the first law to the device illustrated in Figure 4.9 introduces several features not involved in the first law treatment of closed systems. These are:

1. The internal energy (U or u) that appears in equations such as (1.4) and (3.2) is supplemented by the kinetic energy of the moving fluid, $\frac{1}{2}\dot{m}\mathbf{V}^2$. Changes in gravitational potential energy between the inlet and exit lines should also be added to the total energy, but for simplicity, this component of the energy is not treated here.

2. The internal energy convected through the boundaries of the open system are included in writing the first law.
3. In addition to shaft work, pressure-volume work is inherently involved in pushing the fluid into the inlet and out of the exit line. This work component is called the *flow work* $\dot{W}_f$.* The total work performed by the system is $\dot{W} = \dot{W}_s - (\dot{W}_{fi} - \dot{W}_{fe})$ where $\dot{W}_s$ is the rate at which the system performs shaft work, and the terms in parentheses are the inlet and outlet flow work rates. The sign convention is that of the first law for closed systems: work is positive if performed by the system and heat is positive if added to the system.

The flow work terms are of the pressure-volume type because they involve force acting over a distance, or, equivalently, pressure displacing volume. Thus, the rate of performing flow work is the pressure times the volume flow rate, or

$$\dot{W}_{fi} = p_i A_i \mathbf{V}_i = \dot{m} p_i v_i \qquad \text{and} \qquad \dot{W}_{fe} = p_e A_e \mathbf{V}_e = \dot{m} p_e v_e$$

where Equation (4.7) has been used to replace the volume flow rate in terms of the mass flow rate.

The first law for the steady-state open system equates the net rate of energy transport across system boundaries at the flow ports to the net rate of energy input in the form of heat and work exchanged with the surroundings:

$$\dot{m}\left(u_e + \mathbf{V}_e^2/2 - u_i - \mathbf{V}_i^2/2\right) = \dot{Q} - \left[\dot{W}_s - (\dot{m}p_i v_i - \dot{m}p_e v_e)\right]$$

Note that the properties of the fluid inside the system do not enter the statement of the first law. The reason for this absence is the restriction to steady-state operation.

An important simplification of the above equation is obtained by combining the internal energy terms with the flow-work terms. The combination $u + pv$ is the enthalpy of the fluid, h, so the first law becomes:

$$\dot{Q} - \dot{W}_s = \dot{m}\left(h_e + \mathbf{V}_e^2/2 - h_i - \mathbf{V}_i^2/2\right) \tag{4.8}$$

Dividing Equation (4.8) by the mass flow rate converts the heat and shaft work terms to a per-unit-mass basis rather than a per-unit-time basis:

$$q - w_s = h_e + \mathbf{V}_e^2/2 - h_i - \mathbf{V}_i^2/2 \tag{4.9}$$

The units of $\mathbf{V}^2$ are m^2/s^2. Because a Joule is one $kg\text{-}m^2/s^2$, the units of all terms in Equation (4.9) are J/kg.

* Strictly speaking, the work W with a dot over it is a work rate, or power, but to avoid confusion of terminology, it will be referred to as simply work.

4.4.2 The Second Law for Open Systems

Only in some cases does the second law provide an equation that is as universally useful as the first law in the form of Equation (4.9). The reason is that irreversibilities in the device create entropy that cannot be quantitatively determined from thermodynamics. The second law can be applied to the open system in two ways. Steady-state flow through the device is assumed so that the entropy and heat can be expressed on a per-unit-mass basis.

The second law in the form of Equation (1.10) relates the difference in entropy (per unit mass of the flowing fluid) between the outlet and inlet conditions to the heat added to the system:

$$\Delta s = s_e - s_i \geq q / T \tag{4.10}$$

In the event that the temperature varies with location in the device, the right-hand side of Equation (4.10) is replaced by $\sum \delta q_j / T_j$, the summation of increments of heat δq_j added at local fluid temperature T_j. However, the single-temperature version is used here for simplicity.

An alternate method of applying the second law to the open system is via the total entropy change version of Equation (1.13). This involves the entropy changes of the surroundings, $\Delta s_{surr} = -q/T_{surr}$, where the minus sign appears because heat addition to the system (q) is heat removal from the surroundings. Using the equality in Equation (4.10) for Δs of the system (the moving unit mass of fluid), the total system + surroundings entropy change is:

$$\Delta s_{total} = \Delta s + \Delta s_{surr} = s_e - s_i - q/T_{surr} \geq 0 \tag{4.11}$$

The distinction between Equations (4.10) and (4.11) lies in the level of irreversibility present. If the process is internally reversible (Section 1.7), the equality in Equation (4.10) applies. However, internal reversibility does not guarantee total reversibility; if the heat q is transferred over a nonzero ΔT, an external irreversibility is present. This is revealed by substituting (from Equation [4.10]) q/T for $s_e - s_I$ in Equation (4.11):

$$\Delta s_{total} = \frac{q}{T} - \frac{q}{T_{surr}} \geq 0$$

Only if the process is both internally reversible *and* externally reversible ($T = T_{surr}$) is total reversibility achieved.

Essentially all common applications of the entropy balance are for devices such as pumps, turbines, nozzles, valves and in-line flow components. A subset of these in-line flow devices operates in a nearly adiabatic manner ($q = 0$). If they also function reversibly, then Equation (4.10) becomes:

$$s_e = s_i \tag{4.12}$$

Equations (4.11) and (4.12) are the second law analogs of the first law expressed by Equation (4.9). To the extent that the device can be assumed to operate reversibly, the second law provides another equation for thermodynamic analysis of the open system.

4.4.3 Reversible Work of a Flow System

If a flow device operates reversibly, the minimum work needed for the task can be determined by application of the first and second laws. The first law gives:

$$h_e - h_i = q - w_{s,rev}$$

where $w_{s,rev}$ is the reversible shaft work. The differential of Equation (1.16a), $dh = Tds + vdp$, can be integrated to yield:

$$h_e - h_i = \int_{s_i}^{s_e} Tds + \int_{p_i}^{p_e} vdp = q + \int_{p_i}^{p_e} vdp$$

Because the process is reversible, the Tds integral is the heat added during the process.

From the above two equations, the reversible work is:

$$w_{s,rev} = -\int_{p_i}^{p_e} vdp \tag{4.13}$$

This formula is the open-system analog of the pV work integral in a closed system (Equation [3.1]). The work done by the 100% efficient turbine can be calculated from Equation (4.13) by expressing v as a function of p at constant s using the steam tables. However, if the fluid is a liquid, the specific volume v is approximately independent of pressure and the reversible work simplifies to:

$$w_{s,rev} = -v(p_e - p_i) \tag{4.13a}$$

4.5 PRACTICAL POWER CYCLES

Most real power cycles consist of a working fluid that flows through various devices that change the state of the fluid in a manner that mimics the Carnot cycle. Few power-producing systems are of the single-device type, such as the piston/cylinder used to demonstrate the Carnot cycle in Section 4.3. All practical power cycles contain four distinct steps, or rather two pairs of steps. Each pair of steps is conducted as an iso process. The cycles differ in the thermodynamic properties that are held constant in the pairs of steps. Table 4.1 lists the principal power cycles of practical significance, along with the Carnot cycle for comparison.

TABLE 4.1
Power Cycles

Cycle	Component Steps	Efficiency	Reference
Carnot	2 isentropic; 2 isothermal	$1 - T_L/T_H$	Potter and Somerton (1993, 101); Abbott and Van Ness (1989, 37)
Rankine (water)	2 isentropic; 2 isobaric	—	Potter and Somerton (1993, 149); Van Ness (1983, Chap. 4)
Brayton (gas)	2 isentropic; 2 isobaric	$1-(p_L / p_H)^{(\gamma-1)/\gamma}$	Potter and Somerton (1993, 201)
Otto (gas)	2 isentropic, 2 isochoric	$1-(V_L / V_H)^{\gamma-1}$	Abbott and Van Ness (1989, 38)
Stirling	2 isochoric; 2 isothermal	$1 - T_L/T_H$	Potter and Somerton (1993, 199)
Ericsson	2 isobaric, 2 isothermal	$1 - T_L/T_H$	Potter and Somerton (1993, 199)

The Rankine and Brayton cycles differ in the nature of the circulating fluid. The Rankine cycle (Figure 4.3) is close to the Carnot cycle because water is used as the working medium, so the isobaric steps that involve condensation or vaporization of water also occur isothermally. Cycles that include a pair of isothermal steps (Stirling and Ericsson) have the same theoretical efficiency as the Carnot cycle. The Otto cycle utilizes a gas as a working fluid, and is the basis of the internal combustion engine. The Stirling and Ericsson cycles also normally use gases, but Problem 4.3 analyzes an Ericsson cycle using water/steam as the working fluid.

Equation (4.9) is the starting point for the following analyses of power cycles, beginning with the steam power plant shown in Figure 4.3.

4.5.1 The Rankine Cycle: A Steam Power Plant

This cycle consists of four components:

1. A means of producing heat, typically a hydrocarbon-burning furnace or a nuclear reactor. The actual device that receives this energy and transmits it to the circulating fluid is a boiler or a steam generator, respectively.
2. An apparatus that converts the energy in the flowing steam to shaft work. This is invariably a turbine (Figure 4.10).
3. A condenser that converts the exhaust steam from the turbine to subcooled liquid.
4. A pump that pressurizes the water to the value at which the boiler and turbine operate.

The Rankine cycle is best understood by example. The four components are shown in the flow chart in Figure 4.11 along with the quantities that need to be specified to undertake the analysis. Figure 4.12 is the T–v projection of the equation of state of water (see Figure 2.6) upon which is drawn a process diagram of the cycle. The high and low temperatures of the cycle are labeled T_H and T_L, respectively,

500 MW Steam Turbine

Turbine Rotor

FIGURE 4.10 A steam turbine.

in order to compare the efficiency to that of a hypothetical Carnot engine operating between these temperatures.

High-pressure steam (state 1). The enthalpy and entropy of the superheated steam entering the turbine are obtained from Table 2.3 and the pressure and temperature at this location.

Turbine exhaust (state 2). Here water is a two-phase mixture with the enthalpy and entropy of the saturated liquid and the saturated vapor determined by the specified pressure and Table 2.2. The given steam quality provides the mixture enthalpy via Equation (2.22) and the analogous equation for the mixture entropy.

Condensed water (state 3). The pressure and temperature determine the remaining properties of the condensed liquid from table 2.4.

Compressed water (state 4). The enthalpy at the exit of the condenser is obtained from extrapolation of the data in table 2.4 to low pressure.

With the above information in hand, the heat and work associated with each component can be determined by application of the first law in the form of Equation (4.9).

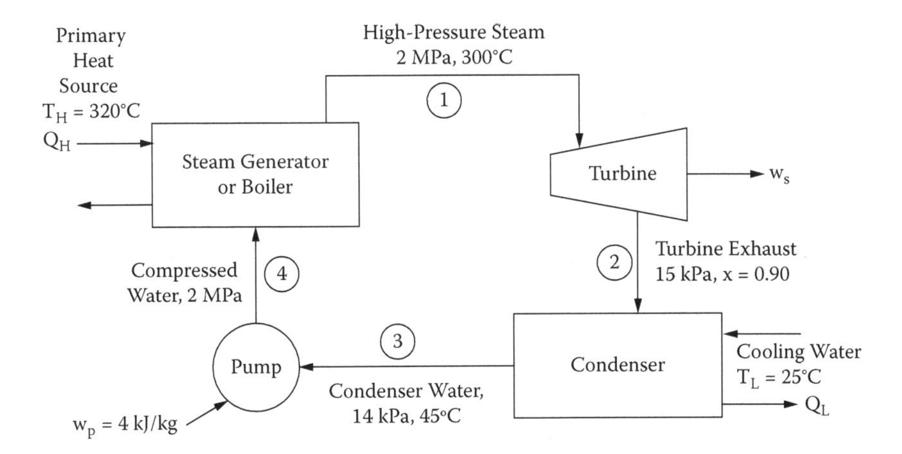

FIGURE 4.11 Flow chart of a Rankine-cycle power plant.

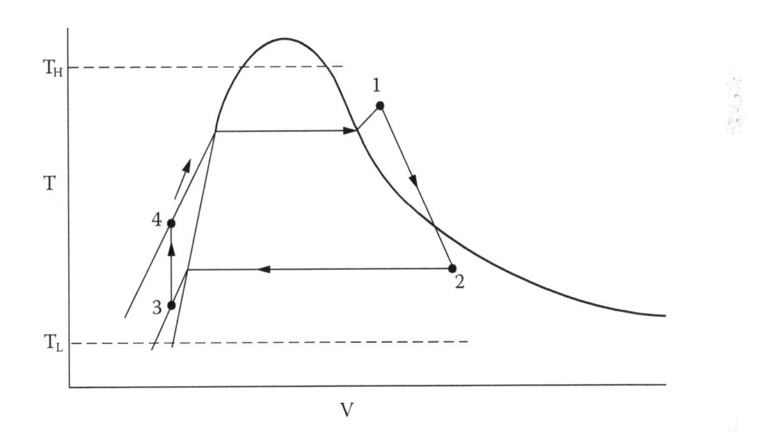

FIGURE 4.12 The Rankine cycle process diagram superimposed on the EOS of water.

In all cases, the kinetic energy terms are neglected because V^2 is small compared to the enthalpy terms. The results are summarized in Table 4.2.

The numbers in ordinary type are specifications, which also neglect pressure losses in the boiler and condenser. Values in bold type are obtained from the steam tables using the input specifications. The underlined numbers are determined by application of the first law to the values of the previous two categories. Specifically:

Turbine. The heat loss from a large turbine is negligible compared to the shaft work produced. Therefore, this component can be assumed to operate adiabatically, or $q_{1\text{-}2} = 0$. With these eliminations, Equation (4.9) reduces to: $w_{1\text{-}2} = h_1 - h_2$, the shaft work produced by the turbine.

Condenser. The condenser is responsible for supplying the reject heat of the cycle to the low-temperature reservoir. The reject heat is obtained by condensing the exhaust steam from the turbine and subcooling the liquid.

TABLE 4.2
First Law/EOS Analysis of a Rankine Cycle

Component	Site	p (kPa)	T (°C)	x	h (kJ/kg)	s (kJ/kg)	Work (kJ/kg)	Heat (kJ/kg)
	1.	2000	300	1.0	***3025***	***6.77***		
Turbine							*661*	0
	2.	15	***54***	0.9	***2364***	***7.29***		
Condenser							0	*2178*
	3.	15	45	0	***186***	***0.643***		
Pump							4	0
	4.	2000	*45*	0	*190*	***0.637***		
Boiler							0	*2835*
	1.	2000	300	1.0	***3025***	***6.77***		

(This is necessary because the pump cannot handle a two-phase mixture.) The first law for the condenser is $q_{2\text{-}3} = q_L = h_2 - h_3$.

Pump. Cycle specifications include the work required to operate the pump. The first law utilizes this number to determine the enthalpy at the pump outlet $h_4 = h_3 + w_{3\text{-}4}$.

Boiler. This component receives the heat input to the cycle from the hot reservoir. Inside this unit, the inlet subcooled liquid is heated to saturation, completely vaporized, and the steam is superheated. The first law for the boiler is $q_{4\text{-}1} = q_H = h_1 - h_4$.

4.5.1.1 Cycle Efficiency

The net (shaft) work produced by the cycle is:

$$w_S = w_{2\text{-}3} - w_{4\text{-}1} = 658 \text{ kJ/kg}$$

and the efficiency of the Rankine cycle, as defined by Equation (4.4), is:

$$\eta_R = \frac{w_S}{q_H} = \frac{657.7}{2851.9} = 0.23\ (23\%)$$

4.5.1.2 Use of the Second Law to Calculate Efficiencies of Work-Producing and Work-Consuming Devices

Even in cases where an adiabatic flow component such as a pump or a turbine is not reversible, Equation (4.12) provides a means of calculating its efficiency, or the ratio of the actual work produced or consumed to that for perfectly reversible operation.

Turbine. The first step is to determine the exit steam quality if the turbine were reversible. If this were so, the entropy of the inlet superheated steam (location 1) should be equal to the entropy of the exhaust steam, or $(s_2)_{rev} = s_1 = 6.77$ J/kg-K. Together with the prescribed 15 kPa outlet pressure and the associated saturated liquid/vapor entropies, the outlet quality would be:

$$x_{rev} = \frac{(s_2)_{rev} - s_f}{s_g - s_f} = \frac{6.766 - 0.755}{8.009 - 0.755} = 0.83$$

A reversible turbine with the same inlet steam conditions and the same exhaust pressure condenses 7% more steam than does the actual turbine (Figure 4.11). In the reversible situation, the exit enthalpy is evaluated from x_{rev} and the enthalpies of the saturated phases at location 2:

$$(h_2)_{rev} = h_f + x_{rev}h_{fg} = 225.9 + 0.83 \times 2373.1 = 2192 \text{ kJ/kg}$$

and the reversible work is:

$$(w_{1\text{-}2})_{rev} = h_1 - (h_2)_{rev} = 3025 - 2192 = 833 \text{ kJ/mole}$$

Using the actual work calculated previously in conjunction with the above reversible work gives the efficiency of the turbine:

$$\text{turbine efficiency} = \frac{w_{1-2}}{(w_{1-2})_{rev}} = \frac{661}{833} = 0.79 \quad (79\%)$$

Additional problems involving steam turbines in tandem with other components of a power generation system are Problems 4.6 and 4.13.

Pump. The efficiency of the pump in Figure 4.11 is calculated using Equation (4.13a). For compressed liquid water, $v \sim 0.001$ m^3/kg. The pump's inlet pressure is 15 kPa and the outlet pressure is 2 MPa. Using these values in Equation (4.13a) yields:

$$(w_{3\text{-}4})_{rev} = 0.001(2000 - 15) = 2 \text{ kJ/kg}$$

The actual work required for pump operation is 4 kJ/kg, so the pump efficiency is 50%.

4.5.1.3 Total Entropy Increase

The turbine and pump efficiency calculations indicate the presence of irreversibilities in the Rankine power cycle. The entropy increase of the cycle (the system) and the reservoirs (the surroundings) are expressed by:

$$\Delta S_{tot} = \Delta S_{fluid} + \Delta S_{pump} + \Delta S_{turbine} + \Delta S_{thermal\ reservoirs} \tag{4.14}$$

Because the working fluid (water) cycles through four states but returns to the starting state,

$$\Delta S_{fluid} = 0$$

The component irreversibilities are obtained from Table 4.2:

$$\Delta S_{pump} = 0.643 - 0.637 = 0.006$$

$$\Delta S_{turbine} = 7.29 - 6.77 = 0.52$$

External irreversibilities due to heat exchange with the hot and cold reservoirs contribute:

$$\Delta S_{thermal\ reservoirs} = -\frac{Q_H}{T_H} + \frac{Q_L}{T_L} = -\frac{2835}{593} + \frac{2178}{298} = 2.53$$

This portion assumes that the temperature of cooling water for the condenser is 25°C and the high-temperature supply from the boiler/steam generator is 320°C.

Substituting the component entropy gains into Equation (4.14) gives the entropy production per kg of water circulated:

$$\Delta S_{tot} = 0 + 0.01 + 0.52 + 2.53 = 3.06 \text{ kJ/kg-K}$$

The largest contribution is due to heat transfer over nonzero temperature differences between the cycle components and their associated thermal reservoirs.

4.5.2 The Brayton Cycle: A Gas Turbine

The Brayton cycle utilizes a gas (e.g., air, He) as the working fluid which, unlike the water Rankine cycle, is directly heated by the primary energy source. The latter can be either a combustor wherein hydrocarbon fuel burns in air, or a nuclear reactor, in which fission energy is transformed into heat that is removed by the flowing coolant gas. These two variations are shown in Figure 4.13 along with the *T-S* process diagram for each.

The combustion version is not a true cycle in that the coolant air is not recycled to the inlet of the compressor, which draws in fresh air. With the nuclear reactor heat source, the coolant is high-purity helium, which is recycled to the compressor after passing through a cooler. In the combustion version, the hot exhaust simply dissipates heat to the atmosphere. In both cases, states 1 and 2 are at a high pressure p_H and states 3 and 4 are at a low pressure p_L. The output of the turbine (W_T) is divided between the net work produced by the cycle (W) and the work required to operate the compressor (W_C), or $W = W_T - W_C$. The first law for the cycle is $W = Q_H - Q_L$, where Q_H is the energy delivered to the gas and Q_L is the heat rejected to the low-temperature reservoir.

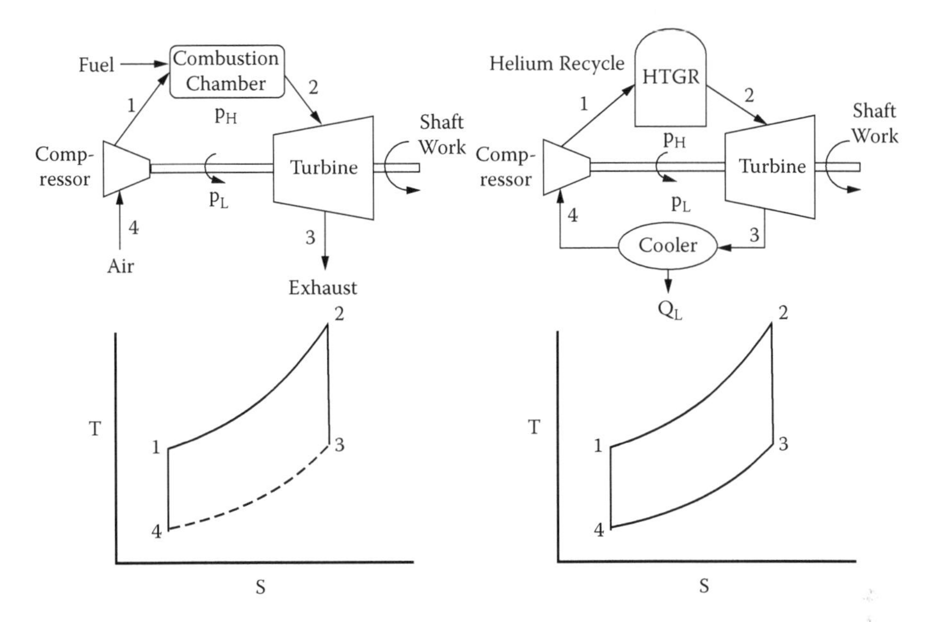

FIGURE 4.13 Two types of Brayton cycles. Left: combustion heat source. Right: nuclear heat source (HTGR means high-temperature gas-cooled reactor).

Focusing on the right-hand diagram of Figure 4.13, the cycle efficiency is:

$$\eta_B = \frac{W}{Q_H} = \frac{Q_H - Q_L}{Q_H} = 1 - \frac{Q_L}{Q_H} \tag{4.15}$$

Flow through the nuclear reactor and the cooler is (approximately) isobaric, so for an ideal gas,

$$Q_L = C_P(T_3 - T_4) \qquad \text{and} \qquad Q_H = C_P(T_2 - T_1)$$

Substituting these into Equation (4.15) yields:

$$\eta_B = 1 - \frac{T_3 - T_4}{T_2 - T_1} = \frac{T_4}{T_1}\frac{T_3/T_4 - 1}{T_2/T_1 - 1}$$

Assuming that flow through the compressor and the turbine is isentropic, Equation (3.14) gives:

$$\frac{T_3}{T_2} = \frac{T_4}{T_1} = \left(\frac{p_L}{p_H}\right)^{\frac{\gamma-1}{\gamma}}$$

which leads to: $T_3/T_4 = T_2/T_1$ and to the final result:

$$\eta_B = 1 - \left(p_L / p_H\right)^{\frac{\gamma-1}{\gamma}} \tag{4.16}$$

Contrary to the Carnot cycle, the Brayton cycle efficiency is determined by the high and low pressures but is independent of the temperature of the low-temperature thermal reservoir (there is no high-temperature thermal reservoir).

4.5.3 The Refrigeration Cycle (Heat Pump)

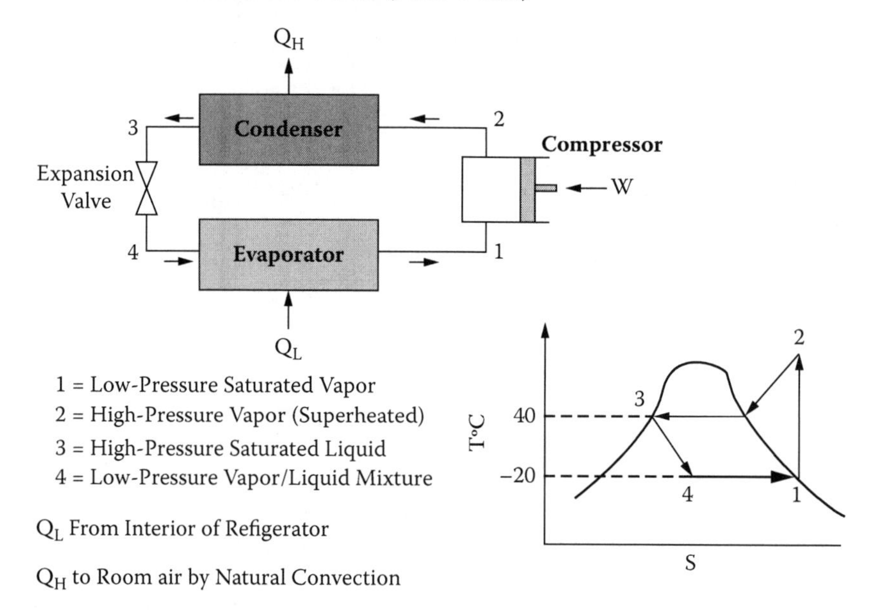

FIGURE 4.14 A refrigeration cycle.

This cycle utilizes a condensable fluid such as ammonia (NH_3) or freon and resembles a reverse Rankine cycle. A typical refrigeration cycle is shown in Figure 4.14. The heat extracted from a low-temperature reservoir converts a two-phase mixture to saturated vapor, which is then compressed adiabatically and (nearly) reversibly. On the *T-s* diagram, the line from state 1 to state 2 is shown as isentropic. The condenser rejects heat to a high-temperature reservoir and in the process converts the superheated vapor to saturated liquid. The final unit is an expansion valve which, by reducing the pressure of the flow, evaporates some of the liquid to produce the two-phase mixture for the cycle to repeat. This component is highly irreversible, which accounts for the tilt of the 3–4 arrow in the *T-s* diagram sketched above.

A numerical example of the conditions at various states is shown in Table 4.3. The underlined numbers are unit specifications and correspond to the values on the process/EOS diagram in Figure 4.14. The horizontal lines point to values obtained from the freon equivalent of the steam tables; for example, at –20°C, the saturation pressure is 1.5 atm and the enthalpy and entropy of the saturated vapor are 179 kJ/kg

TABLE 4.3
A Freon Refrigeration Cycle

State	T °C	p, atm	h, kJ/kg	s J/kg-K	quality
1	20	1.5	179	0.71	1
2	51	9.6	211	0.71	1
3	40	9.6	75	-	0
4	-20	1.5	75	-	0.35

Specifications:

1. The condenser operates at 40°C
2. The evaporator " " -20°C
3. State 1 is saturated vapor (x = 1)
4. State 3 is saturated liquid (x = 0)

From "Freon tables" → XXX

Process Linkage ↓

and 0.71 kJ/kg-K, respectively. The vertical arrows are the result of process restraints (constant pressure, enthalpy, or entropy). Thus, the arrows between states 2 and 3 approximate an isobar. Similarly, the enthalpies at states 3 and 4 are equal because of the first law for a flow device (the expansion valve) with no heat or work exchanged with the surroundings. Finally, the compressor is assumed to be both adiabatic and reversible, so the entropy at states 1 and 2 are the same. T_2 is fixed by intersection of the isobar originating at state 3 and the isentrope starting from state 1.

Additional applications of the first law yield the following information:

$$\text{Condenser: } Q_H = h_2 - h_3 = 136 \text{ kJ/kg}$$

$$\text{Evaporator: } Q_L = h_1 - h_4 = 104 \text{ kJ/kg}$$

$$\text{Compressor (shaft) work} = Q_H - Q_L = h_2 - h_1 = 32 \text{ kJ/kg}$$

Instead of an efficiency, the quality of the refrigeration operation is described by the *coefficient of performance*:

$$\text{COP} = Q_L/W = 104/32 = 3.2$$

The power requirements for a typical household refrigerator with this COP, a cooling capacity (Q_L) of 3.1 kW and a freon flow rate of 0.03 kg/s is 3.1/3.2 ~ 1 kW.

4.6 FLOW DEVICES WITHOUT SHAFT WORK

Numerous categories of common industrial fluid-flow equipment operate in essentially adiabatic fashion and do not exchange work with their surroundings. Throttling

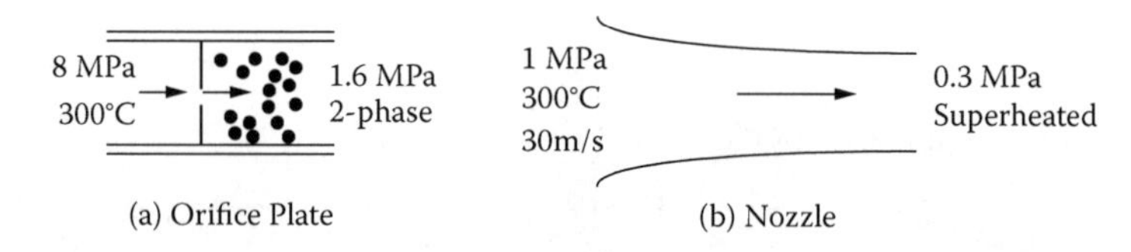

FIGURE 4.15 Flow devices that are both adiabatic and do not perform or accept work.

devices such as orifice plates or their more complicated cousins, valves, provide a pressure drop without changing velocity. Nozzles, on the other hand, are designed to speed up the flow of a gas. These devices are represented in Figure 4.15.

The two classes of flow devices differ in two principal ways: fluid velocity changes and reversibility. In both cases, the open system consists of the section of pipe containing the constriction and imaginary meshes perpendicular to the flow direction that are sufficiently far upstream and downstream to escape perturbations of the flow by the device. The inlet fluid condition is completely fixed (thermodynamic state and velocity) and the outlet condition is partially specified. The first law, and if applicable, the form of the second law for reversible processes, are used to calculate the remaining downstream conditions. The mass flow rate is constant in time, so the analysis is on the basis of a unit of flowing mass.

4.6.1 Orifice or Valve

For an orifice or a valve, the first law reduces to:

$$h_e = h_i \tag{4.17}$$

Example: What is the outlet steam quality in Figure 4.15(a)? The assumption that condensation occurs in the outlet stream needs to be verified. Using the upstream pressure and temperature in the superheated steam Table 2.3, the entrance and exit enthalpies are: $h_i = h_e = 2785$ kJ/kg. This enthalpy is slightly less than the enthalpy of the saturated vapor at the downstream pressure of 1.6 MPa. Therefore the outlet water is in the two-phase region and the temperature, obtained from Table 2.2, is $T_e = 201$°C. The quality of the downstream steam is:

$$x_e = \frac{h_e - h_f}{h_{fg}} = \frac{2785 - 859}{1935} = 0.995$$

Suppose, however, that the downstream pressure is specified as 1.0 MPa. In this case, h_e is greater than the enthalpy of the saturated vapor, and the downstream steam remains superheated. To determine the final state of the gas, steam table A.3 is entered with the combination p_e =1.0 MPa, h_e =2785 kJ/kg, which yields a downstream temperature of 183°C.

That flow through an orifice is irreversible seems intuitive because of fluid turbulence (or at least laminar friction) created by the abrupt reduction in flow cross section. This qualitative assessment can be verified by comparing the entropies before and after passage through the orifice. At 8 MPa and 300°C, table A.3 gives $s_i = 5.79$ kJ/kg-K. For the downstream condition ($p_e = 1.0$ MPa, $h_e = 2785$ kJ/kg), $s_e = 6.60$ kJ/kg-K. Thus, $s_e > s_i$, proving that gas flow through an orifice is indeed an irreversible process.

4.6.2 Nozzle

The first law applied to a nozzle is:

$$h_i + \tfrac{1}{2}V_i^2 = h_e + \tfrac{1}{2}V_e^2 \tag{4.18}$$

Because of the smooth shape of the walls, the flow is approximately reversible. Because the system is also adiabatic, the second law provides the additional relation:

$$s_i = s_e \tag{4.19}$$

In order to relate the constant-entropy condition of this particular type of open system to the changes in pressure and temperature of the ideal gas, Equation (3.14) is employed. Even though this equation was derived for changes in a closed system, it is applicable to open systems as well. The reason is that this equation involves only thermodynamic properties, and so depends only on the initial and final states of the fluid, not on the process that caused the change. Another way of viewing isentropic flow is to imagine that the flow consists of small packets of fluid acting as closed systems undergoing reversible adiabatic expansion or compression as they move from the inlet to the outlet.

Once T_e is determined from Equation (3.14), the first law can be applied to complete the solution of the problem.

Example: What are the exit temperature and velocity of steam flowing through the nozzle of Figure 4.15(b)? Steam is assumed to behave as an ideal gas. The specific heat of steam (C_P) is 33 J/mole-K (Figure 2.4) or 1833 J/kg-K. The heat capacity ratio $C_P/C_V = 1.337$.

Applying Equation (3.14) yields:

$$T_e = T_i\left(\frac{p_e}{p_i}\right)^{\frac{\gamma-1}{\gamma}} = 573\left(\frac{0.3}{1.0}\right)^{0.252} = 423\ K\ \ (150°\text{C})$$

Since the enthalpy change $h_e - h_i = C_P(T_e - T_i)$ for an ideal gas, Equation (4.18) can be solved for the exit velocity:

$$V_e = \left[V_i^2 + 2C_P\left(T_i - T_e\right)\right]^{1/2} = \left[30^2 + 2\times 1833(573-423\right]^{1/2} = 740\ \text{m/s}$$

The area reduction in the converging nozzle needed to generate this increase in steam velocity is calculated using Equation (4.7) for steady-state flow:

$$\frac{A_i V_i}{v_i} = \frac{A_e V_e}{v_E}$$

The specific volume ratio is obtained from the ideal gas law:

$$\frac{v_e}{v_i} = \frac{T_e}{T_i}\frac{p_i}{p_e} = \frac{423}{573}\frac{1.0}{0.3} = 2.46$$

and

$$\frac{A_e}{A_i} = \frac{v_e}{v_i}\frac{V_i}{V_e} = 2.46 \times \frac{30}{740} = 0.10$$

Other applications of the first and second laws to flow devices are found in Problems 4.7, 4.8, and 4.13.

4.7 SUMMARY: PROPERTIES OF FLOW DEVICES

Table 4.4 collects the essential features of each of the open systems analyzed in this chapter.

TABLE 4.4
Properties of Flow Devices

Device	Δ(KE)	Heat	Work	Reversible
Pump	No	No	Yes (in)	Yes (approx.)
Turbine	No	No	Yes (out)	Yes (approx.)
Condenser/boiler	No	Yes	No	No
Valve	No	No	No	No
Nozzle	Yes	No	No	Yes

PROBLEMS

4.1 A cyclical engine receives 325 kJ of heat from a reservoir at 1000 K, delivers 200 kJ of work and rejects 125 kJ of heat to a reservoir at 400 K. Does this engine violate either the first law or the second law?

4.2 Consider a Carnot cycle of 25% efficiency using water as the working fluid. Heat transfer to the engine takes place at 300°C, during which the water changes from saturated liquid to saturated vapor. Similarly, heat rejection occurs isothermally as the vapor condenses.

(a) Sketch this Carnot cycle superimposed on the *T-s* diagram of water (below).
(b) What is the temperature of the heat sink?

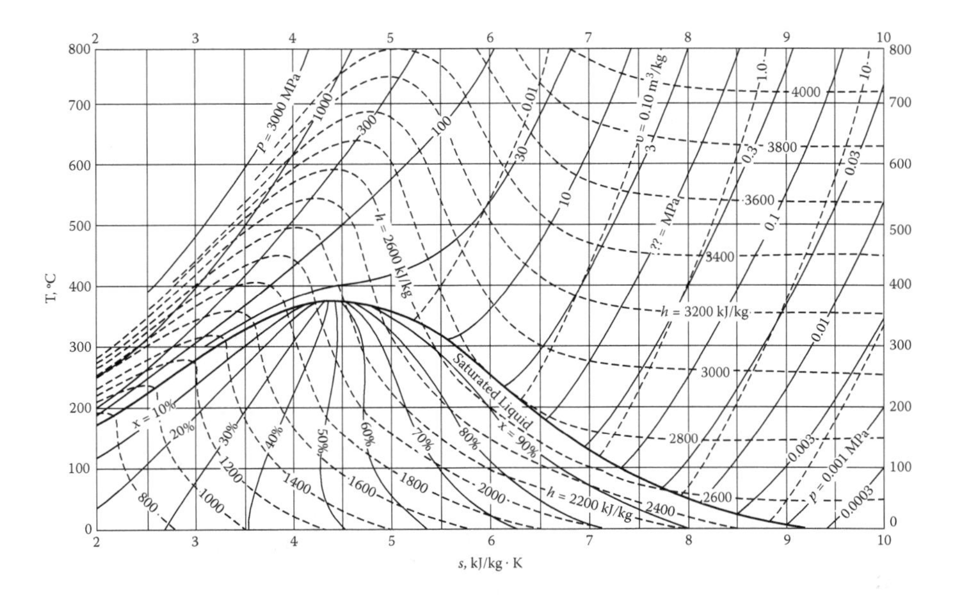

(c) What is the steam quality at the beginning and end of the heat-rejection step?

(d) What is the work done by the engine (per kilogram of water)?

4.3 Starting from state 1 in the two-phase region, 1 kg of water undergoes the cycle shown below. Steps 1-2 and 3-4 are isobaric and reversible and steps 2-3 and 4-1 are adiabatic and reversible. The system remains in the two-phase region throughout the cycle.

(a) Prepare a table with the following column headings: State, Temperature, Pressure, Specific Volume, and Quality. Fill in this table for the four states shown in the diagram.

(b) What is the efficiency of this cycle?

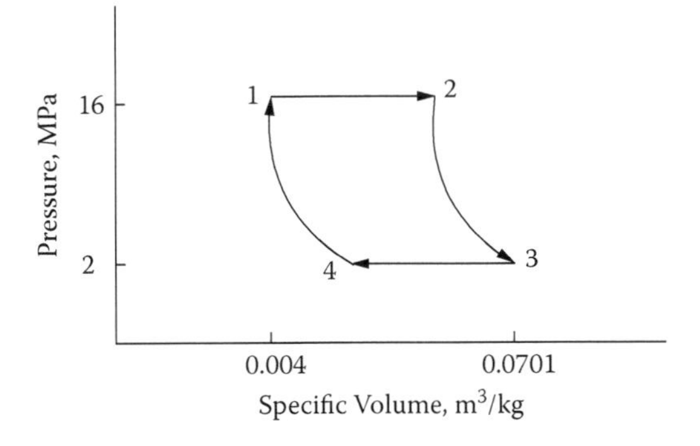

4.4 The sketch below shows two identical, interacting Carnot engines. The first operates between a reservoir at 600 K and rejects heat at a temperature T to the second cycle, which rejects its heat to a reservoir at 300 K. If the two cycles have the same efficiency, what is the intermediate temperature T?

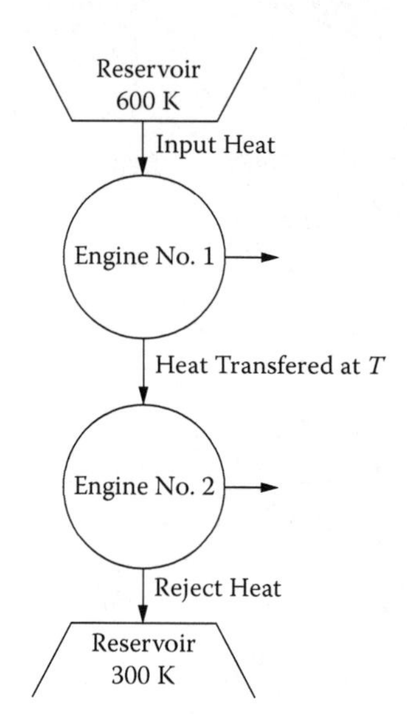

4.5 A Carnot engine operating in outer space receives heat from a nuclear power source at T_H. The engine rejects heat by thermal radiation at a rate given by $\dot{Q}_L = AKT_L^4$, where A is the area of the radiating surface and K is a constant. T_L is the temperature of the radiator, which is considered to be the low-temperature reservoir for the engine. Show that if the power is fixed at $\dot{W}$, the area of the radiator is a minimum when $T_L/T_H = 3/4$. What is the minimum radiator area in terms of $\dot{W}$, K, and T_H?

4.6 A 100% efficient, adiabatic turbine in a power cycle of the type shown in Figure 4.3 is fed with superheated steam at 325°C and 10 MPa and produces an exhaust of 90% quality. Determine:

(a) The exit pressure and temperature.
(b) The volume fraction of steam in the exit gas.
(c) The work produced per unit mass of inlet steam.
(d) The condenser in the power cycle is cooled by water at 20°C. What is the cycle efficiency compared to the maximum possible value? Neglect pressure drops in the circuit (except for the pump and turbine) and assume that the pump is reversible and that heat transfer in the boiler is reversible.

4.7 A nuclear rocket is propelled by lithium vapor in a stream of helium. A 50-50 mole percent mixture is contained in the coolant vessel. The gas mixture is heated to 3000 K and 10 MPa in a nuclear reactor and then expanded through an adiabatic nozzle. The exhaust temperature is 500 K.

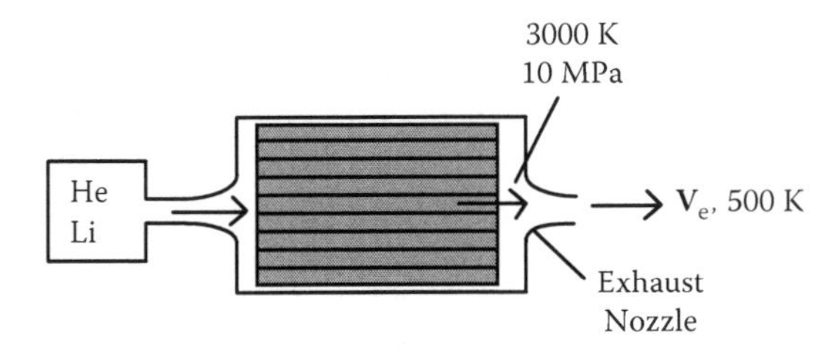

PE: verify horiontal lines in figure.

Calculate the nozzle exhaust velocity (V_e) for the following conditions during the expansion:

(a) The lithium remains gaseous throughout the expansion even though it becomes supercooled during the process. Helium and lithium can be treated as ideal monatomic gases.
(b) During the expansion, lithium remains gaseous until the temperature has dropped to 1600 K. At this point, it completely condenses and remains liquid thereafter. The enthalpy change of liquid lithium from 500 to 1600 K is 22.9 kJ/mole and the heat of vaporization at 1600 K is 147.5 kJ/mole.

4.8 Superheated steam enters an insulated pipe attached to a large reservoir at 3 MPa and 350°C. In the pipe, the pressure is 1.6 MPa and the velocity is 550 m/s. The steam mass flow rate is 0.5 kg/s. Calculate:

(a) The exit quality.
(b) The exit temperature.
(c) The pipe cross-sectional area.

4.9 A helium gas turbine operates isentropically with an inlet pressure of 3 MPa and an exhaust pressure of 0.1 MPa. The inlet temperature is 800 K.

(a) Although the process is isentropic and the gas is ideal, Equation (3.16) does not apply in this case. Explain why.
(b) How much shaft work can be obtained per mole of gas flowing through the turbine?
(c) Modify Equation (3.19) so that it is similar in form to the shaft work equation derived in Part (b). What is the difference in these two work formulas?

4.10 A helium gas turbine operates adiabatically with the following specifications: inlet temperature and pressure: 1100 K, 3 MPa. Outlet pressure: 0.5 MPa. When built and operated, the outlet temperature is 700 K. What is the efficiency of the turbine?

4.11 A Carnot engine with air as the working fluid operates on the *p-v* cycle shown in Figure 4.2. If the high-temperature reservoir is maintained at 200°C and the conditions at state 4 are 200 kPa and 0.3 m^3/kg, what is the efficiency of the engine?

4.12 A 1000 MW (electric) power plant uses steam as the working fluid and the condenser is cooled by river water (see sketch). The maximum steam temperature is 550°C and the condenser pressure is 10 kPa. Estimate the minimum temperature rise of the river downstream.

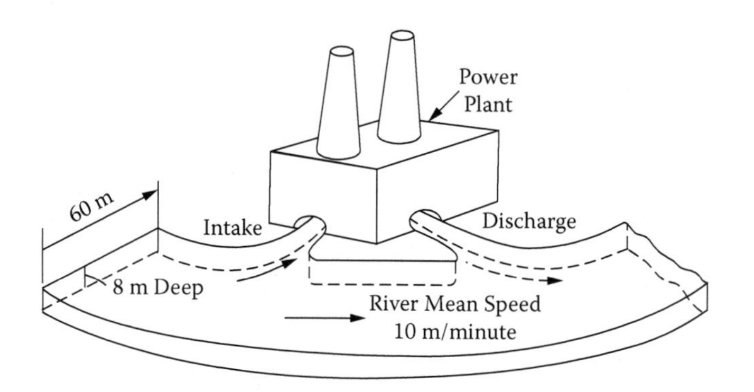

4.13 A geothermal supply of hot water at 500 kPa and 150°C is fed to an insulated evaporator at the rate of 1.5 kg/s. The evaporator contains vapor and liquid water and operates at 200 kPa. Saturated liquid is drained from the bottom of the evaporator and the saturated vapor is fed to a turbine. The turbine has an efficiency of 70% at an exit pressure of 15 kPa.

(a) Draw a flow diagram of the evaporator/turbine combination and label all streams entering or leaving the units with their thermodynamic properties(e.g., pressure, temperature, enthalpy, entropy).
(b) Calculate the flow rates of liquid and steam leaving the flash evaporator.
(c) Determine the quality of the steam exiting the turbine. (Hint: first determine the work produced by the turbine if it were isentropic, or 100% efficient.)
(d) Determine the rate of entropy production, $\dot{S}_{irr}$, in the combined evaporator/turbine.

4.14 A 100% efficient adiabatic gas turbine normally operates with an inlet gas at 4 MPa pressure and 1000°C. To reduce turbine power, a throttle valve in the line before the turbine inlet is partially closed. This reduces the inlet gas pressure to 3 MPa. The exhaust pressure is 1 MPa in both cases. The gas is ideal and has a heat capacity ratio of 1.4. At what percentage of full power does the turbine operate in the throttled condition?

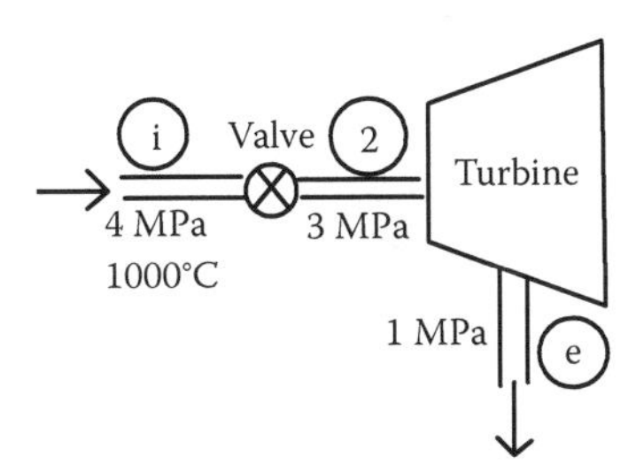

4.15 A block of metal at temperature T_o is cooled by a heat reservoir at T_c, which remains constant during the process. Cooling of the block from T_o to T_c is accompanied by a quantity Q_H of heat. If the block and the reservoir had been connected by a reversible heat engine, the quantity of heat absorbed by the reservoir during the block cooling process would have been reduced to Q_C. What is the ratio Q_C/Q_H in terms of T_o and T_c?

4.16 The mixing equipment shown below preheats 45°C water (input 1) using superheated steam at 250°C (input 2). The unit is adiabatic and operates at 600 kPa. Assuming that the two water streams are saturated, calculate the temperature of the heated water (output 3).

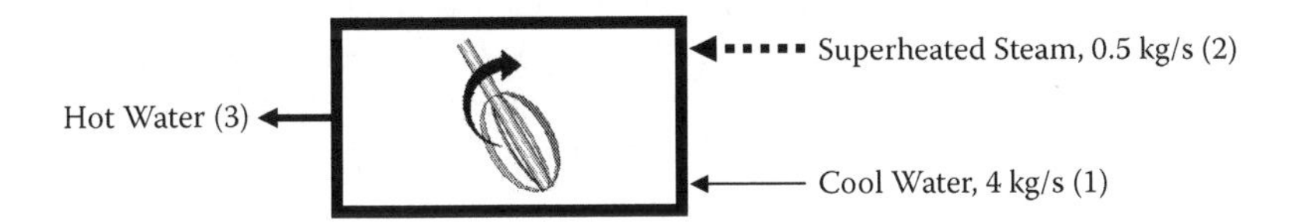

4.17 Consider the Carnot cycles for the open system (Figures 4.4 and 4.5) and the closed system (Figures 4.2 and 4.3).

(a) How do the calculations of the net cycle work differ?
(b) Why are the 1 → 2 and the 3 → 4 processes different?
(c) What do the 2 → 3 and the 4 → 1 lines in the two plots have in common?
(d) What is different about the fluids that the 2 → 3 and the 4 → 1 lines in the two plots represent?
(e) Why can't an ideal gas be used in the Carnot-type flow cycle?

4.18 A reversible, isothermal gas turbine is connected to two closed gas reservoirs. Initially, the upper reservoir contains n_T moles of an ideal gas and the lower reservoir is a vacuum. The volumes of the upper and lower reservoirs are V_U and V_L, respectively. The entire apparatus remains at a constant temperature T during the process. The process consists of flow of gas from the upper reservoir to the lower reservoir through the turbine until the pressures equalize.

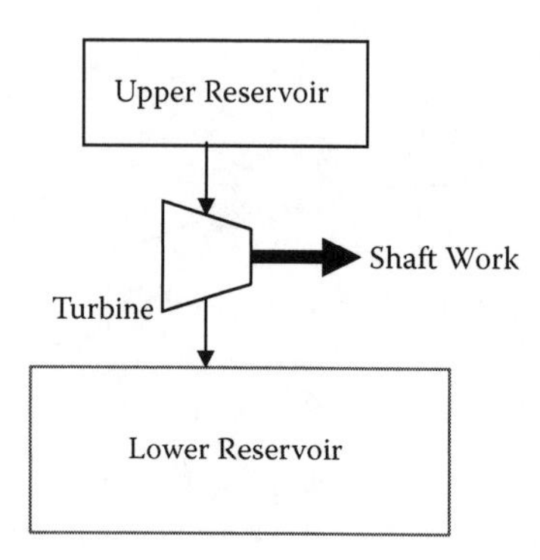

(a) How much work is performed? (Hint: establish a relation between the differential work δW_S and the differential number of moles dn_U flowing through the turbine. This relation will contain the ratio of the pressures in the two reservoirs, p_U/p_L, which must be related to the current number of moles of gas in the upper vessel, n_U.)

(b) How much heat is absorbed by the turbine in the process? (*Part (a) need not be solved to answer this part.*)

4.19 A reversible helium gas turbine operates with inlet and outlet pressures of 4 and 0.5 MPa, respectively. The inlet gas temperature is 1000°C. However, the turbine is not adiabatic, which causes the temperature of the gas moving through the unit to decrease faster than in the absence of heat loss. The pressure–temperature relationship of the gas flowing through the turbine is:

$$\frac{T}{T_i} = \left(\frac{p}{p_i}\right)^{0.5}$$

Using the second law only, determine the fraction of the enthalpy loss of the gas between the outlet and inlet of the turbine that escapes as heat to the environment.

4.20 A helium gas turbine operates isentropically with an inlet pressure of 2 MPa and temperature of 300°C. If the unit produces 3 kJ of shaft work per mole of He, what is the outlet pressure?

REFERENCES

Abbott, M. and H. C. Van Ness. 1989. *Theory and Problems in Engineering Thermodynamics*. New York: McGraw-Hill.

Potter, M. and C. Somerton. 1993. *Theory and Problems in Engineering Thermodynamics*. New York: McGraw-Hill.

Van Ness, H. C. 1983. *Understanding Thermodynamics*. Mineola, NY: Dover.

5 Phase Equilibria in One-Component Systems

5.1 INTRODUCTION

Just as the thermodynamic properties of single-phase, one-component gases, liquids and solids are of considerable practical importance, so too are the properties of their two-phase mixtures. In Chapter 2, the properties of ideal and real one-component materials were presented. Because of its unique standing in our environment, Chapter 2 also reviewed the thermodynamics of liquid-vapor mixtures of water. Few other substances have been as exhaustively studied to warrant extensive thermodynamic compilations analogous to the steam tables. For most materials, the equivalent of the steam tables must be contained in a few properties (e.g., temperatures, enthalpies and entropies of phase changes, specific heats, and coefficients of thermal expansion and compressibility). Thermodynamics provides the relationships between these basic quantities and useful properties such as vapor pressures and phase diagrams.

This chapter treats a broad range of two-phase mixtures of simple substances (or one-component materials), including solid as well as liquid and vapor phases. Simple substances do not present the complications arising from composition as a variable. Phase equilibria in two-component systems will be analyzed in Chapter 7. For one-component systems, pressure and temperature are the only variables.

Questions dealing with a simple substance that are explored in this chapter include:

- What *p*-*T* conditions permit the coexistence of two phases?
- Upon increasing temperature, when does one phase disappear and the other form?
- How does the free energy provide the basic criterion of phase equilibrium?

The change of one phase to another is termed a *phase transition*. The following four transitions are commonly encountered:

- Liquid–gas (vaporization)
- Solid–gas (sublimation)
- Solid–liquid (melting, or fusion)
- Solid I–solid II (allotropy)

Solids and liquids are collectively termed *condensed phases* chiefly because of their high density and incompressibility compared to those of gases.

In analyzing the p-T characteristics of phase transitions, the meaning of the term "pressure" needs to be clarified. For transitions between condensed phases, pressure means the total applied pressure, and is designated by p. The total pressure can be controlled by an inert gas in contact with the condensed phases. Alternatively, pressure can be generated by a piston acting on the mixture, without a gas phase present. In dealing with phase transitions involving the gaseous form of the substance (i.e., vaporization and sublimation), the pressure generally means the *vapor pressure*, also called the *saturation pressure*, denoted by p_{sat}. This is a property of the substance, as opposed to the total pressure p, which is applied to the system. Being a thermodynamic property, the vapor pressure of a condensed phase is a function of both temperature and total pressure. However, the effect of p on p_{sat} is small, and the vapor pressure is usually considered to be a function of temperature only. Thus, the vapor pressure of liquid water at 100°C is 1 atm whether the gas phase is pure steam or contains in addition a 10 atm partial pressure of an inert gas such as air (actually, the vapor pressure rises to 1.005 atm when the total pressure is 10 atm).

The concept of a characteristic pressure also applies to transitions between condensed phases, where it is denoted by p_{tr}. For example, the pressure at which graphite transforms to diamond is a property of carbon. However, only when one phase is gaseous are the two pressures p and p_{sat} distinct quantities. When both phases are condensed, the applied pressure and the transition pressure are identical.

A similar distinction is made between the imposed temperature T and the characteristic temperature of a phase transition. These are denoted by T_{sat} if one phase is gaseous or T_{tr} if both are condensed phases.

5.2 EQUILIBRIUM BETWEEN TWO PHASES

The equilibrium criterion of minimum free energy (Section 1.11) can be applied to any of the phase transitions described in the previous section. At fixed pressure and temperature, let the system contain n_I moles of phase I and n_{II} moles of phase II, with molar free energies of g_I and g_{II}, respectively. The total free energy of the two-phase mixture is:

$$G = n_I g_I + n_{II} g_{II} \tag{5.1}$$

The requirement of equilibrium is that G remain unchanged (at its minimum value) for any variations in the state of the system. Because p and T are fixed, so are the molar free energies of the individual phases, g_I and g_{II}. The only possible change is the conversion of some of one phase to the other. Because the system is closed, an increment dn_I of phase I implies an equal and opposite change $dn_{II} = -dn_I$ in phase II. The equilibrium criterion of Equation (1.20a) yields:

$$dG_{T,p} = g_I dn_I + g_{II} dn_{II} = (g_I - g_{II}) dn_I = 0$$

or

$$g_I = g_{II} \tag{5.2}$$

This is an expression of *chemical equilibrium*. It complements the conditions of thermal equilibrium ($T_I = T_{II}$) and mechanical equilibrium ($p_I = p_{II}$). Because the free energy is defined by $g = h - Ts$, another form of Equation (5.2) is:

$$\Delta g_{tr} = g_{II} - g_I = \Delta h_{tr} - T_{tr}\Delta s_{tr} = 0 \tag{5.3}$$

where the subscript *tr* means transition from one phase to the other. $\Delta h_{tr} = h_{II} - h_I$ and $\Delta s_{tr} = s_{II} - s_I$ are, respectively, the enthalpy and entropy differences of the substance in the two states at the temperature T_{tr} of the transition. Equation (5.3) provides a relation between the two property changes:

$$\Delta s_{tr} = \frac{\Delta h_{tr}}{T_{tr}} \tag{5.4}$$

If one phase is a gas, Equation (5.4) is written as:

$$\Delta s_{sat} = \frac{\Delta h_{sat}}{T_{sat}} \tag{5.4a}$$

By common usage, the subscript "sat" is replaced by "vap" for a liquid (denoting vaporization) or "sub" for sublimation of a solid.

Equation (5.4a) was applied in Section 3.5.2 for the specific case of water vaporization. It is used in Problem 5.6 as part of a calculation of the absolute entropy of a substance.

By common convention, the phase transition denoted by the generic subscripts *tr* in Equation (5.4) or *sat* in Equations (5.4) are written in the form of a simple chemical reaction with the higher-enthalpy phase on the right-hand side. Thus, vaporization of a liquid is written in chemical reaction terminology as:

$$Lrg = Vap$$

The equal sign denotes equilibrium between liquid and vapor. With this convention, Δh_{vap} is a positive quantity. Transitions between condensed phases are written as:

$$Solid = Liquid \qquad \text{or} \qquad SolI = SolII$$

And the enthalpy differences Δh_M for melting or Δh_{tr} for crystal-structure changes are positive.

The consequence of the above convention is that the entropy of the transition must also be positive, because temperature cannot be negative. Thus, the entropy of the vapor is always greater than that of the coexisting liquid state, corresponding to the relative degrees of disorder of these two phases. Similarly, Δs_M for melting and Δs_{tr} for solid/solid transitions are also positive quantities.

The equilibrium condition of Equation (5.3) can be regarded as a balance between the positive enthalpy change of the transition, which tends to favor the more tightly bound state (i.e., the liquid), and the positive entropy change that acts to drive the system towards the state of greater disorder (i.e., the vapor). These competing tendencies just balance at a unique temperature, T_{tr}, where the free energies of the two phases are equal.

5.3 THE CLAPYRON EQUATION

The phase rule (Equation [1.21]) allows one degree of freedom for a two-phase, one-component system. As the temperature is changed, the pressure must also adjust in order to maintain both phases in equilibrium. The criterion of Equation (5.2) determines the p-T relationship for coexistence of two-phases.

If Equation (5.2) is satisfied at a particular combination of T and p, and the temperature is incremented by dT, the pressure must change by dp to maintain both phases in equilibrium. As a result of these changes, the molar free energies of the two phases change by dg_I and dg_{II}. For equilibrium at the new conditions, the molar free energies must also be equal, or $g_I + dg_I = g_{II} + dg_{II}$. Because $g_I = g_{II}$ at the initial state, then $dg_I = dg_{II}$. These free energy increments can be expressed in terms of other properties using the fundamental differential of Equation (1.18a):

$$dg_I = v_I dp - s_I dT \qquad \text{and} \qquad dg_{II} = v_{II} dp - s_{II} dT$$

Since the left-hand sides of the above equations are equal, equating the right-hand sides yields:

$$\frac{dp}{dT} = \frac{s_{II} - s_I}{v_{II} - v_I} = \frac{\Delta s_{tr}}{\Delta v_{tr}}$$

Making use of Equation (5.4) gives the final form of the *Clapyron equation*:

$$\frac{dp}{dT} = \frac{\Delta h_{tr}}{T \Delta v_{tr}} \tag{5.5}$$

The subscripts on p and T in the above equation have been omitted because the designations vary with the application. In vaporization or sublimation, for example, T is the imposed property, and the resulting property is the vapor pressure, p_{sat}. For melting of a solid, on the other hand, pressure is the imposed property and the resulting property is the melting temperature. These two applications are developed below.

5.4 VAPORIZATION (OR SUBLIMATION)

Application of the Clapyron equation to vapor-liquid equilibria is identical to that for vapor-solid equilibrium, so only the former is presented. When one of the phases is a vapor, its molar volume is so much larger than that of the condensed phase that the latter can be neglected in Δv_{vap}. In addition, assuming the vapor to behave ideally is generally adequate. With these two approximations the volume change on vaporization is:

$$\Delta v_{tr} = \Delta v_{vap} = v_g - v_L \cong v_g \cong RT/p_{sat}$$

where p_{sat} is the vapor pressure, or saturation pressure, at temperature T.

Substituting this equation into Equation (5.5) and identifying Δh_{tr} with the enthalpy of vaporization Δh_{vap} yields:

$$\frac{d \ln p_{sat}}{dT} = \frac{\Delta h_{vap}}{RT^2}$$

or

$$\frac{d \ln p_{sat}}{d(1/T)} = -\frac{\Delta h_{vap}}{R} \tag{5.6}$$

This relation is the *Clausius-Clapyron equation.* Its integration requires knowledge of the temperature dependence of the enthalpy of vaporization. Assuming that the heat capacities of the liquid and vapor are independent of temperature, the enthalpy of vaporization is given by a formula analogous to the enthalpy of melting (Equation. [3.2]):

$$\Delta h_{vap}(T) = \Delta h_{vap}(T_{ref}) + \Delta C_P(T - T_{ref}) \tag{5.7}$$

where $\Delta C_P = C_{pg} - C_{PL}$ and T_{ref} is an arbitrary reference temperature.

To assess the importance of the second term in Equation (5.7), consider water between 0 and 100°C. The reference state is chosen as the normal boiling point (i.e., where the vapor pressure is 1 atm, or $T_R = 373$ K). At this temperature, $\Delta h_{vap}(373) =$ 41.1 kJ/mole. The heat capacities of liquid water and water vapor are 75 and 33 J/mole-K, respectively. Substituting these values into Equation (5.7) for $T = 273$ K shows that Δh_{vap} changes by about 10% over this 100-degree temperature range. For most other higher-melting inorganic materials (metals and ceramics), the heat of vaporization is much larger than that of water and the second term on the right in Equation (5.7) is only a few percent of the first term. Thus, over a substantial temperature interval, the temperature dependence of Δh_{vap} can be neglected and Equation (5.6) can be integrated from the reference state at T_{ref} to temperature T, resulting in:

$$\ln\left(\frac{p_{sat}}{p_{sat,ref}}\right) = \frac{\Delta h_{vap}}{R}\left(\frac{1}{T_{ref}} - \frac{1}{T}\right) \tag{5.8}$$

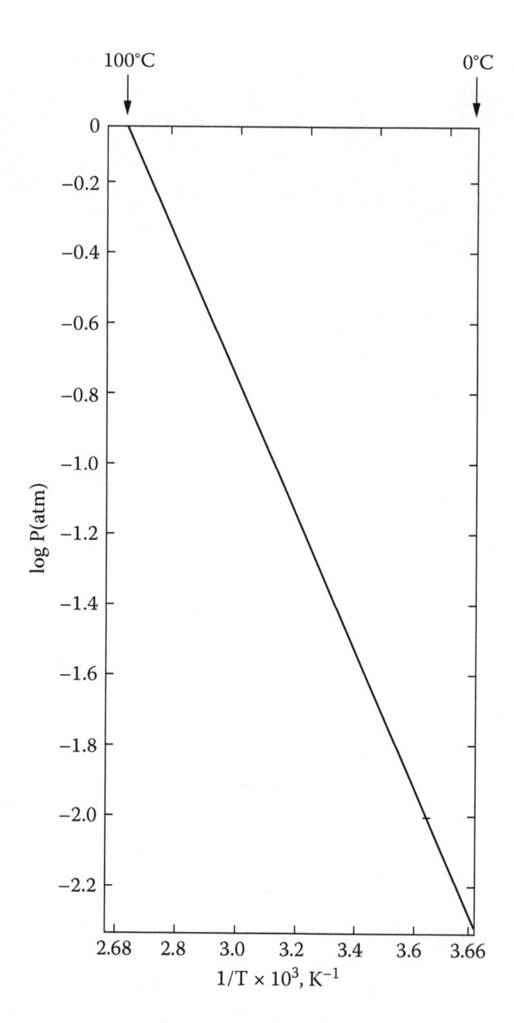

FIGURE 5.1 The vapor pressure of water as a function of temperature.

Problems 5.3, 5.6 and 5.7 utilize this form of the Clausius-Clapyron equation in typical applications. Equation (5.8) is often written in the more compact form that avoids the need to specify a reference state:

$$\ln p_{sat} = A - B/T \tag{5.9}$$

p_{sat} is expressed in atmospheres and T in Kelvins. A and B are constants.

A plot of the logarithm of the vapor pressure against the reciprocal of the absolute temperature should be a straight line with a slope of $-\Delta h_{vap}/R$. Figure 5.1 shows such a plot for water. The line is nearly straight, but the approximately 10% change in slope over the 100°C temperature range is discernible. Ignoring this slope change, the coefficient values for water are:

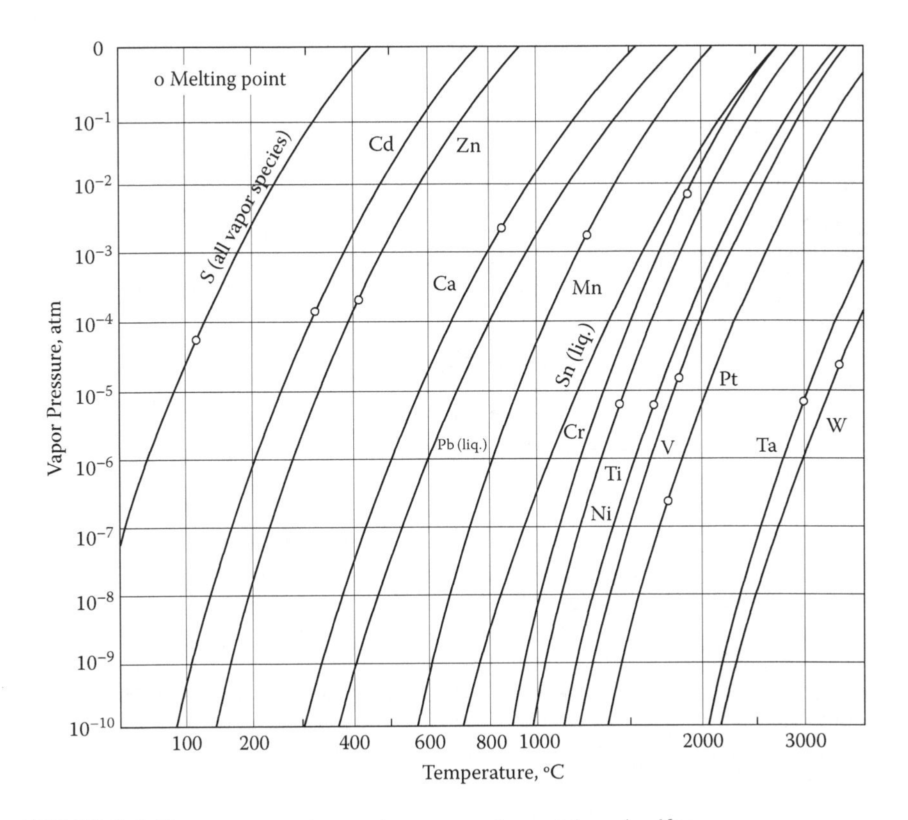

FIGURE 5.2 Vapor pressure–temperature curves for metals and sulfur.

$$A = \Delta h_{vap} / RT_{ref} + \ln p_{sat,ref} = 14.3$$
$$B = \Delta h_{vap} / R = 5300 \text{ K} \tag{5.10}$$

For most other materials, the approximate version of Equation (5.8) or (5.9) is adequate. These vapor pressure–temperature relations are specified by either of two pairs of constants:

For Equation (5.8): Δh_{vap} and $p_{sat,ref}$, (in atm), the vapor pressure at reference temperature T_{ref}
For Equation (5.9): The constants A and B, with p_{sat} usually expressed atmospheres

Equations identical in form to (5.8) and (5.9) apply to sublimation, with Δh_{sub} in place of Δh_{vap}. Problem 5.3 provides an exercise in simultaneous application of the sublimation and vaporization equilibrium pressure equations.

The very large range of vapor pressures of the elements is illustrated in Figure 5.2. For clarity of presentation, the abcissa is temperature rather than $1/T$ (as suggested in Equation [5.9]). However, if these curves were plotted in the form of Equation (5.9), they would be converted to straight lines.

When the ΔC_P term in Equation (5.7) is too large to be neglected, integration of Equation (5.6) leads to an additional term in Equation (5.9) (see Problem 5.1). Other applications of the more accurate vapor-pressure formula are given in Problems 5.2, 5.8, and 5.9 and in Chapter 11. The temperature dependence of the enthalpy of vaporization caused by the ΔC_P term in Equation (5.7) also complicates analysis of processes in which a condensed phase (i.e., a liquid or solid) undergoes both a phase transition and a temperature change. Problem 5.8 illustrates this situation.

5.5 PSYCHROMETRY (GAS–VAPOR MIXTURES)

The gas–vapor mixture of greatest practical importance is the air-water system.* The thermodynamic behavior of this mixture affects the weather and engineered devices such as air conditioners. In these applications, air can be treated as an inert ideal gas.

Water vapor can be considered to behave ideally because its concentration in air is low. In this two-component gas mixture, air is treated as a single species. The gas and the vapor obey Dalton's rule (Section 7.2):

$$p_w = x_w p \qquad p_a = x_a p \tag{5.11}$$

where the subscripts w and a denote water vapor and air (or other inert gas). p_w and p_a are the partial pressures of water vapor and air, respectively, and $p = p_w + p_a$ is the total pressure. Psychrometry problems frequently deal with water vapor in equilibrium with liquid water. In the two-phase case (at equilibrium), the gas phase is said to be saturated, and $p_w = p_{sat,w}$ is a property of water; if no liquid water is present, p_w is an adjustable parameter like p_a. The relative humidity ϕ quantitatively characterizes the water content of air as the ratio of the partial pressure of water to its vapor pressure:

$$\phi = p_w / p_{sat,w} \tag{5.12}$$

Because the temperature in practical applications is generally between 0 and 100°C, the representation of $p_{sat,w}$ by the Clausius-Clapyron equation is simpler and sufficiently accurate. In this temperature range, the vapor pressure of water is adequately represented by Equations (5.9) and (5.10):

$$p_{sat,w} = 1.6 \times 10^6 e^{-5300/T} \tag{5.13}$$

Like all thermodynamic properties of a pure substance, the vapor pressure is in principle a function of two variables. Equation (5.13) shows only the temperature dependence. There is also a dependence of $p_{sat,w}$ on total pressure, but this effect is very small (see Problem 5.6).

Equation (5.13) gives the partial pressure of water vapor in saturated air at a specified temperature. The inverse representation of saturation gives the temperature at which saturation occurs in air with a specified water vapor partial pressure. This

* Potter and Somerton (1993, 252) give a short discussion of gas-vapor mixtures.

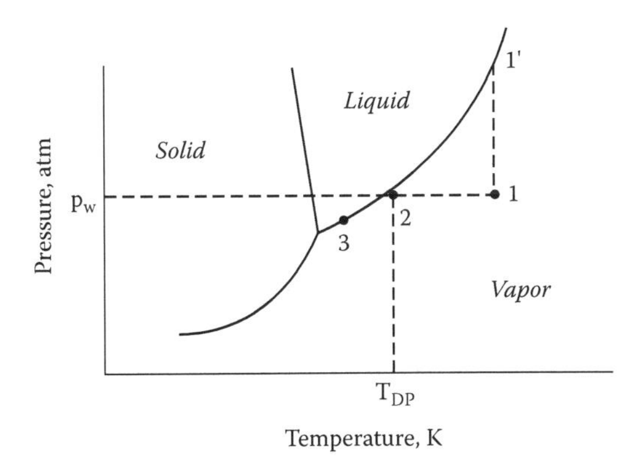

FIGURE 5.3 The *p-T* diagram of water showing one unsaturated state (No. 1) and two saturated states (Nos. 2 and 3).

temperature is called the *dew point* because the phenomenon describes the first appearance of dew on grass as the night air cools. The dew point is obtained from Equation (5.13) by replacing $p_{sat,w}$ by p_w and T by T_{DP}, the dew point:

$$T_{DP} = \frac{5300}{\ln(1.6 \times 10^6 / p_w)} \tag{5.14}$$

The onset of liquid water condensation is illustrated in the *p-T* diagram of Figure 5.3. The unsaturated air in state 1 has a relative humidity given by $p_w/p_{sat,w}(T_1)$, where p_w is specified. $p_{sat,w}(T_1)$ is located at point 1′ on the liquid-vapor coexistence curve. When this gas is cooled at constant total pressure, both p_w and p_a remain constant until the saturation curve at state 2 is reached. The temperature at state 2 is the dew point given by Equation (5.14). Upon further cooling from state 2 to state 3, water condenses from the gas, p_w (now equal to $p_{sat,w}$) decreases as determined by Equation (5.13), and p_a increases to keep the total pressure constant. The following example illustrates this process in detail.

Example: Unsaturated air is cooled from state 1 to state 3 in a piston/cylinder container maintained at a constant pressure of 1 atm.

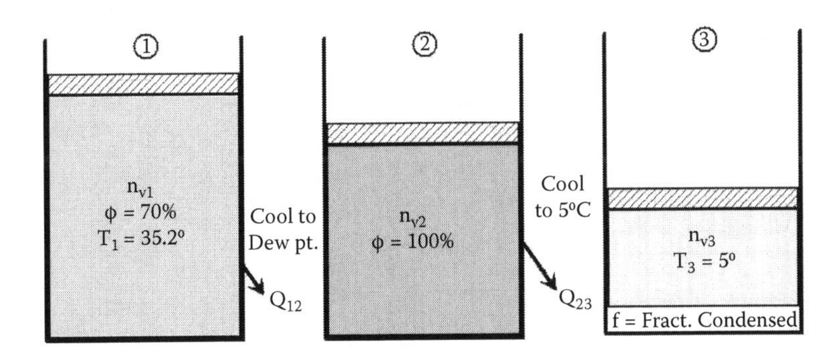

Calculate:

(a) The dew point of the gas in state 1.
(b) The fraction of water condensed at state 3.
(c) The heat removed from the container per mole of air during cooling from state 1 to state 2 and from state 2 to state 3.

(a) The saturation pressure associated with the air temperature of state 1 is on the liquid-vapor curve in Figure 5.3 (point 1′) directly above point 1. As calculated from Equation (5.13) for T_1 = 35.2°C (308.2 K), p_{sat1} = 0.0568 atm. The relative humidity is 70%, so Equation (5.12) gives p_{w1} = 0.0398 atm. Because the air is just saturated in state 2, the water partial pressure does not change on cooling from 1 to 2 and $p_{sat2} = p_{w2} = p_{w1}$. The dew point of air with this partial pressure of water lies on the saturation curve at state 2 in Figure 5.3. From Equation (5.14), $T_2 = T_{DP}$ = 301.9 K (28.9°C).

(b) The fraction of the water vapor in the gas phase in state 1 (or in state 2) that condenses when state 3 is attained is determined by combining Dalton's rule and the vapor pressure curve. The number of moles of water (n_w) per mole of air (n_a) in the gas in states 1 and 2 is determined from Equation (5.11):

$$\frac{p_w}{p_a} = \frac{x_{w2}}{x_a} = \frac{n_{w2}/n_{tot}}{n_a/n_{tot}} \quad \text{or} \quad \frac{n_{w2}}{n_a} = \frac{p_{sat2}}{p - p_{sat2}} = \frac{0.0398}{0.96} = 0.041$$

where p = 1 atm is the total pressure. In state 3, Equation (5.13) gives p_{sat3} = 0.0088 atm. The vapor-to-air mole ratio is

$$\frac{n_{w3}}{n_a} = \frac{p_{sat3}}{p - p_{sat3}} = \frac{0.0088}{0.99} = 0.0089$$

Since all of the air remains in the gas phase, n_a is constant. The fraction of the initial water condensed in state 3 is:

$$f = \frac{n_{w2} - n_{w3}}{n_{w2}} = \frac{0.041 - 0.0089}{0.041} = 0.79$$

(c) Because the cooling process is isobaric, the heat removed is equal to the enthalpy decrease (Section 3.4). From state 1 to state 2, only sensible heat is extracted from the gas mixture. The specific heats at constant pressure (C_P) of water vapor and air are 33 and 29 J/mole-K, respectively, so the mixture heat capacity is:

$$n_{tot}C_P = n_{w2}C_{Pw} + n_aC_{Pa} = 0.041 \times 33 + 1 \times 29 = 30.3 \text{ J/K-mole air}$$

The heat removed per mole of air is:

$$Q_{12} = H_2 - H_1 = n_{tot}C_P(T_1 - T_{DP}) = 30.3 \times (35.2 - 28.9) = 191 \text{ J/mole air}$$

Even though liquid water continuously condenses during cooling from state 2 to state 3, to permit calculation, heat removal is broken into two distinct but hypothetical steps as shown in the diagram.

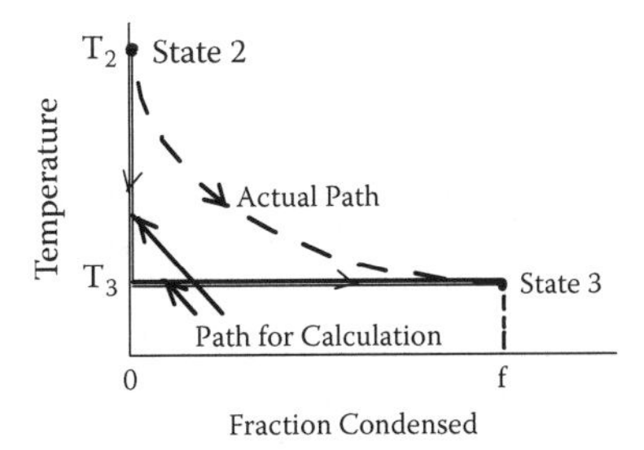

First, the mixture is cooled from T_2 to T_3 without condensation. The heat effect of this step is removal of sensible heat. Second, the required quantity of water is condensed at temperature T_3, during which the enthalpy of vaporization is released.

From Equation (5.10): $\Delta h_{vap} = 5300 \times 8.314 = 44{,}000$ J/mole and:

$$Q_{23} = H_2 - H_3 = n_{tot}C_P(T_2 - T_3) + f(n_{v2}/n_a)\Delta h_{vap} = 725 + 1425 = 2150 \text{ J/mole air}$$

Problems 8.2, 8.12, 8.13, 8.15 and 8.21 explore other aspects of the above condensation process.

5.6 ONE-COMPONENT PHASE DIAGRAMS

The most familiar graphical representation of the phase relationships of pure substances is the p-T diagram, as illustrated for water and carbon dioxide in Figure 5.4.

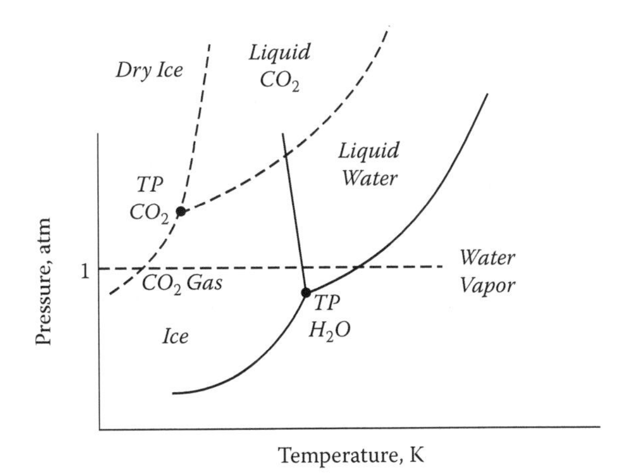

FIGURE 5.4 p-T phase diagrams for water and carbon dioxide.

p-T diagrams are comprised of three lines that intersect at the triple point. In the areas representing the solid, liquid and vapor phases, a single phase is stable over a range of temperature and pressure. The line separating the liquid and vapor regions is the vapor pressure equation given by Equation (5.9). A curve of the same mathematical form but with different coefficients (A and B) separates the solid and vapor zones; this is the sublimation pressure curve of the substance.

If the triple point pressure is less than 1 atm, as is the case for water, the dotted line on the diagram at 1 atm passes through all three phases. If the triple-point pressure for the substance is above 1 atm, the liquid phase does not appear along a horizontal line at 1 atm. The reason that carbon dioxide is called "dry ice" is because it sublimes rather than melts at atmospheric pressure. The triple point pressure of CO_2 is 5.2 atm. Liquid CO_2 exists only at pressures exceeding this value.

The two straight lines emanating from the triple points separate coexisting solid and liquid phases. The line for CO_2 has a positive slope, which is found for most substances. The melting line for water is unusual in that it has a negative slope. The explanation for this behavior is given below.

In *p-T* diagrams, there is an important distinction between the solid/vapor and liquid/vapor curves and the remainder of the diagram, including the solid/liquid line. The meaning of the ordinate "pressure" for the solid/vapor and liquid/vapor lines is "saturation pressure," p_{sat}, whereas for the latter, "pressure" means "total pressure," p. The first is a property of the substance but the latter is an imposed condition.

Problems 5.2, 5.7, and 5.13 explore other aspects of the triple point.

5.7 THE EFFECT OF PRESSURE ON THERMODYNAMIC PROPERTIES

The effect of total pressure on properties of pure substances is of interest from more than a theoretical point of view. For example,

- Solid–liquid and solid I–solid II transformations are of great practical importance in understanding phase transformations of minerals driven by the enormous pressures deep below the earth's surface.
- High pressures can accentuate the nonideal behavior of gases.
- In a more frivolous vein, because pressure on ice reduces its melting point, ice-skate blades glide on a thin film of liquid water created by the force exerted by the blade on the ice.

5.7.1 Solid–Liquid (Melting) and Solid I–Solid II Transformations

The direction and magnitude of the pressure effect on phase transformation equilibria are determined from the Clapyron equation. Inverting Equation (5.5) and regarding the temperature as the dependent property and total pressure as the adjustable property, the variation of the transition point T_{tr} with pressure is given by:

$$\frac{dT_{tr}}{dp} = \frac{T_{tr}\Delta v_{tr}}{\Delta h_{tr}} \tag{5.15}$$

TABLE 5.1
Pressure Effect on Solid–Liquid and Solid I–Solid II Transitions

Species	Transition	T_{tr}, K	Δh_{tr}, J/mole	Δv_{tr}, m³/mole × 10⁶	dT_{tr}/dp, K/MPa
H_2O	Ice → L	273	5960	−1.62	−0.074
Fe	$\alpha \rightarrow \gamma$	1185	900	−0.03	−0.04
Fe	$\gamma \rightarrow$ L	1811	13800	+0.25	+0.03

where Δh_{tr} is the enthalpy change of phase transformation and $\Delta v_{tr} = v_{II} - v_I$ is the corresponding volume change. Table 5.1 summarizes applications of Equation (5.15) to melting of ice and iron and to the α-γ solid–solid transition in iron. For most materials, Δv_M is positive—phase I (solid) is denser than phase II (liquid). Ice/water is the notable exception to this general rule because the liquid is denser than the solid; application of 13.5 MPa lowers the melting point by 1°C.

In iron, the stable low-temperature phase (the α phase) has a body-centered-cubic (bcc) structure that transforms to the face-centered-cubic (fcc) γ phase at 1185 K. Since the fcc crystal is close-packed whereas the bcc lattice is more open, the specific volume change of the $\alpha \rightarrow \gamma$ transition, $\Delta v_{\alpha\text{-}\gamma}$, is negative. In addition, $\Delta h_{\alpha\text{-}\gamma}$ is positive, so Equation (5.15) shows that the transition temperature decreases with increasing pressure. The $\gamma \rightarrow$ L phase-change (i.e., melting) temperature increases with increasing pressure because, according to Equation (5.15), this is the behavior expected when both Δv_{tr} and Δh_{tr} are positive. For either the solid–solid phase change or melting, the pressure effect is small because the volume changes associated with the transitions are small.

5.7.2 Transformation of Graphite to Diamond

Another notable solid–solid equilibrium is the graphite-to-diamond transition in the element carbon. Graphite is fairly common in the Earth's crust but the rarity of diamond is the origin of its value. Under normal terrestrial conditions (300 K, 1 atm) the two forms of carbon are not in equilibrium and so, thermodynamically speaking, only one form should exist. The stable form is the one with the lowest Gibbs free energy. At 300 K, the enthalpy difference between diamond and graphite is $\Delta h_{d\text{-}g} = 1900$ J/mole, with diamond less stable than graphite in this regard. Being a highly ordered structure, diamond has a molar entropy lower than that of graphite, and $\Delta s_{d\text{-}g} = -3.3$ J/mole-K (see Figure 3.6). This difference also favors the stability of graphite. The combination of the enthalpy and entropy effects produces a free-energy difference of:

$$\Delta g_{d\text{-}g} = g_{diamond} - g_{graphite} = \Delta h_{d\text{-}g} - T\Delta s_{d\text{-}g} = 1900 - 300(-3.3) = 2880 \text{ J/mole}$$

Since the phase with the lowest free energy (graphite) is stable, diamond is a metastable phase. It exists only because the kinetics of transformation to graphite is extremely slow at ambient temperature. How then does diamond form?

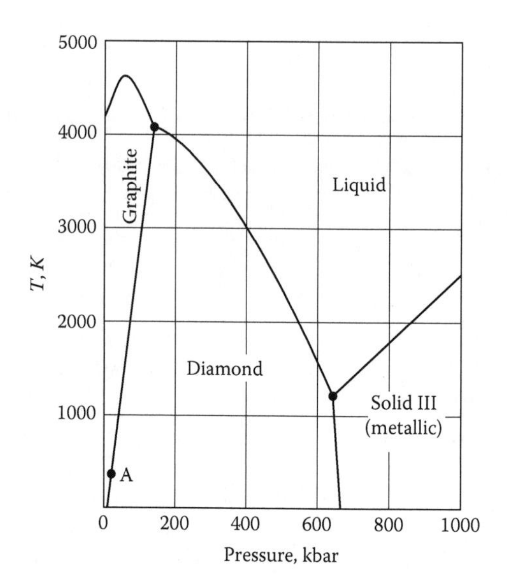

FIGURE 5.5 The equilibrium phase diagram for carbon.

The phase diagram of carbon is shown in Figure 5.5. At low pressure, graphite is stable at all temperatures up to the melting point. In order to transform graphite into diamond at constant temperature, the pressure must be very high. This is how, deep under the Earth, diamond was created.

There is a considerable industrial market for synthetic diamonds, so a practical method of effecting the $g \to d$ transition is highly desirable. The pressure required for the transformation can be read from the phase diagram of Figure 5.5, or it can be calculated by applying the fundamental differential $dg = vdp$ at constant temperature to both phases and taking the difference:

$$\left(\frac{\partial \Delta g_{d-g}}{\partial p}\right)_T = \Delta v_{d-g}$$

Example: At what pressure does graphite transform to diamond at room temperature?

Diamond has a higher density (and consequently a lower molar volume) than graphite, and $\Delta v_{d\text{-}g} = -2 \times 10^{-6}$ m^3/mole. Assuming that this quantity is independent of pressure, which is equivalent to assuming that the compressibilities of the two allotropes are the same, the above equation can be integrated to obtain the pressure at which the free energies of the two forms are the same:

$$\Delta g_{d\text{-}g} = 0 = \Delta g_{d\text{-}g}(1 \text{ atm}) + (p_{d\text{-}g} - 10^5)\Delta v_{d\text{-}g}$$

where $p_{d\text{-}g}$ is the transformation pressure (in Pa) at 300 K, which is the temperature at which $\Delta g_{d\text{-}g}(1 \text{ atm}) = 2880$ J/mole and $\Delta v_{d\text{-}g} = -2 \times 10^{-6}$ m^3/mole are specified. Using these values in the above equation yields:

$$p_{d-g} = 10^5 - \frac{\Delta g_{d-g}(1\ atm)}{\Delta v_{d-g}} = 10^5 - \frac{2880}{-2\times 10^{-6}} = 1.45\times 10^9\ \text{Pa} = 14,500\ \text{atm} = 14.5\ \text{kbar}$$

At 300 K, this pressure falls on the line in Figure 5.5 separating graphite and diamond (point A).

This elementary thermodynamic analysis has provided a fairly accurate estimate of the pressure needed to transform graphite into diamond at room temperature. Problem 5.12 presents an alternative approach to the same type of transformation that appears in a geologic setting.

Other applications of Equation (5.15) are considered in Problems 5.4, 5.9 and 5.10.

5.7.3 Solid–Vapor and Liquid–Vapor: Vapor Pressures

The starting point for the analysis of the effect of total pressure on the saturation pressure of two-phase systems in which one phase is a vapor is the fundamental differential of Equation (1.18a). At constant temperature, this equation reduces to $dg = vdp$. Applied to an ideal gas (or vapor), this becomes:

$$\left(\frac{\partial g_g}{\partial p}\right)_T = v_g = \frac{RT}{p}$$

The partial pressure of a vapor in equilibrium with the liquid is p_{sat}. Making this substitution for p in the above equation and integrating from a reference pressure $p_{sat,ref}$ at temperature T where $g_g = g_{g,ref}$ yields:

$$g_g = g_{g,ref} + RT\ln\left(\frac{p_{sat}}{p_{sat,ref}}\right) \tag{5.16a}$$

$p_{sat,ref}$ is the saturation pressure in the absence of any external agent (such as an inert gas) to cause the total pressure to differ from the saturation pressure.

Pressure in excess of the saturation pressure alters the free energy of the condensed phase by:

$$\left(\frac{\partial g_c}{\partial p}\right)_T = v_c$$

When integrated from $p_{sat,ref}$ to some higher pressure p, neglecting the pressure effect on the molar volume of the condensed phase, the result is:

$$g_c = g_{c,ref} + v_c(p - p_{sat,ref}) \tag{5.16b}$$

Although the total pressure p enters directly into the free energy of the condensed phase, it only indirectly affects g_g via its effect on p_{sat}. Since the vapor and the condensed phase are always in equilibrium ($g_{c,ref} = g_{c,ref}$ and $g_g = g_c$), equating the right-hand sides of Equations (5.16a) and (5.16b) yields:

$$\frac{p_{sat}}{p_{sat,ref}} = \exp\left[\frac{v_c(p - p_{sat,ref})}{RT}\right] \tag{5.17}$$

The right-hand side of this equation is called the *Poynting factor* after the British physicist who derived it at the turn of the twentieth century.

Example: What is the vapor pressure of water at 500 atm if the reference state is the normal boiling point?

This choice of reference state implies $T = 373$ K and $p_{sat,ref} = 1$ atm. The molar volume of water at this state (from Table 2.2) is $v_L = 1.88 \times 10^{-5}$ m^3/mole. With these values, and converting the pressure difference to 5×10^7 Pa, Equation (5.17) gives a Poynting factor of 1.35, or $p_{sat} = 1.35$ atm.

5.7.4 Triple Point (of Water)

The triple point is the pressure-temperature combination at which the solid-vapor and liquid-vapor lines in Figure 5.4 intersect. For water, the total pressure over the two condensed phases at this point is their common vapor pressure (0.00611 atm). However, if the total pressure on a liquid–solid phase mixture is increased, the triple point migrates from its low-pressure value. To quantify this effect, the differential forms of the vapor-pressure functions of the solid and liquid phases $p_{sat,S}(p,T)$ and $p_{sat,L}(p,T)$, are used:

$$dp_{sat,S} = \left(\frac{\partial p_{sat,S}}{\partial T}\right)_p dT + \left(\frac{\partial p_{sat,S}}{\partial p}\right)_T dp \tag{5.18S}$$

and

$$dp_{sat,L} = \left(\frac{\partial p_{sat,L}}{\partial T}\right)_p dT + \left(\frac{\partial p_{sat,L}}{\partial p}\right)_T dp \tag{5.18L}$$

The temperature effect is given by the Clausius-Clapyron equation, Equation (5.6):

$$\left(\frac{\partial p_{sat,S}}{\partial T}\right)_p = \frac{p_{sat,S}\Delta h_{sub}}{RT^2} \quad \text{and} \quad \left(\frac{\partial p_{sat,L}}{\partial T}\right)_p = \frac{p_{sat,L}\Delta h_{vap}}{RT^2} \tag{5.19}$$

Taking the derivative of Equation (5.17) with respect to p (at constant T) gives:

$$\left(\frac{\partial p_{sat}}{\partial p}\right)_T = \frac{v_c p_{sat,ref}}{RT}\exp\left[\frac{v_c(p - p_{sat,ref})}{RT}\right] \cong \frac{v_c p_{sat,ref}}{RT} \tag{5.20}$$

which assumes that the Poynting factor is close to unity (as it is because v_c is very small). Applying Equation (5.20) to the two condensed phases yields:

$$\left(\frac{\partial p_{sat,s}}{\partial p}\right)_T = \frac{v_s p_{sat,ref}}{RT} \quad \text{and} \quad \left(\frac{\partial p_{sat,L}}{\partial p}\right)_T = \frac{v_L p_{sat,ref}}{RT} \tag{5.21}$$

Substituting Equations (5.19) and (5.21) into Equations (5.18S) and (5.18L) results in:

$$dp_{sat,s} = \frac{p_{sat,s}\Delta h_{sub}}{RT^2}dT + \frac{v_s p_{sat,s}}{RT}dp \tag{5.22S}$$

$$dp_{sat,L} = \frac{p_{sat,L}\Delta h_{vap}}{RT^2}dT + \frac{v_L p_{sat,L}}{RT}dp \tag{5.22L}$$

Taking the reference state as the normal triple point (T^o_{TP} = 273.1 K, p^o_{TP} = 611 Pa), integration of these two equations gives:

$$\ln\left(\frac{p_{sat,s}}{p^o_{TP}}\right) = \frac{\Delta h_{sub}}{RT^o_{TP}}\left(1 - \frac{T^o_{TP}}{T_{TP}}\right) + \frac{v_s p^o_{TP}}{RT_{TP}}\left(\frac{p}{p^o_{TP}} - 1\right) \tag{5.23S}$$

$$\ln\left(\frac{p_{sat,L}}{p^o_{TP}}\right) = \frac{\Delta h_{vap}}{RT^o_{TP}}\left(1 - \frac{T^o_{TP}}{T_{TP}}\right) + \frac{v_L p^o_{TP}}{RT^o_{TP}}\left(\frac{p}{p^o_{TP}} - 1\right) \tag{5.23L}$$

As the pressure is raised, the triple point remains the intersection of the $p_{sat,s}$ versus T and the $p_{sat,L}$ versus T curves. Setting $p_{sat.s} = p_{sat,L} = p_{TP}$ in the above equations and equating the two right-hand sides yields the following equation giving T_{TP} as a function of p:

$$19.8 \times \left(1 - \frac{T^o_{TP}}{T_{TP}}\right) + 4.85 \times 10^{-6}\left(\frac{p}{p^o_{TP}} - 1\right) = 22.4 \times \left(1 - \frac{T^o_{TP}}{T_{TP}}\right) + 5.29 \times 10^{-6}\left(\frac{p}{p^o_{TP}} - 1\right)$$

Numerical values for the coefficients in Equations (5.23S) and (5.23L) for water have been inserted. This equation is solved for T_{TP} as a function of p and the corresponding p_{TP} calculated from either Equation (5.23S) or Equation (5.23L). The results are shown in Figure 5.6 The vapor-liquid and vapor-solid lines are drawn for inert-gas pressures of 0 and 100 atm. At 500 atm, the triple-point temperature has decreased by approximately 4 K and the triple-point pressure has increased by approximately 12%.*

* This value is slightly in error due to the neglect of the exponential term in Equation (5.20).

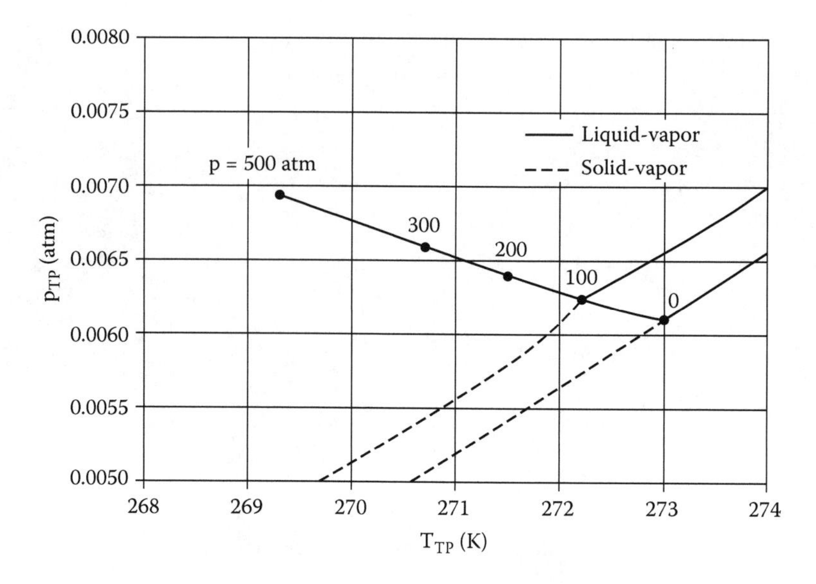

FIGURE 5.6 Effect of total (applied) pressure on the triple point of water (curve with points).

5.7.5 Summary of Pressure Effects

In addition to the influence of total pressure on the thermodynamic properties derived in the preceding section, several other similar formulae are given in other chapters. These are collected in Table 5.2.

TABLE 5.2
Pressure Effects on Thermodynamic Properties

Property	X	$(\partial X/\partial p)_T$	Refer to Equations
Specific volume	v	$-v\beta$	1.2, 2.19
Free energy	g	v	1.18a
Enthalpy	h	$-T(\partial v/\partial T)_p + v$	6.22a
Specific heat	C_P	$-T(\partial^2 v/\partial T^2)_p$	6.27
Melting temp.	T_M	$T_M \Delta v_M/\Delta h_M$	5.15
Vapor pressure	p_{sat}	$\frac{v_c p_{sat,ref}}{RT}\exp\left[\frac{v_c(p-p_{sat,ref})}{RT}\right]$	5.20
Triple point	T_{TP}, p_{TP}	$\ln\left(\frac{p_{sat,s}}{p^o_{TP}}\right)=\frac{\Delta h_{sub}}{RT^o_{TP}}\left(1-\frac{T^o_{TP}}{T_{TP}}\right)+\frac{v_s p_{TP}}{RT_{TP}}\left(\frac{p}{p^o_{TP}}-1\right)$	(5.23S), (5.23L)
		$\ln\left(\frac{p_{sat,L}}{p^o_{TP}}\right)=\frac{\Delta h_{vap}}{RT^o_{TP}}\left(1-\frac{T^o_{TP}}{T_{TP}}\right)+\frac{v_L p^o_{TP}}{RT^o_{TP}}\left(\frac{p}{p^o_{TP}}-1\right)$	
		$p_{sat,s}=p_{sat,L}$	

PROBLEMS

5.1 In the text, Equation (5.6) was integrated assuming that $\Delta h_{vap} = h_g - h_L$ is constant. More generally, Δh_{vap} varies with temperature.

(a) Using Equation (5.7) in the Clausius-Clapyron equation, show that

$$\ln p_{sat} = A - B/T + D\ln T$$

and identify A, B, and D (take the normal boiling point as the reference state).

(b) Assuming that the liquid and vapor are monatomic and $C_{VL} = 3R$, where R is the gas constant, what is the value of the coefficient D?

(c) For water, use your best estimates of C_{Pg} and C_{PL} to determine D.

5.2 The vapor pressures (in atmospheres) of the liquid and solid phases of a particular metal are given by the equation:

$$\ln p_{sat} = A - B/T + D\ln T$$

where T is in Kelvin. The coefficients in this equation are:

Transition	A	B(K)	D
Solid–gas	19.25	15773	–0.755
Liquid–gas	21.79	15246	–1.255

Using the results of Problem 5.1a, identify the coefficients B and D. Then calculate:

(a) The normal boiling point (i.e., the temperature at which $p_{sat} = 1$ atm).

(b) The triple point temperature.

(c) The heat of vaporization at the normal boiling point.

(d) The difference between the heat capacities of the liquid and solid, $C_{PL} - C_{Ps}$.

(e) The heat of fusion at the triple point.

(f) The slope of the liquid-solid boundary line at the triple point. The densities of the liquid and solid phases are 5.66 and 7.14 g/cm^3 and the atomic weight of the metal is 65.4.

5.3 The vapor pressures of solid and liquid CO_2 are expressed by:

$$\ln p_{sat,S} = 15.0 - 3116/T \quad \ln p_{sat,L} = A_L - B_L/T$$

The heat of fusion of CO_2 is 8.3 kJ/mole and the triple point temperature is 217 K.

(a) Determine the numerical values of A_L and B_L.
(b) What is the stable condensed phase of CO_2 at 1 atm pressure?
(c) At what temperature is the vapor pressure equal to 1 atm?

5.4 The triple point of water is defined as 0.01°C and 611 Pa. Show why the freezing point of water at 1 atm pressure is 0°C.

5.5 Table 2.2 contains entries for v_f, v_g,...s_f, s_g, but none for g_f and g_g.

(a) What would be the most important characteristic of g_f and g_g if they were displayed in the tables?
(b) How would g_f and g_g be determined from the entries in the existing tables?

5.6 The specific heat of solid copper (C_P) is shown on the plot below. Other properties are:

Atomic weight: 63.5
Melting point: 1356 K
Heat of sublimation: 329 kJ/mole (independent of temperature)
Vapor pressure at the melting point: 0.0505 Pa

At 1200 K, determine the following properties of the saturated vapor:

(a) The saturation pressure
(b) The absolute entropy of the vapor. Use $C_P = 2.48 \times 10^{-7}T^3$ cal/g-K from 0–30 K, and $C_P = 0.085 + 2.7 \times 10^{-5}T$ cal/g-K from 300 K to the melting point. In the intermediate range, $C_P(T)$ must be obtained from the graph. Convert units to J/g-K.

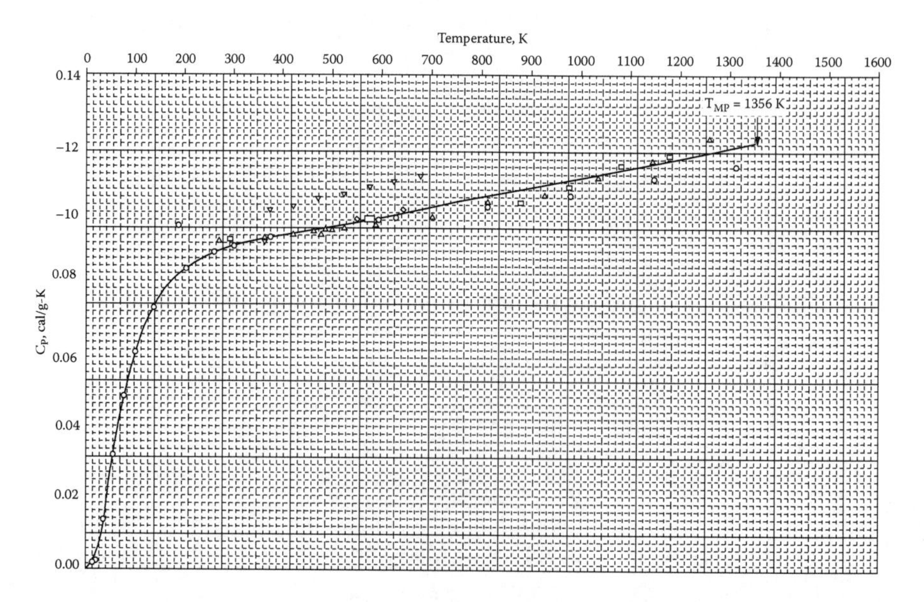

5.7 Using the triple-point data for water from Table 2.5 and assuming the heat of sublimation of ice to be constant, estimate the water vapor pressure over ice at –40°C. Compare to the value given in Table 2.5.

5.8 A compressed liquid at T_1 and p_1 is heated at constant pressure by addition of a quantity q of heat per mole. In the final state, two phases are present at the temperature T_2 fixed by the pressure. The vapor and liquid heat capacities are C_{Pg} and C_{PL} and the enthalpy of vaporization at T_1 is ΔH_{V1}.

Derive the equation for the fraction of liquid vaporized in this process by either the top (A-B) or bottom (D-E) two-step routes shown below:

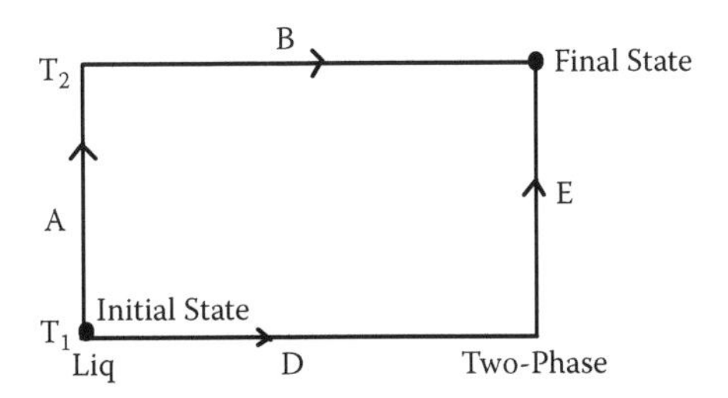

5.9 The melting point of iodine is 113°C. The vapor pressures of the solid and liquid states are given by $\ln p_{sat} = A - B/T + D\ln T$, where:

$B_s = 8240$, $B_L = 7381$, $D_s = -2.51$, $D_L = 5.18$, $A_s = 34.16$, and $A_L = 47.83$

(a) Calculate the triple point temperature and pressure of iodine. Why does this result differ from the reported melting temperature?
(b) Consider iodine vapor initially at 0.04 atm and 150°C. For the following processes, determine which condensed phase first appears and at what p and T. Sketch the p-T projection for iodine and identify on it the paths of the two processes. Assume that the vapor obeys the ideal gas law.
(c) The initial vapor is isothermally compressed.
(d) The initial vapor is cooled at constant volume.

5.10 At 800 K and 1 atm pressure, the Gibbs free energy of a liquid is 1 kJ/mole greater than that of the solid phase. The heat of fusion of the substance is 25 kJ/mole and the heat capacities of the liquid and solid are equal.

(a) What is the melting point of the substance at 1 atm pressure?
(b) If the molar volume of the liquid is 1 cm^3/mole smaller than that of the solid, at what pressure does the substance melt at a temperature of 800 K?

5.11 It is an axiom of thermodynamics that the change in a property during a process from state 1 to state 2 is independent of the path followed. This is to be demonstrated for the following process: 1 mole of a superheated vapor at temperature T_1 is cooled at constant pressure to the saturation temperature T_2 and further cooled to condense n_L moles of liquid. The enthalpy difference $h_2 - h_1$ is to be calculated for this path and for an alternate (hypothetical) path. The second path is condensing n_L moles of liquid from the vapor at a constant temperature T_1 then cooling the two-phase mixture to T_2.

(a) Draw a diagram like the one in Problem 5.8 showing these two paths.
(b) Derive the equations for $h_2 - h_1$ for the two paths and show that they are identical. Assume that the heat capacities of the vapor and the liquid are temperature independent and that the heat of vaporization varies with T according to Equation (5.7).

5.12 In the Earth's crust, $CaCO_3$ occurs in two crystalline modifications, calcite (C) and aragonite (A). At very high pressures (as occurs deep below the surface), the effect of total pressure on the Gibbs free energies of these two mineral forms must be considered. The free energy change corresponding to the conversion A → C is given by:

$$g_C - g_A = -50 + 9 \times 10^{-3}T + 0.07(p - 1) \text{ J/mole}$$

where T is in Kelvins and p is in atmospheres.

(a) Identify each of the numerical quantities in the above equation in terms of thermodynamic quantities characterizing the transformation from A to C. For the last term, use the equation on the bottom of page 152 of the text, with $\Delta s_{A\text{-}C}$ independent of p.
(b) Which phase is stable at the Earth's surface?
(c) The Earth's temperature increases linearly with depth with a gradient of 1 K/m. A lithostatic pressure gradient is also created by a rock density of 5 g/cm³. At what depth does the form stable at the surface transform to the other crystal form? Neglect the effect of temperature on $\Delta h_{A\text{-}C}$.

5.13 The line containing point A in the carbon p-T diagram (Figure 5.5) shows that the pressure required for the graphite-to-diamond transition increases with temperature. At 14.5 kbar and 300 K, compare the slope $dp_{g\text{-}d}/dT$ obtained from the plot and that obtained from the Clapyron equation. Assume that $\Delta v_{g\text{-}d} = -2 \times 10^{-6}$ m³/mole is independent of temperature and pressure. Account for the effect of pressure on $\Delta h_{g\text{-}d}$ using the equation on the bottom of page 152. Neglect the effect of pressure on the entropy of transition. At 1 atm, $\Delta h_{g\text{-}d}$ = 1900 J/mole.

5.14 The constant-pressure cooling of an air-water vapor mixture based on Figure 5.3, was worked out in Section 5.5. Calculate the system volume at the beginning and end of each step. Assume that the system contains one mole of air and that the total pressure is 1 atm.

5.15 What is the percentage change in the vapor pressure of water from the steam tables value at 25°C when liquid water is in equilibrium with water vapor in air at a total pressure of 1 atm?

5.16 A sealed, rigid container of volume V is partially filled with a mass m of liquid water. The remainder of the volume is filled with air. Initially, the system temperature is T_o, its pressure is p_o and the air has zero relative humidity. A quantity Q of heat is added to the contents of the vessel.

(a) Derive the equations from which the final temperature T, pressure p, and steam quality x can be derived. The vapor pressure of water is given by Equation (5.13). Assume that the specific heats of air and liquid water are known. The internal energy change upon vaporization can be approximated by Δh_{vap}.
(b) Solve for T, p, and x using the following input values: V = 1 liter; m = 200 g; T = 25°C; p, = 1 atm; Q = 50.8 J. Estimate the heat capacities. How is the input heat partitioned between sensible heat to the moist air and the heat for vaporization of water?
(c) Solve the problem using the steam tables instead of the equations developed in part (a). Use the properties given in (b).

5.17 A condenser operates as a steady-flow device, receiving an input vapor with an inert gas (air) molar flow rate $\dot{n}_I$ at a temperature T_o and a relative humidity ϕ_o. The condenser extracts heat from the gas at a rate $\dot{Q}$ and the exit gas is at temperature T_e. The unit operates at a constant total pressure p.

(a) Derive the equations for: (1) the unit's heat flux and the exit temperature; (2) the fraction of the water vapor in the inlet gas condensed in the unit. The saturation pressure curve is given by Equation (5.13).
(b) Solve the equations derived in part (a) for the heat flux $\dot{Q}$ for the following conditions:

$$\dot{n}_I = 0.2 \text{ moles/s};\ \phi_o = 85\%;\ T_o = 105°\text{C};\ p = 2 \text{ atm};\ T_e = 75°\text{C}.$$

The heat capacities are to be taken from Figure 2.3.

5.18 An air-water vapor mixture is contained in a closed volume initially at 1 atm total pressure and 7°C. The relative humidity of the mixture is 50%. The vessel is heated at constant volume to 25.7°C. Calculate the following:

(a) The dew point of the air in the vessel before heating (state 1).
(b) The relative humidity and dew point of the mixture after heating (state 2).
(c) The heat absorbed per mole of gas.

The specific heats at constant pressure for air and water vapor are 3.5R and 4R, respectively. The gas components can be considered to be ideal.

5.19 Saturated air initially at 1 atm total pressure and 40°C is heated. Calculate the relative humidity at 50°C if the heating is conducted at:

(a) Constant volume
(b) Constant pressure

Hint: do not use the steam tables for this problem. Instead, assume that air and water vapor behave ideally.

5.20 At the end of the day, the temperature is 30°C and the relative humidity is 30%. By 3:00 a.m., the temperature has dropped to 20°C. What is the relative humidity at this time?

5.21 The triple point of CO_2 is –57°C, 5 atm. At –78°C, the vapor pressure of the solid is 1 atm. The enthalpy of melting is 8300 J/mole. Determine:

(a) The enthalpy of vaporization, Δh_{vap}.
(b) The vapor pressure of the liquid at –23°C.

5.22 An insulated, constant-pressure vessel contains moist air at 1 atm, 21°C and 50% relative humidity. 0.003 moles of liquid water at 15°C are injected into the vessel. Find the final temperature and relative humidity.

REFERENCE

Potter, M. and C. Somerton. 1993. *Theory and Problems in Engineering Thermodynamics.* New York: McGraw-Hill.

6 The Mathematics of Thermodynamics

6.1 MATHEMATICAL PRELUDE

Of all the thermodynamic properties that have been introduced in the preceding chapters, only a few can be measured directly by laboratory experiments. Pressure, temperature, and volume are obviously among the measurable properties. However, there is no instrument to measure entropy or any of the properties related to energy (u, h, f and g). Except for entropy, which according to the third law, is zero at 0 K, these quantities cannot be assigned absolute values; only changes in them as a result of a process have quantitative meaning. All of the thermodynamic information concerning simple substances (i.e., one-component systems) is based upon measurements of the equation of state $v(p,T)$, the temperature-dependent heat capacity $C_P(T)$ at a particular pressure, and the enthalpy changes associated with phase transitions.

One of the significant achievements of classical thermodynamics is to connect various properties, so that only a few measurements are needed for a complete description of a substance. For example, the difference $C_P - C_V$ and the effect of pressure on the enthalpy can be obtained from the p-v-T equation of state of the substance. For the ideal gas and the approximate model of condensed phases presented in Chapter 2, there was no need for a formalized approach to these relations.

For an accurate description of nonideal gases and condensed phases under extreme conditions, relationships between thermodynamic properties need to be developed. Establishing these connections is the purpose of the present chapter. First, certain mathematical fundamentals need to be reviewed. These are based on the fact that all thermodynamic properties of a simple substance are smoothly varying functions of any two.

6.1.1 Exact Differentials

Let x and y be two specified properties. Denoting one of the dependent properties by z, the relation between the three is written as $z(x,y)$. The *total differential* of z is:

$$dz = \left(\frac{\partial z}{\partial x}\right)_y dx + \left(\frac{\partial z}{\partial y}\right)_x dy = M(x,y)dx + N(x,y)dy \tag{6.1}$$

The partial derivatives depend on x and y as expressed by the functions M and N in the second equality of Equation (6.1):

$$M(x,y) = \left(\frac{\partial z}{\partial x}\right)_y \qquad N(x,y) = \left(\frac{\partial z}{\partial y}\right)_x \tag{6.2}$$

Total differentials have the mathematical characteristic of being *exact* or *inexact*. The distinction is determined by comparison of the mixed second derivatives of z:

$$\left(\frac{\partial M}{\partial y}\right)_x = \left[\frac{\partial}{\partial y}\left(\frac{\partial z}{\partial x}\right)_y\right]_x \qquad \left(\frac{\partial N}{\partial x}\right)_y = \left[\frac{\partial}{\partial x}\left(\frac{\partial z}{\partial y}\right)_x\right]_y \tag{6.3}$$

The total differential is exact if the order of forming the mixed second derivative is immaterial, or if:

$$\left(\frac{\partial N}{\partial x}\right)_y = \left(\frac{\partial M}{\partial y}\right)_x \tag{6.4}$$

A mathematical corollary of the exactness property of a total differential of a dependent variable z is that its integral $z_2 - z_1$ from state 1 to state 2 [i.e., for the process from (x_1,y_1) to (x_2,y_2)] is independent of the path (or constraint) $y = F(x)$. For example, z is volume and x and y denote p and T, the path could be a combination of an isothermal step and an isobaric step. Or, x and y could vary in a continuous fashion as shown by the curve $y = F(x)$ on the x - y plane in Figure 6.1. This curve generates a corresponding path on the $z(x,y)$ surface, as shown in the diagram.

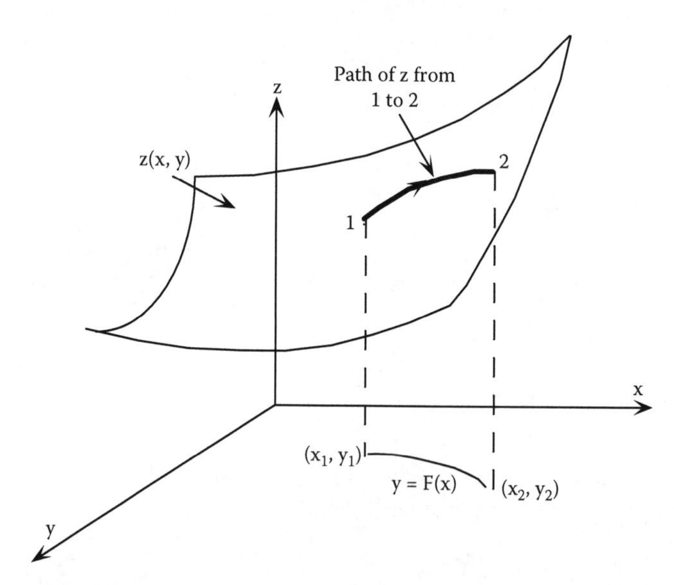

FIGURE 6.1 The surface $z(x,y)$ and a path $y = F(x)$ from state 1 to state 2.

If y is eliminated from Equation (6.1) using $y = F(x)$, the total differential can be integrated from state 1 to state 2 to yield:

$$z_2 - z_1 = \int_{x_1}^{x_2} \left\{ M\left[x, F(x)\right] + N\left[x, F(x)\right] \frac{dF}{dx} \right\} dx \tag{6.5}$$

If the total differential is exact, Equation (6.4) is satisfied *and* Equation (6.5) gives the same value of $z_2 - z_1$ for all paths $F(x)$. Problem 6.1 provides a purely mathematical exercise to help understand the distinction between exact and inexact differentials.

The relevance of the above mathematical prelude to thermodynamic calculations stems from the independence of thermodynamic property changes on the path of a process. This characteristic implies that all thermodynamic functions must satisfy Equation (6.4), whether M and N are partial derivatives of particular properties (as in Equation [6.2]) or, more generally, when they are *any* functions of thermodynamic properties.

6.1.2 The "Divide-and-Hold-Constant" Method

The total differential of Equation (6.1) can be manipulated to produce relationships between partial derivatives of x, y, and z, or other variables. The technique can be described as the "divide-and-hold-constant" method. It works as follows: Equation (6.1) is divided by a differential (not necessarily dx, dy or dz) and another (dx, dy or dz) is set equal to zero, thus holding constant the variable associated with it. For example, suppose Equation (6.1) is divided by dy with z held constant (i.e., $dz = 0$). The result is the following relation:

$$0 = \left(\frac{\partial z}{\partial x}\right)_y \left(\frac{\partial x}{\partial y}\right)_z + \left(\frac{\partial z}{\partial y}\right)_x$$

or, because the partial derivatives can be inverted, the above equation becomes what is called the *cyclic transformation*:

$$\left(\frac{\partial x}{\partial y}\right)_z \left(\frac{\partial y}{\partial z}\right)_x \left(\frac{\partial z}{\partial x}\right)_y = -1 \tag{6.6}$$

In another example of the "divide-and-hold-constant" method, Equation (6.1) is divided by a fourth variable w with x held constant to yield:

$$\left(\frac{\partial z}{\partial w}\right)_x = \left(\frac{\partial z}{\partial y}\right)_x \left(\frac{\partial y}{\partial w}\right)_x \tag{6.7}$$

which is the *chain rule* for partial derivatives.

It is obvious that there are so many combinations of variables to which the "divide-and-hold-constant" method can be applied that attempting to construct an exhaustive catalog of relations such as Equations (6.6) and (6.7) would be fruitless. The practical approach is to apply the method to suit the needs of particular problems.

Example: If a fluid is heated in a constant-volume container from T to $T + \Delta T$, what is the pressure rise Δp?

For this problem, the starting function is the equation of state $v(T,p)$, for which the total differential is:

$$dv = \left(\frac{\partial v}{\partial T}\right)_p dT + \left(\frac{\partial v}{\partial p}\right)_T dp$$

Dividing by dT and holding v constant produces the analog of Equation (6.6), written as:

$$\left(\frac{\partial p}{\partial T}\right)_v = -\frac{(\partial v / \partial T)_p}{(\partial v / \partial p)_T} = \frac{\alpha}{\beta} \tag{6.8}$$

where α and β are the coefficients of thermal expansion and isothermal compressibility, respectively (see Equation [1.2]). If these coefficients are substantially constant over the range of T and p involved, Equation (6.8) can be directly integrated to yield:

$$\Delta p = \frac{\alpha}{\beta} \Delta T$$

6.2 MAXWELL RELATIONS AND OTHER USEFUL FORMULAS

The fundamental differentials described in Section 1.10 are of the form of Equation (6.1). They provide the starting point for obtaining many useful thermodynamic relations. The fundamental differentials are represented as exact differentials of $u(s,v)$, $h(s,p)$, $f(T,v)$, and $g(T,p)$:

$$du = \left(\frac{\partial u}{\partial s}\right)_v ds + \left(\frac{\partial u}{\partial v}\right)_s dv = Tds - pdv \tag{6.9}$$

$$dh = \left(\frac{\partial h}{\partial s}\right)_p ds + \left(\frac{\partial h}{\partial p}\right)_s dp = Tds + vdp \tag{6.10}$$

$$df = \left(\frac{\partial f}{\partial T}\right)_v dT + \left(\frac{\partial f}{\partial v}\right)_T dv = -sdT - pdv \tag{6.11}$$

$$dg = \left(\frac{\partial g}{\partial T}\right)_p dT + \left(\frac{\partial g}{\partial p}\right)_T dp = -sdT + vdp \tag{6.12}$$

Note that each of the energy-like properties has a pair of "natural" variables associated with it. The variables in each pair are those that appear as differentials in the above equations. However, these associations are not immutable; it is possible, for example, to assume u to be a function of p and T and write the total differential du based on these variables instead of the "natural" pair s and v.

Equating the coefficients of the differentials in the second equalities of Equations (6.9) to (6.12) produces eight relations:

$$\left(\frac{\partial u}{\partial s}\right)_v = \left(\frac{\partial h}{\partial s}\right)_p = T \tag{6.13}$$

$$\left(\frac{\partial u}{\partial v}\right)_s = \left(\frac{\partial f}{\partial v}\right)_T = -p \tag{6.14}$$

$$\left(\frac{\partial h}{\partial p}\right)_s = \left(\frac{\partial g}{\partial p}\right)_T = v \tag{6.15}$$

$$\left(\frac{\partial f}{\partial T}\right)_v = \left(\frac{\partial g}{\partial T}\right)_p = -s \tag{6.16}$$

A most useful foursome of thermodynamic relations is obtained by noting that the form of the fundamental differentials of Equations (6.9) to (6.12) is the same as the total differential of Equation (6.1). Consisting solely of thermodynamic properties, the fundamental differentials are exact (in the mathematical sense defined earlier), and Equation (6.4) applies to them. For example, for the fundamental differential $du = Tds - pdv$, the correspondence with the general formula sets $M = T$, $N = -p$, $x = s$ and $y = v$. With these variables, Equation (6.4) gives:

$$\left(\frac{\partial T}{\partial v}\right)_s = -\left(\frac{\partial p}{\partial s}\right)_v \tag{6.17}$$

The remaining fundamental differentials in Equations (6.10) to (6.12) yield:

$$\left(\frac{\partial T}{\partial p}\right)_s = \left(\frac{\partial v}{\partial s}\right)_p \tag{6.18}$$

$$\left(\frac{\partial s}{\partial v}\right)_T = \left(\frac{\partial p}{\partial T}\right)_v \tag{6.19}$$

$$\left(\frac{\partial s}{\partial p}\right)_T = -\left(\frac{\partial v}{\partial T}\right)_p \tag{6.20}$$

Equations (6.17) to (6.20) are collectively known as the *Maxwell relations*. In the following sections, these relations, together with the fundamental differentials and Equations (6.13) to (6.16), will be utilized to derive a number of useful relationships between the thermodynamic properties of simple substances.

6.3 THERMODYNAMIC RELATIONS FOR NONIDEAL BEHAVIOR

In Chapters 2 and 3, numerous property relations were presented for ideal gases and idealized solids. The latter are characterized by constant coefficients of thermal expansion and compressibility and obey the equation of state given by Equation (2.18). For these substances,

- The specific heats (and hence the internal energy and enthalpy) are functions of temperature but are independent of pressure or specific volume.
- The entropy of the ideal gas varies with T and v (or p) according to Equations (3.9) and (3.10). The entropy of the idealized solid is much more sensitive to temperature than to pressure.
- For an ideal gas, the difference between C_P and C_V is equal to the gas constant. For the ideal solid, $C_P = C_V$.

Although the simplified representations above are reasonably applicable to most materials at moderate pressures and temperatures, more exact descriptions are occasionally needed. The objective of this section is to express the deviations from ideality in terms of the equation of state of the substance and its heat capacity.

6.3.1 INTERNAL ENERGY AND ENTHALPY

Consider the internal energy to be a function of temperature and specific volume, or $u(T,v)$. The total differential is:

$$du = \left(\frac{\partial u}{\partial T}\right)_v dT + \left(\frac{\partial u}{\partial v}\right)_T dv$$

The coefficient of dT is by definition the heat capacity at constant volume, C_V. The coefficient of dv is obtained from the fundamental differential $du = Tds - pdv$ by dividing by dv and holding T constant:

$$\left(\frac{\partial u}{\partial v}\right)_T = T\left(\frac{\partial s}{\partial v}\right)_T - p = T\left(\frac{\partial p}{\partial T}\right)_v - p \tag{6.21a}$$

where the partial derivative involving the entropy has been replaced by the Maxwell relation of Equation (6.19). The final result for du is:

$$du = C_V dT + \left[T\left(\frac{\partial p}{\partial T}\right)_v - p\right]dv \tag{6.21b}$$

This form is applicable to nonideal gases for which an equation of state $p(T,v)$ is available. For solids, replacement of $(\partial p/\partial T)_v$ by α/β (see Equation [6.8]) provides a more useful form.

The relation analogous to Equation (6.21a) for the enthalpy is obtained from the total differential of $h(T,p)$ and following a procedure similar to that used above for the internal energy. The fundamental differential $dh = Tds + vdp$ is divided by dp at constant T to arrive at $(\partial h/\partial p)_T$ in terms of $(\partial s/\partial p)_T$, which is then eliminated by the Maxwell relation given by Equation (6.20). The result is:

$$\left(\frac{\partial h}{\partial p}\right)_T = T\left(\frac{\partial s}{\partial p}\right)_T + v = -T\left(\frac{\partial v}{\partial T}\right)_p + v \tag{6.22a}$$

The total differential of h is:

$$dh = C_P dT + \left[v - T\left(\frac{\partial v}{\partial T}\right)_p\right]dp \tag{6.22b}$$

For solids, αv is substituted for the partial derivative.

For an ideal gas, the bracketed terms in Equations (6.21b) and (6.22b) are identically zero. For solids, the second terms on the right-hand sides of these equations are essential if the process is isothermal and involves changes of v or p.

Other derivatives of the internal energy and enthalpy can be obtained by manipulations of the fundamental differentials of du and dh different from those employed above to give Equations (6.21a) and (6.22a). Problem 6.2 applies this method to the derivatives $(\partial u/\partial p)_T$ and $(\partial u/\partial T)_p$, wherein the "off-natural" variable p replaces the "natural" variable v associated with u (see Section 6.2). Problem 6.3 shows that $(\partial u/\partial p)_T$ can be obtained in three different ways (in thermodynamics, there is often more than one way to skin a cat).

For an ideal gas, internal energy and enthalpy are independent of specific volume or pressure. For a nonideal gas such as one obeying the Van der Waals equation of state, both u and h depend on v, as shown in Problem 6.4.

6.3.2 Entropy

For the entropy expressed as $s(T,v)$, the total differential is:

$$ds = \left(\frac{\partial s}{\partial T}\right)_v dT + \left(\frac{\partial s}{\partial v}\right)_T dv \tag{6.23a}$$

The coefficient of dT is C_V/T, as can be demonstrated from $du = Tds - pdv$ by dividing by dT while holding v constant. The coefficient of dv is eliminated using the Maxwell relation Equation (6.19), leading to the final result:

$$ds = \frac{C_V}{T} dT + \left(\frac{\partial p}{\partial T}\right)_v dv \tag{6.23b}$$

Starting from $s(T,p)$ yields an alternative entropy differential:

$$ds = \left(\frac{\partial s}{\partial T}\right)_p dT + \left(\frac{\partial s}{\partial p}\right)_T dp \tag{6.24a}$$

The fundamental differential $dh = Tds + vdp$ gives C_P/T as the coefficient of dT and use of the Maxwell relation Equation (6.20) gives the coefficient of dp. The end result is:

$$ds = \frac{C_P}{T} dT - \left(\frac{\partial v}{\partial T}\right)_p dp \tag{6.24b}$$

When integrated for an ideal gas with constant heat capacity, Equations (6.23b) and (6.24b) reduce to Equations (3.9) and (3.10), respectively. The first pair of equations apply to any one-component substance, but to integrate them, a path $v = F(T)$ or $p = G(T)$ must be specified. The resulting change in entropy, however, is independent of the path chosen.

For solids, the coefficients of dv in Equation (6.23b) and dp in Equation (6.24b) are best replaced by α/β and αv, respectively.

Equations (6.23b) and (6.24b) are sometimes called the "*ds*" equations. There is a third variant in which the independent variables are p and v. This equation is derived in Problem 6.6.

6.3.3 Heat Capacities

Subtracting Equation (6.23b) from Equation (6.24b) gives:

$$(C_P - C_V)dT = T\left(\frac{\partial v}{\partial T}\right)_p dp + T\left(\frac{\partial p}{\partial T}\right)_v dv$$

Dividing by dv and holding p constant gives:

$$(C_P - C_V)\left(\frac{\partial T}{\partial v}\right)_p = T\left(\frac{\partial p}{\partial T}\right)_v$$

Inverting the partial derivative on the left-hand side yields:

$$C_P - C_V = T\left(\frac{\partial v}{\partial T}\right)_p\left(\frac{\partial p}{\partial T}\right)_v \tag{6.25a}$$

For an ideal gas, the product of T and the two partial derivatives is equal to the gas constant. For nonideal gases, on the other hand, the difference between the two heat capacities can differ significantly from R (see Problem 6.8).

For condensed phases, the first partial derivative in Equation (6.25a) is replaced by αv and the second by Equation (6.8), yielding:

$$C_P - C_V = \alpha^2 Tv/\beta \tag{6.25b}$$

For solids or liquids with a small coefficient of thermal expansion, the heat capacities are approximately equal, as shown by the following example.

Example: $C_P - C_V$ for liquid water at 20°C and 10 MPa.

The pertinent properties are: $\alpha = 2.0 \times 10^{-4}$ K^{-1}; $\beta = 4.4 \times 10^{-4}$ MPa^{-1}; $v = 1.04 \times 10^{-3}$ m^3/kg.

With these values, Equation (6.25b) gives:

$$C_P - C_V = 0.5 \text{ J/mole-K} = 0.06R$$

By way of comparison, C_P of water is 75 J/mole-K or 9.0R. The difference in the heat capacities of this substance is clearly negligible.

Equation (6.23b) provides the starting points for calculating the effect of specific volume on C_V. As a thermodynamic property, ds must be an exact differential, so Equation (6.4) applies to the coefficients of dT and dv in Equation (6.23b). The terms in Equation (6.1) can be paired with the corresponding terms in Equation (6.23b) by: $x = T$, $y = p$, $M = C_V/T$, $N = (\partial p/\partial T)_v$. Using these identifications in Equation (6.4) yields:

$$\left(\frac{\partial C_V}{\partial v}\right)_T = T\left(\frac{\partial^2 p}{\partial T^2}\right)_v = T\frac{\partial}{\partial T}\left(\frac{\alpha}{\beta}\right)_v \tag{6.26}$$

The second equality utilizes Equation (6.8).

Analogous identification of the coefficients of dT and dp in Equation (6.24b) with M and N in Equation (6.4) yields (Problem 6.12):

$$\left(\frac{\partial C_P}{\partial p}\right)_T = -T\left(\frac{\partial^2 v}{\partial T^2}\right)_p = -T\frac{\partial}{\partial T}(\alpha v)_p \tag{6.27}$$

For ideal gases, the second derivatives in Equations (6.26) and (6.27) are zero, and the heat capacities are independent of pressure or specific volume.

Example: What change in C_P is caused by increasing the pressure of water at 20°C by 10 MPa?

Assuming α to be constant over the pressure range considered, Equation (6.27) simplifies to:

$$\left(\frac{\partial C_P}{\partial p}\right)_T \cong -\alpha^2 T v = -(2\times10^{-4})^2(293)(1.04\times10^{-3}) = -1.2\times10^{-8}\ \frac{\text{J/mole-K}}{\text{N/m}^2}$$

For a pressure increase of 10 MPa (10^7 n/m²), C_P decreases from 75 to 74.9 J/mole-K.

Although the derivatives of C_P and C_V with respect to pressure can be determined, it is not possible to relate $(\partial C_V/\partial T)_V$ or $(\partial C_P/\partial T)_p$ to the EOS of the substance. These partial derivatives are available only from data such as shown in Figures 2.4 or 2.6.

The following example demonstrates the method of integrating the differentials represented by Equations (6.21) to (6.24) between two states of different temperature and pressure.

Example:

Nitrogen obeys the Van der Waals equation of state and serves as the vehicle for highlighting the importance of nonideality corrections if the pressures are sufficiently high.

The path of the process is $1 \rightarrow A \rightarrow 2$ shown in the figure below. The enthalpy difference $h_2 - h_1$ is to be calculated from the following two-step process: an isobaric step from 1 to A followed by an isothermal step from A to 2.

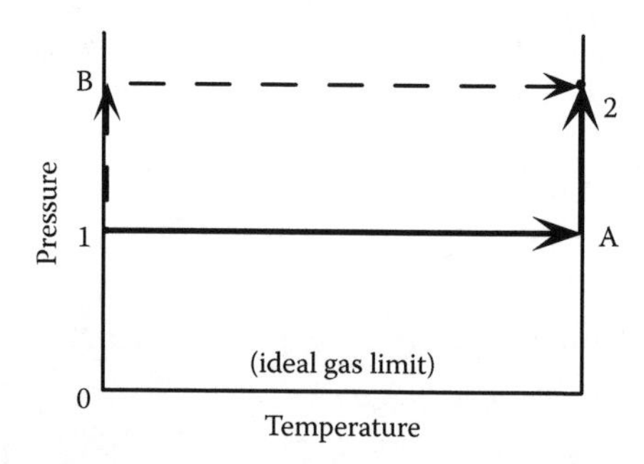

The conditions at states 1 and 2 are: T_1 = 200 K, p_1 = 0.1 MPa and T_2 = 300 K, p_2 = 10 MPa.

Level 0 represents the ideal gas limit, where the heat capacity is C_{P0} = 29.1 J/mole-K.

The equation of state of the gas is given by the Van der Waals EOS, Equation (2.5):

$$v = \frac{RT}{p} + b - \frac{a}{RT}$$

where a = 0.14 J-m^3/mole^2 and $b = 3.9 \times 10^{-5}$ m^3/mole. The heat capacity at pressure p_1 is calculated by substituting the EOS into the first equality of Equation (6.27) and integrating from $p_0 \sim 0$ where $C_P = C_{P0}$ to p_1, the common pressure of states 1 and A:

$$C_{P1} = C_{P0} + \frac{2ap_1}{RT^2}$$

This result is used to determine the enthalpy change for the temperature increase from state 1 to intermediate state A:

$$h_A - h_1 = \int_{T_1}^{T_2} C_{P1}dT = \left(C_{P0} + \frac{2ap_1}{RT_1T_2}\right)(T_2 - T_1) = (29.1 + 0.06) \times 100 = 2916\frac{\text{J}}{\text{mole}}$$

The pressure correction to C_{P0} amounts to less than 0.2% of the ideal gas heat capacity.

For the isothermal step from state A to the final state 2, the second term in Equation (6.22b) is integrated using v given by the Van der Waals EOS:

$$h_2 - h_A = \int_{p_1}^{p_2}\left[v - T\left(\frac{\partial v}{\partial T}\right)_p\right]dp = -\left(\frac{2a}{RT_2} - b\right)(p_2 - p_1) = -736\frac{\text{J}}{\text{mole}}$$

The total enthalpy change, $h_2 - h_1$ = 2916 – 736 = 2180 J/mole is approximately 25% smaller than the ideal gas value of 2910 J/mole.

The choice of the path 1→A→2 to perform the calculation is purely a matter of calculational convenience. Any other path from state 1 to state 2, such as 1→B→2, would produce the same value of $h_2 - h_1$ (see Problem 6.9).

6.4 NONIDEAL GASES WITH SPECIAL PROCESS RESTRAINTS

Section 6.3 treated the effects of changes in p, T and v on the dependent variable properties u, h and s. The partial derivatives that appear in these relations (Equations [6.21] to [6.24]) involved only the EOS variables p, T, and v. In the present section, the roles of the independent and dependent variables are reversed. Processes in which s, u, and h are held constant are analyzed to determine the p-v-T behavior of a nonideal gas.

6.4.1 Isentropic Process

Isentropic expansion of an ideal gas was treated in Section 3.5. Here, the same process is analyzed without the restriction of ideality. Equation (6.23b) is divided by dv while holding s constant, which produces the relation:

$$\left(\frac{\partial T}{\partial v}\right)_s = -\frac{T}{C_V}\left(\frac{\partial p}{\partial T}\right)_v \tag{6.28}$$

To illustrate the effect of gas nonideality on property changes during an isentropic expansion, the right-hand side of Equation (6.28) is evaluated for a Van der Waals gas obeying the equation of state given by Equation (2.3):

$$p = \frac{RT}{v-b} - \frac{a}{v^2}$$

Substituting the above EOS into Equation (6.28) yields:

$$\left(\frac{\partial T}{\partial v}\right)_s = -\frac{RT}{C_V(v-b)}$$

In the limit of low pressure, the specific heats are independent of p (or v) and their difference is the gas constant. The nonideal gas effect on C_V is given by the first equality of Equation (6.26). For the Van der Waals EOS $(\partial^2 p/\partial T^2)_v = 0$, so C_V is not a function of v. Substituting $R = C_{P0} - C_{V0}$, setting $C_V = C_{V0}$ and using the symbol γ for C_{P0}/C_{V0}, the above equation becomes:

$$\left(\frac{\partial T}{\partial v}\right)_s = -(\gamma - 1)\frac{T}{(v-b)}$$

Integrating between states 1 and 2 gives:

$$T_2 = T_1\left(\frac{v_1 - b}{v_2 - b}\right)^{\gamma-1} \tag{6.29}$$

If the gas were ideal, the parameter b would vanish from Equation (6.29) and Equation (3.14) would be recovered.

Example: N_2 ($\gamma = 1.4$) is compressed isentropically from 0.1 MPa and 20°C (state 1) to 10 MPa (state 2). Find the final temperature if (a) nonideality is characterized by the Van der Waals EOS with $a = 0.14$ J-m^3/mole^2 and $b = 3.9 \times 10^{-5}$ m^3/mole and (b) ideality is assumed.

(a) In state 1, N_2 is very nearly ideal, so $v_1 = 2.44 \times 10^{-2}$ m^3/mole. However, v_2 depends on T_2, which is also an unknown. The Van der Waals EOS [Equation (2.3)] is written for the conditions of state 2, RT_2 is divided by RT_1 and the result equated to the right-hand side of Equation (6.29). The result is:

$$\frac{T_2}{T_1} = \frac{v_2 - b}{RT_1}\left(p_2 + \frac{a}{v_2^2}\right) = \left(\frac{v_1 - b}{v_2 - b}\right)^{\gamma-1}$$

Solving the second equality numerically yields $v_2 = 9.4 \times 10^{-4}$ m^3/mole and the first equality gives $T_2 = 1098$ K.

(b) If the gas were assumed to behave ideally, the simpler formulas of Equations (3.13) and (3.14) would yield $(v_2)_{id} = 9.1 \times 10^{-4}$ m^3/mole and $(T_2)_{id} = 1092$ K. For the process conditions used in this example, the nonideality corrections are modest.

Problems 6.5 and 6.6 provide additional exercises involving isentropic processes of nonideal gases.

6.4.2 Joule Expansion (Constant Internal Energy Process)

One of the simplest yet most significant experiments in the long history of thermodynamics was performed in 1843 by James Joule. The experimental setup consisted of two vessels connected by a valve and immersed in a water bath with a thermometer to detect any temperature changes arising from cooling or heating effects in the two vessels. Initially, one vessel contained air at high pressure and the other was evacuated. When the valve connecting them was opened and the gas expanded to fill both vessels equally (doubled in volume), no increase in the temperature of the water bath was noted. Since the gas in the pair of vessels constitutes a system of constant internal energy,* Joule concluded that the internal energy of the air does not depend on its specific volume. This key feature of ideal gases was stated without proof in Section 2.4.1.

Joule's experiment can be modified to more clearly reveal the thermal effects of expansion of a nonideal gas when its internal energy is held constant. Or, how is the partial derivative $(\partial T/\partial v)_u$ related to the equation of state of the gas? The modification consists of eliminating the water bath, insulating the two vessels against heat flow from outside, and placing the thermometer directly in the first vessel. For simplicity, the valve is replaced by a membrane that is ruptured to initiate gas expansion. The apparatus is shown in Figure 6.2.

* The water bath serves only to measure the temperature change of the gas following opening of the valve.

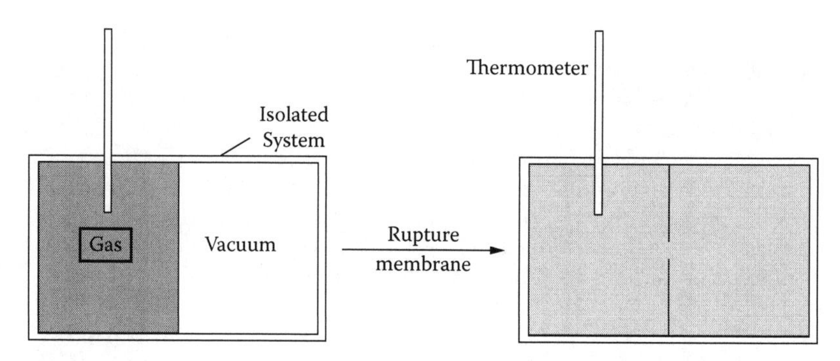

FIGURE 6.2 Modified Joule apparatus.

To relate the temperature difference between the two states shown in Figure 6.2 to the equation of state, Equation (6.21b) is divided by dv with u held constant. This yields the desired coefficient:

$$\left(\frac{\partial T}{\partial v}\right)_u = \frac{1}{C_V}\left[p - T\left(\frac{\partial p}{\partial T}\right)_v\right] \tag{6.30}$$

For an ideal gas, it is readily seen that the right hand side of this equation is zero, which is the theoretical basis for Joule's observation of the lack of temperature change on expansion in his experiment. Thus, the experiment proved that the heat capacity of ideal gases does not depend on specific volume.

If an equation of state appropriate to a nonideal gas is used in the bracketed term in Equation (6.30), the temperature change on expansion can be predicted. If the initial pressure is sufficiently high and the initial temperature low enough, the nonideal effect will be measurable.

Example: What is the change in temperature of nitrogen when the specific volume initially at 20°C and 10 MPa undergoes the process shown in Figure 6.2?

Noting that $C_V = C_{V0} = 7/2R = 29.1$ J/mole-K, and using Equation (2.3) for the equation of state in Equation (6.30) yields:

$$\left(\frac{\partial T}{\partial v}\right)_u = -\frac{a}{C_{V0}v^2}$$

Integrating from v_1 to $2v_1$ gives a temperature change of:

$$\Delta T = -\frac{a}{2C_{V0}v_1} = -\frac{ap_1}{2C_{V0}RT_1} = -\frac{(0.14)(10^7)}{(2)(29.1)(8.314)(293)} = -10°\text{C}$$

It is clear from this result that measurement of the gas temperature decrease in a Joule expansion provides a means of experimentally determining the coefficient **a** in the Van der Waals equation.

6.4.3 The Joule-Thompson Coefficient (Constant Enthalpy Process)

Equation (4.9) shows that flow of a fluid through a device that exchanges neither heat nor work with the surroundings and involves negligible kinetic energy changes proceeds without change in enthalpy. Valves, porous plugs and orifices inserted into flow lines are examples of devices through which the flow is isenthalpic. Even though the enthalpies upstream and downstream of such devices are equal, the pressures are not. In fact, the main practical purpose of these devices is to produce an abrupt reduction in pressure, and for refrigerators, of temperature as well. For this reason, they are called *throttling devices.*

If the fluid is a gas, the change in temperature across the device may be positive, negative, or zero, depending on the equation of state and the upstream temperature. The partial derivative $(\partial T/\partial p)_h$ representing this process is called the *Joule-Thompson coefficient.* If the gas is ideal, no change in temperature occurs because the enthalpy is constant.

The relation between the Joule-Thompson coefficient and the equation of state is obtained from Equation (6.22b) by dividing by dp while holding h constant:

$$\left(\frac{\partial T}{\partial p}\right)_h = \frac{1}{C_P}\left[T\left(\frac{\partial v}{\partial T}\right)_p - v\right] \tag{6.31}$$

This equation is the enthalpy analog of Equation (6.30). It is obtained from the latter by exchanging p for v and h for u.

Example: If N_2 at 20°C is reduced in pressure in a throttling device from 10 MPa to 0.1 MPa, what is the temperature change?

The pressure dependence of C_P of a nonideal gas is given by Equation (6.27). However, this effect is of second order, and it suffices to approximate C_P by the ideal gas value: $C_P = C_{P0} = 9/2R = 36.4$ J/mole-K. Substituting the approximate form of the Van der Waals EOS, Equation (2.5), into Equation (6.31) and using the constants for N_2 yields:

$$\left(\frac{\partial T}{\partial p}\right)_h = \frac{1}{C_P}\left[\frac{2a}{RT} - b\right] = 2\times 10^{-6}\,\frac{\text{K}}{\text{Pa}} = 2\,\frac{\text{K}}{\text{MPa}}$$

The 9.9 MPa pressure drop over the throttling device is accompanied by a 20°C temperature decrease. Repeating this test at another temperature provides enough data to determine the constants a and b in the Van der Waals EOS.

PROBLEMS

6.1 Consider the following total differentials of the functions $w(x,y)$ and $z(x,y)$:

$$dw = y(3x^2 + y^2)dx + x(x^2 + 2y^2)dy$$

and

$$dz = y(3x^2 + y)dx + x(x^2 + 2y)dy$$

Determine:

(a) Whether either or both is exact in the sense of Equation (6.4).
(b) Whether either or both is exact in the sense of Equation (6.5). Test this condition by integrating over two paths: Path A: $y = F(x) = x$ and Path B: $y = F(x) = x^2$. The initial and final states are $x = 0$ and $x = 1$.
(c) If either total differential satisfies the (a) and (b) exactness conditions (to be exact, it must satisfy both), determine the function $w(x,y)$ and/or $z(x,y)$ from which the total differentials arise.

6.2 (a) Starting from the fundamental differentials for du and dh, derive expressions for $(\partial u/\partial p)_T$ and $(\partial u/\partial T)_p$ in terms of C_P and the equation of state variables α, β, and v. Use Maxwell's equations where needed.
(b) For a solid, what temperature change (at 1 atm) produces the same change in internal energy as a pressure increases from 1 atm to 100 atm? The specific heat is 400 J/kg-K, $T = 300$ K and the p-v-T properties of the substance are: density = 8.8 g/cm^3; $\alpha = 5 \times 10^{-5}$ K^{-1}; $\beta = 9 \times 10^{-6}$ MPa^{-1}. Assume that the density is independent of p and T.
(c) Express $(\partial h/\partial v)_T$ in terms of the EOS variables α and β.

6.3 One kilogram of a solid is compressed reversibly from p_1 to p_2 at a fixed T. Calculate:

(a) The work required.
(b) The heat exchanged in maintaining constant T.
(c) The change in the internal energy.

6.4 A gas obeying the Van der Waals EOS is compressed from specific volume v_1 to v_2 at constant temperature. Derive the equations for the changes in u, h, and s for this process. What do these reduce to for an ideal gas?

6.5 A gas obeying the Van der Waals equation of state is initially at pressure $p_1 = 10$ MPa and specific volume $v_1 = 2.5 \times 10^{-4}$ m^3/mole. It undergoes an isentropic expansion to $p_2 = 0.1$ MPa.

(a) What is the equation giving the final specific volume, v_2?
(b) For N_2, calculate the ratio v_2/v_1 by:

1. The full Van der Waals formula (i.e., the equation derived in Part [a]).
2. The Van der Waals formula with the constant $a = 0$.
3. The ideal gas law (both a and b equal to zero).

6.6 (a) Use the chain rule to show that:

$$\left(\frac{\partial T}{\partial p}\right)_s = -\frac{(\partial s / \partial p)_T}{(\partial s / \partial T)_p}$$

(b) Express the partial derivatives on the right-hand side of this equation in terms of C_P and the EOS variables p, v, and T.

6.7 The differentials of the entropy given by Equations (6.23b) and (6.24b) have a third variant:

$$ds = \frac{C_V}{T}\left(\frac{\partial T}{\partial p}\right)_v dp + \frac{C_P}{T}\left(\frac{\partial T}{\partial v}\right)_p dv$$

Prove this relationship. (Hint: start with s as a function of p and v, form its differential and use the chain rule to convert the partial derivative coefficients to the desired forms.)

6.8 What is the difference between C_P and C_V (in units of R) for N_2 at 300 K and 10 MPa? Nitrogen at these conditions behaves as a Van der Waals gas with constants given by Equation (2.6).

6.9 Show that the equation for the enthalpy change from state 1 to state 2 in the diagram on the bottom of page 174 calculated by the 1-B-2 path is the same as was determined for the 1-A-2 path.

6.10 Show that:

$$\left(\frac{\partial p}{\partial v}\right)_S = -\frac{C_P}{C_V \beta v}$$

where β is the coefficient of compressibility.
Follow these steps:

1. To obtain an expression for $(\partial p/\partial v)_S$:
   a. Take the differential of $s(p,v)$.
   b. Use the divide-and-hold-constant rule.
2. To introduce C_P and C_V, use the chain rule (Equation [6.7]).
3. To introduce β, use the cyclic transformation (Equation[6.6]).

7 Gas Mixtures and Nonaqueous Solutions

7.1 MIXTURES AND SOLUTIONS DEFINED; MEASURES OF COMPOSITION

Up to this stage, only systems containing one component have been analyzed. These simple substances can exist (at equilibrium) in up to three phases. This limit is imposed by the phase rule (Equation [1.21]), $F = C + 2 - P$ where C is the number of components, P the number of phases, and F the number of degrees of freedom, or adjustable thermodynamic properties.

The objective of this chapter is to understand the thermodynamics of single-phase ($P = 1$), two-component ($C = 2$) systems. The characteristics of mixtures of ideal gases and the simplest models of nonideal binary solid or nonaqueous liquid solutions are treated. Temperature and total pressure are specified.

7.1.1 Mixtures versus Solutions

Before starting, clarification of the terms "mixture" and "solution" is in order. These terms are nearly, but not quite, synonymous. A *solution* unequivocally refers to a homogeneous system of two or more components. This term is applied to liquids and solids, but not to gases. Salt dissolved in water is an aqueous *solution* of NaCl; a gold-silver alloy is a solid *solution* of these two elements. However, air is a *mixture* of oxygen and nitrogen (plus minor species), not a solution of O_2 in N_2. A multiphase system of a single component is referred to as a *mixture* of phases. Reference to a mixture of steam and liquid water is acceptable terminology, but a salt solution would never be called a mixture. These semantic distinctions between mixtures and solutions are usually clear from the context in which they are used.

7.1.2 Composition

Composition is not a variable in one-component systems. A two-component system possesses a single composition variable. In general, a C–component system is characterized by $C - 1$ independent compositions.

Several measures of composition (or concentrations) are available. The most commonly used is the *mole fraction*, which for component i is defined by:

$$x_i = \frac{n_i}{n} \qquad \text{where} \qquad n = \sum n_i \tag{7.1}$$

n is the total moles in the system and n_i is the number of moles of component *i*. By definition, the sum of the mole fractions is unity. The mole fraction unit is used for gas mixtures, nonaqueous liquid solutions and solid solutions. In aqueous solutions, these units are generally avoided in favor of volumetric units such as moles or mass of *i* per unit volume of solution.

7.2 IDEAL GAS MIXTURES

In an ideal gas mixture, there are no intermolecular interactions in either the pure components or in the mixture.

The mixing process starts from prescribed quantities of the pure gases at specified temperature and pressure. The mixture is at the same *T* and *p*. Because the mixture is ideal, this process does not release or absorb heat from the surroundings and the combined volume is the sum of the individual volumes of the pure constituents:

$$V = \sum n_i v_i \qquad \text{or} \qquad v = \frac{V}{n} = \sum x_i v_i \tag{7.2a}$$

Absent *pV* work and heat exchange with the surroundings, the first law requires that the internal energy of the mixture be equal to the sum of the values of the pure components, or:

$$U = \sum n_i u_i \qquad \text{or} \qquad u = \frac{U}{n} = \sum x_i u_i \tag{7.2b}$$

where u_i is the molar internal energy of component *i* and *u* is the internal energy per mole of mixture or solution. Because mixing occurs at constant pressure and there is no change in system volume, a similar equation applies to the enthalpy:

$$H = \sum n_i h_i \qquad \text{or} \qquad h = \frac{H}{n} = \sum x_i h_i \tag{7.2c}$$

Since the specific heats C_V and C_P are temperature derivatives of *u* and *h*, respectively, the same mixing rules apply:

$$C_V = \sum x_i C_{Vi} \qquad \text{and} \qquad C_P = \sum x_i C_{Pi} \tag{7.2d}$$

where C_{Vi} and C_{Pi} are the molar specific heats of pure component *i*.

The entropy of an ideal mixture or solution is not as transparent as are *U* and *H*. This issue is treated in Section 7.2.2.

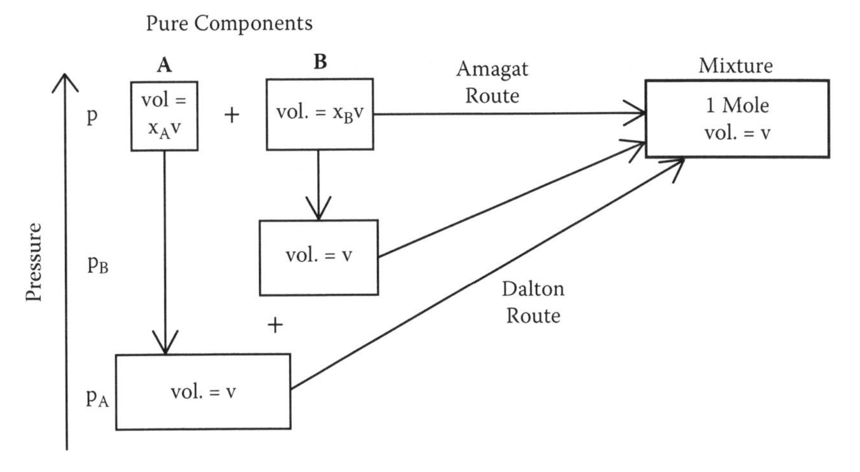

FIGURE 7.1 Two methods of mixing x_A moles of pure A with x_B moles of pure B to form 1 mole of an ideal gas mixture. The temperature is constant.

7.2.1 Dalton's Rule

Two routes for forming 1 mole of an ideal gas mixture from x_A and x_B moles of pure components A and B are shown in Figure 7.1. Both mixing processes are conducted isothermally at temperature T. x_A and x_B are the mole fractions of A and B in the mixture, so $x_A + x_B = 1$. In the diagram, the molar volume is $v = RT/p$, where p is the pressure of the initial pure components and of the final mixture.

The initial states of A and B at pressure p obey the ideal gas law:

$$p(x_A v) = x_A RT \qquad p(x_B v) = x_B RT$$

(the mole fractions appear on both sides of these equations to facilitate subsequent explanations). The two pure gases are mixed in two ways.

The upper path in Figure 7.1, called the *Amagat route*, simply mixes the two pure components at constant pressure p. The lower path, called the *Dalton route*, is more circuitous but produces the same end result. In this route, the pure components are first isothermally expanded from their initial volumes to the final volume v of the mixture. When expanded to volume v, the pressures of the pure components are reduced according to the ideal gas law:

$$p_A v = x_A RT \qquad p_B v = x_B RT$$

Comparing the left-hand sides of the above pairs of equations shows that the *partial pressures* are:

$$p_A = x_A p \qquad p_B = x_B p \tag{7.3}$$

When the two expanded volumes of A and B are mixed by the Dalton route in Figure 7.1, the pressures p_A and p_B are retained in the mixture. This is the origin of the term "partial" pressures. Adding the two equations in (7.3) yields:

$$p_A + p_B = p \tag{7.4}$$

The partial pressures play dual roles as measures of mixture composition and as contributions to the total gas pressure. Equations (7.3) and (7.4) are collectively known as *Dalton's rule.*

The components of an ideal gas mixture individually obey the ideal gas law. Each constituent species independently occupies the same volume at the same temperature as the others. The equation of state for species i in a general multi-component gas mixture is:

$$p_i V = n_i RT \tag{7.5}$$

The mixture obeys the ideal gas law at the specified total pressure:

$$pV = nRT \tag{7.6}$$

7.2.2 Entropy of Mixing

In isothermal, isobaric mixing of ideal substances, all thermodynamic properties remain unchanged:

$$\Delta V_{mix} = 0 \qquad \Delta U_{mix} = 0 \qquad \Delta H_{mix} = 0 \tag{7.7}$$

except for the entropy. Despite the absence of intermolecular interactions in ideal gases, mixing of pure components increases the entropy because the mixture is in a more random state than the separated pure components; mixing is an irreversible process that requires work to undo. The irreversible nature of mixing resides in the Amagat route in Figure 7.1, which entails an entropy increase ΔS_{mix}. The utility of the Dalton route in this figure is that it provides a way of calculating ΔS_{mix}. Figure 7.2 shows the entropy changes for the various steps in the two routes. Because the change in a thermodynamic property is path independent, these entropy changes are related by:

$$\Delta S_{mix} = \Delta S_A + \Delta S_B + \Delta S' \tag{7.8}$$

The first two terms on the right represent isothermal pressure changes for which the entropy changes are given by Equation (3.10):

$$\Delta S_A = -Rx_A \ln(p_A/p) \qquad \Delta S_B = -Rx_B \ln(p_B/p) \tag{7.9}$$

The entropy change on mixing by the Dalton path, $\Delta S'$, is zero.

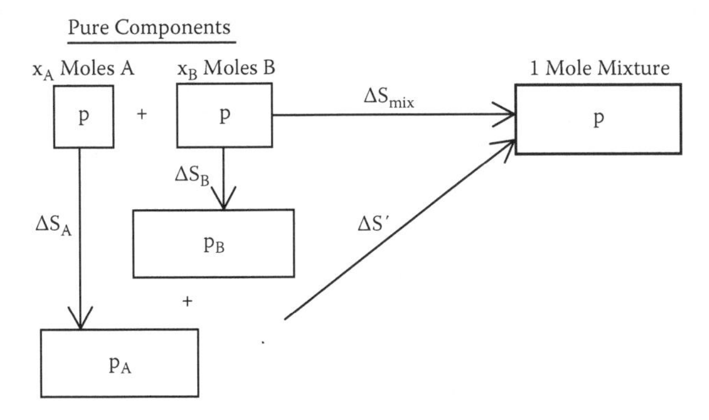

FIGURE 7.2 Entropy changes associated with the Amagat and Dalton routes for mixing pure components A and B.

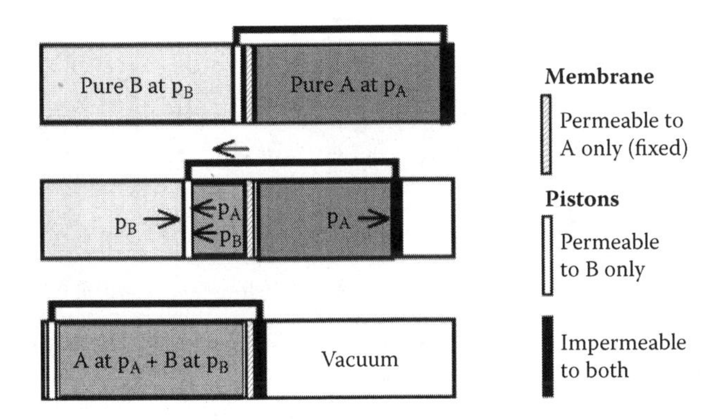

FIGURE 7.3 Apparatus for mixing ideal gases without an entropy change.

Proof of this assertion is based on the cylinder-piston apparatus shown in Figure 7.3. The pure components at their reduced pressures are initially contained in compartments separated by a fixed membrane that is permeable to component A but that will not pass component B (top view). These two compartments are equivalent to the two lower boxes in Figure 7.2. They both have volume v.

The apparatus in Figure 7.3 also contains moveable pistons connected by a rigid rod. The piston on the left permits component B to pass but is a barrier to component A. The piston on the right is impermeable to both gases. In the initial position of the rod, A is kept inthe right-hand chamber by the left piston and B is confined to the left chamber by the fixed central membrane.

The pistons are frictionless and can move from right to left without requiring work to be done on them; the net force on the connected pistons is zero because p_B acts equally on both faces of the B-permeable membrane and p_A acting on the right-hand face of the B-permeable membrane is balanced by the same pressure acting on the lefthand face of the impermeable membrane. The lower plus sign in Figure 7.2 represents the mixing process depicted in the middle view of Figure 7.3.

As the moveable pair of connected pistons moves to the left, component B flows without resistance into the mixture region and component A does likewise through the fixed central membrane. There is no internal energy change during mixing of the gases since they are ideal. Since $W = 0$ and $\Delta U = 0$, the first law requires that $Q = 0$. Since the process is reversible, the second law yields $\Delta S' = Q/T = 0$.

Using this result along with Equation (7.9) in Equation (7.8) and expressing the pressure ratios by Dalton's rule, Equation (7.3), yields:

$$\Delta s_{mix} = -R(x_A \ln x_A + x_B \ln x_B) \tag{7.10}$$

Equation (7.10) applies to one mole of a binary gas mixture. For gas mixtures containing n_i moles of each component, the general form of the entropy of mixing formula is:

$$\Delta S_{mix} = -R\sum n_i \ln x_i \tag{7.11}$$

For nonideal gases or mixing processes that are neither isothermal nor isobaric, or for initial states that are mixtures rather than pure components, Equation (7.11) is but one component of the entropy change. These effects are illustrated in the following examples.

Example: Calculate the changes in enthalpy and entropy when two moles of helium (species 1) at $T_1 = 100°C$ and $p_1 = 1$ atm are mixed with one mole of nitrogen (species 2) at $T_2 = 200°C$ and $p_2 = 0.5$ atm. Mixing takes place adiabatically and the mixture volume is the same as the sum of the volumes of the initial pure species. This process represents opening of a valve between two insulated tanks of the gases and allowing their contents sufficient time to mix to a uniform state.

First, the temperature, pressure, volume, and composition of the mixture are calculated. These properties are designated without a subscript. The composition is characterized by the helium mole fraction $x_{He} = n_{He}/n = 0.667$. The nitrogen mole fraction is $1 - x_{He} = 0.333$. The mixture volume is the sum of the volumes of the two tanks:

$$V = V_1 + V_2 = n_{He}RT_1/p_1 + n_{N2}RT_2/p_2 = 0.141 \text{ m}^3$$

To calculate the mixture temperature, the first law is applied. Because the tanks are insulated ($Q = 0$) and do no work because they do not change volume ($W = 0$), the first law requires the internal energy of the mixture to equal the sum of the internal energies of the pure gases. For ideal gases the internal energy is dependent on temperature only and can be expressed in terms of the heat capacities from Equation (7.2d):

$$C_V(T - T_{ref}) = x_{He}C_{VHe}(T_1 - T_{ref}) + (1 - x_{He})C_{VN2}(T_2 - T_{ref})$$

T_{ref} is an arbitrary reference temperature and $C_{VHe} = (3/2)R$ and $C_{VN2} = (5/2)R$ are the specific heats at constant volume of helium and nitrogen, respectively. Using Equation (7.2d), the specific heat of the mixture is the mole-fraction average of the two pure species: $C_V = (11/6)R$. Solving the above equation for T yields:

$$T = x_{He}\frac{C_{VHe}}{C_V}T_1 + (1 - x_{He})\frac{C_{VN2}}{C_V}T_2 = 146°\text{C}$$

The final pressure is obtained by applying the ideal gas law to the mixture:

$$p = \frac{nRT}{V} = \frac{3 \times 8.314 \times (146 + 273)}{0.141} \times 10^{-5}\ \frac{\text{atm}}{\text{Pa}} = 0.74\ \text{atm}$$

To obtain the enthalpy change upon mixing the pure gases, the specific heats at constant pressure are needed. These are equal to the specific heats at constant volume plus the gas constant: $C_{PHe} = C_{VHe} + R = (5/2)\text{R}$ and $C_{PN2} = C_{VN2} + R = (7/2)\text{R}$. The mixture C_P is (17/6)R.

The enthalpy change for the mixing process is:

$$\Delta H = n\ [C_P(T - T_{ref}) - x_{He}C_{PHe}(T_1 - T_{ref}) - (1 - x_{He})C_{PN2}(T_2 - T_{ref})] = 305\ \text{J}$$

(the terms involving T_{ref} cancel). $n = n_{He} + n_{N2}$

To calculate the entropy change, the pure components must first be brought to the final p and T of the mixture. Using Equation (3.10), this step incurs the following changes in entropy:

$$\Delta S_{He} = n_{He}\left[C_{PHe}\ln\left(\frac{T}{T_1}\right) - R\ln\left(\frac{p}{p_1}\right)\right] = 1.18\text{R}$$

$$\Delta S_{N2} = n_{N2}\left[C_{PN2}\ln\left(\frac{T}{T_2}\right) - R\ln\left(\frac{p}{p_2}\right)\right] = -0.82\text{R}$$

The entropy of mixing at constant temperature and pressure is given by Equation (7.10):

$$\Delta S_{mix} = -R[n_{He}\ln x_{He} + n_{N2}\ln x_{N2})] = 1.91\text{R}$$

The entropy change for the process is:

$$\Delta S = \Delta S_{He} + \Delta S_{N2} + \Delta S_{mix} = 2.27\text{R} = 17.9\ \text{J/K}$$

Example: Two moles of an ideal equimolar A-B gas mixture (designated No. 1) are mixed with three moles of a similar mixture containing 20 mole percent of A (No. 2). Both are at the same temperature. What is the entropy change when these two mixtures are combined to form mixture 3?

The method is to start from the pure components (1.6 moles of A and 3.4 moles of B) and prepare mixtures 1, 2, and 3 from them. The difference in the mixing entropies is the desired result.

In the diagram below, the pure components are shown in the middle two boxes; the final mixture is the large top box. At the bottom are the two starting mixtures.

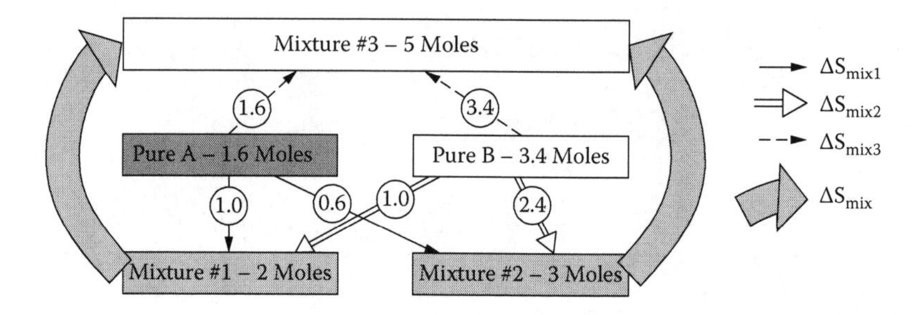

The entropy gains in preparing the two initial solutions from the pure components are:

$$\Delta S_{mix1} = -R[(1.0)\ln(0.5) + (1.0)\ln(0.5)] = 1.39\mathrm{R}$$

$$\Delta S_{mix2} = -R[(0.6)\ln(0.2) + (2.4)\ln(0.8)] = 1.50\mathrm{R}$$

The mole fraction of A in mixture 3 is 0.32 and its mixing entropy is:

$$\Delta S_{mix3} = -R[(1.6)\ln(0.32) + (3.4)\ln(0.68)] = 3.13\mathrm{R}$$

The entropy change due to combining mixture 1 and 2 to form mixture 3 is:

$$\Delta S_{mix} = \Delta S_{mix3} - \Delta S_{mix1} - \Delta S_{mix2} = 0.24\mathrm{R} = 2.0 \text{ J/K}$$

Inclusion of the entropy of mixing in situations with components at different initial conditions (pressure and/or temperature) is the topic of Problem 7.1.

Although the mixing-entropy formula was derived for ideal solutions, it is also present in mixtures of species that behave nonideally (see Problems 7.11 and 7.17). Other exercises in mixing entropy are Problems 7.5/7.6 and 7.18/7.19.

7.3 NONIDEAL LIQUID AND SOLID SOLUTIONS

Although gas mixtures can, for most purposes, be treated as ideal, liquid and solid solutions are generally significantly nonideal. The strong intermolecular interactions that are responsible for the existence of pure condensed phases are also the source of their deviations from ideality when mixed in solutions. A binary solution of A and B is ideal if the average of the A-A and B-B intermolecular forces is just equal to the strength of the A-B interaction (Gaskell, 1981 provides a thorough explanation of this behavior). Nonideal behavior affects all extensive properties of solid and liquid solutions, namely V, U, H, S, F, and G. It is manifest as a departure from linear variation of these properties with changes in the number of moles of one of the components. Linearity in ideal solutions is shown by the formulas for U and H given by Equations (7.2b) and (7.2c).

7.3.1 Partial Molar Properties

For simplicity, the theory of nonideal solutions is presented only for the enthalpy, but analogous formulas apply to all extensive properties. The enthalpy of a nonideal solution is of the same form as that for an ideal solution (Equation [7.2c]). The sole change is replacement of the molar enthalpy of the pure components (h_i) by a quantity called the *partial molar enthalpy*, denoted by $\bar{h}_i$:

$$H = \sum n_i \bar{h}_i \tag{7.12}$$

The partial molar property for species i depends on temperature and on the composition of the solution but the effect of total pressure is negligible. The composition dependence renders $\bar{h}_i$ considerably more difficult to determine than the corresponding molar property of the pure species, h_i.

The physical meaning of $\bar{h}_i$ is best shown by explicitly including the mole numbers of each species in the solution in writing its total enthalpy. Taking the differential of $H(T,p,n_1,n_2,\ldots)$ at constant T and p yields:

$$dH = \sum_i \left(\frac{\partial H}{\partial n_i}\right)_{T,p,n_j} dn_i \qquad \text{or} \qquad dH = \sum_i \bar{h}_i dn_i \tag{7.13}$$

where the partial derivative with respect to n_i is taken with the number of moles of all other components held constant. This partial derivative is the partial molar enthalpy of species i in the solution:

$$\bar{h}_i = \left(\frac{\partial H}{\partial n_i}\right)_{T,p,n_j} \tag{7.14}$$

Equation (7.14) has a definite physical meaning: $\bar{h}_i$ represents the change in the enthalpy of a solution when a small quantity of species i is added while the amounts of all other components are held constant.

Equation (7.13) can be “integrated” in a physical sense by simultaneously adding the pure components to a vessel at rates proportional to their concentrations in the solution (Figure 7.4). This procedure maintains all concentrations constant during the process, and demonstrates that Equation (7.13) is consistent with Equation (7.12).

A very important relation can be deduced from these two equations. The total differential of Equation (7.12) is

$$dH = \sum n_i d\bar{h}_i + \sum \bar{h}_i dn_i$$

which, when combined with Equation (7.13) yields:

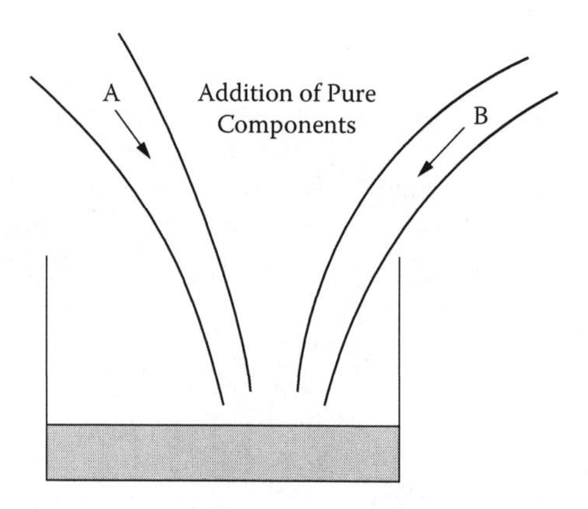

FIGURE 7.4 Mixing to maintain the solution composition constant. The rates of adding A and B are proportional to their concentrations in the solution.

$$\sum_i n_i d\bar{h}_i = 0 \tag{7.15}$$

The most useful forms of the above equations are in terms of mole fractions rather than mole numbers and specialized for an A-B binary solution. Dividing Equation (7.12) by the total moles of A and B gives the molar enthalpy of the solution:

$$h = x_A \bar{h}_A + x_B \bar{h}_B \tag{7.16}$$

Similar treatment of Equation (7.15) yields:

$$x_A d\bar{h}_A + x_B d\bar{h}_B = 0 \tag{7.17}$$

Taking the total differential of Equation (7.16) and taking Equation (7.15) to account produces the mole-fraction analog of Equation (7.13):

$$dh = \bar{h}_A dx_A + \bar{h}_B dx_B \tag{7.18}$$

For a detailed proof of this equation, see problem 7.10.

The importance of Equation (7.17) is that it permits $\bar{h}_B$ to be determined if the variation of $\bar{h}_A$ with composition is known. This relation is obtained by integrating Equation (7.17) from $x_A = 0$ (where $\bar{h}_B = h_B$ because the solution is pure B) to an arbitrary mole fraction of A:

$$\bar{h}_B - h_B = -\int_0^{x_A} \frac{x'_A}{1 - x'_A} \frac{d\bar{h}_A}{dx'_A} dx'_A \tag{7.19}$$

If the dependence of $\bar{h}_A$ with x_A is determined experimentally, Independent measurement of $\bar{h}_B$ is not needed.

On occasion, it is necessary to invert Equation (7.16) to express the partial molar properties in terms of the molar property. For this purpose, Equation (7.18) is divided by dx_A, and since $dx_B = -dx_A$, the result is $dh/dx_A = \bar{h}_A - \bar{h}_B$. Multiplying this equation by x_B and adding the result to Equation (7.16) gives:

$$\bar{h}_A = h + x_B \frac{dh}{dx_A} \tag{7.20a}$$

and a similar approach for component B yields:

$$\bar{h}_B = h + x_A \frac{dh}{dx_B} \tag{7.20b}$$

Partial molar properties cannot be measured directly; only the molar properties are accessible to experiment. However, as shown in Figure 7.5, Equation (7.20) provides the basis for a graphical method for determination of $\bar{h}_A$ and $\bar{h}_B$ from the h versus x_A curve.

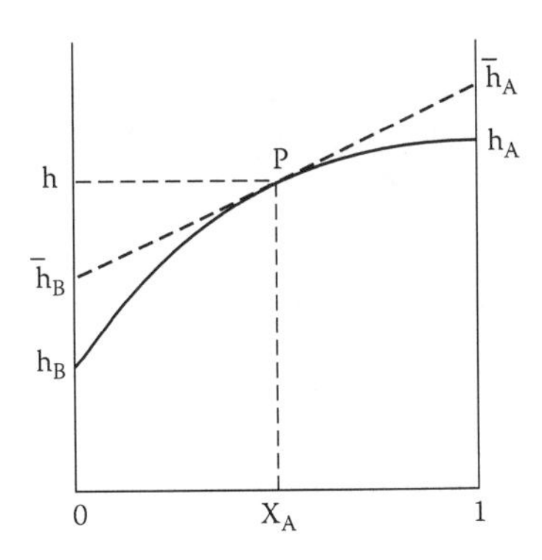

FIGURE 7.5 Graphical method of determining partial molar quantities from molar quantities.

The ordinates at $x_A = 0$ and $x_A = 1$ are the molar enthalpies of the pure components. At each point such as P the tangent line (dashed) is drawn. The intercepts on the two ordinate axes are the partial molar properties (see Problem 7.20 for proof).

7.3.2 Excess Properties

An alternative to the rather abstract partial molar property approach to solution nonideality is the "excess" property method, which describes solutions in terms of their deviations from ideal behavior. Using an A-B binary solution with enthalpy as the generic thermodynamic property, the excess property formulation for the solution enthalpy is:

$$h = x_A h_A + x_B h_B + h^{ex} \tag{7.21}$$

The first two terms on the right represent the combined enthalpies of the pure components and h^{ex} is the *excess enthalpy* and is due solely to nonideality of the solution.

Since the partial molar properties and the excess properties describe the same physical phenomenon, they must be related. The connection between h^{ex} and the partial molar enthalpies is obtained by equating the right-hand sides of Equations (7.16) and (7.21):

$$h^{ex} = x_A(\bar{h}_A - h_A) + x_B(\bar{h}_B - h_B) \tag{7.22}$$

A direct connection between $\bar{h}_A$ and h^{ex} can be obtained by substituting Equation (7.19) into the second term on the right-hand side of Equation (7.22), dividing the resulting equation by x_B, and taking the derivative with respect to x_A. This procedure yields:

$$\bar{h}_A - h_A = (1 - x_A)^2 \frac{d}{dx_A}\left(\frac{h^{ex}}{1 - x_A}\right) \tag{7.23}$$

Knowledge of h^{ex} as a function of x_A gives $\bar{h}_A - h_A$ from Equation (7.23) and Equation (7.22) then yields $\bar{h}_B - h_B$. The utility of this approach is that (at least for the enthalpy) the excess property is amenable to experimental determination.

Example: n_A moles of pure liquid A and n_B moles of pure B, both at temperature T, are mixed in a vessel maintained in a large water bath also held at temperature T. As a result of forming the solution, the temperature of the water bath is observed to increase by ΔT. How is this temperature rise related to the excess enthalpy of solution?

Upon combining the two pure components, the enthalpy of the solution is less than that of the pure components by h^{ex}. Because formation of the solution occurs at constant pressure, the first law (Section 3.4) requires that the enthalpy change appear as heat transferred from the solution to the water bath. This exchange of heat causes the water bath to increase in temperature. With C_{Pw} denoting the specific heat of water and n_w the number of moles, the excess enthalpy of the solution is:

$$(n_A + n_B)\, h^{ex} = -n_w C_{Pw} \Delta T$$

A variant of this example is analyzed in Problem 7.2.

Like partial molar quantities, excess functions apply to solution properties other than enthalpy: volume, internal energy, entropy, Helmholz free energy, and Gibbs free energy. As discussed in the next section, the last of these is particularly important.

7.4 THE CHEMICAL POTENTIAL

The thermodynamic terms heat and work can be viewed as the product of a capacity factor (or quantity of something) and a difference in a potential. Table 7.1 lists several examples of this breakdown of heat/work expressions for mechanical, electrical, thermal, and chemical processes.

TABLE 7.1
Work/Heat as a Capacity Times a Potential Difference

Process	Capacity	Potential	Work/Heat
Lower a weight	Mass	Gravitational	$mg\Delta h$
Pumping a fluid[a]	Volume	Pressure	$V\Delta p$
Electrical	Charge	Electrostatic	$q\Delta\Phi$
Thermal	Energy	Temperature	$nC_P\Delta T$
Chemical[b]	Moles of i	Chemical	$n_i\Delta\mu_i$

[a] See Equation (4.13a); adiabatic process.
[b] Isothermal, isobaric process.

The quantity discussed in this section is the *chemical potential* which is the last entry in Table 7.1.

Although rate processes are not within the purview of thermodynamics, they involve the same potentials as those responsible for producing heat or work. The basic rate laws are of the form: flux = coefficient × potential gradient. Table 7.2 shows the four common rate laws of this type.

TABLE 7.2
Flux as a Kinetic Coefficient Times a Potential Gradient

Rate Process	Rate Coefficient	Potential	Flux	Name of Law
Heat conduction	Thermal conductivity	Temperature	$-k\nabla T$	Fourier's
Momentum transfer (fluid)	Viscosity	Pressure	$-\mu\nabla p$	Newton's
Electricity flow	Electrical conductivity	Electrostatic	$-\kappa\nabla\Phi$	Ohm's
Diffusion	Diffusion coefficient	Chemical	$-(Dc_i/kT)\,\nabla\mu_i$	Fick's

Chemical reactions and interphase mass transfer are also driven by imbalances of the chemical potentials of species in the system. The chemical potential is as important a thermodynamic driving force as are temperature and pressure. This potential drives individual chemical species from one phase to another, from one molecular form to another, or from regions of high concentration to regions of low concentration.

The chemical potential is directly related to the free energy of a system. For a one-component system, the chemical potential is identical to the molar free energy. In solutions or mixtures, the chemical potential is simply another name for the *partial molar free energy*. The discussion in Section 7.3, in which enthalpy was used to illustrate partial molar and excess properties, applies to the Gibbs free energy; one need only replace h everywhere by g.

The reason that the partial molar free energy ($\bar{g}_i$) is accorded the special name "chemical potential" is not only to shorten a cumbersome five-word designation. More important is the role of the chemical potential in phase equilibria and chemical equilibria when the restraints are constant temperature and pressure.

Instead of the symbol $\bar{g}_i$, the chemical potential is designated by μ_i. The connection between the free energy of a system at fixed T and p and the equilibrium state is shown in Figure 1.18. In the remainder of this chapter, the relation between the free energy of a multicomponent system and the chemical potentials of its constituents is developed.

The chemical potential is embedded in the equation for the differential of the free energy of a solution at fixed T and p analogous to Equation (7.13) for the enthalpy:

$$dG = \sum_i \left(\frac{\partial G}{\partial n_i} \right)_{T,p,n_j} dn_i \tag{7.24}$$

The partial derivatives that serve as coefficients of dn_i are the partial molar Gibbs free energies, or the chemical potentials, of each component of the solution:

$$\mu_i = \bar{g}_i = \left(\frac{\partial G}{\partial n_i} \right)_{T,p,n_j} \tag{7.25a}$$

For a one-component system, $G = n_i g_i$, where g_i is the molar free energy. Consequently, for the pure substance, Equation (7.25a) reduces to:

$$\mu_i = g_i \tag{7.25b}$$

Following the lines of the treatment using h and $\bar{h}_i$ in Section 7.3, the following fundamental relations between g and μ_i are obtained:

$$G = \sum_i n_i \mu_i \quad \text{or} \quad g = \sum_i x_i \mu_i \tag{7.26}$$

$$dG = \sum_i \mu_i dn_i \quad \text{or} \quad dg = \sum_i \mu_i dx_i \tag{7.27}$$

$$\sum_i n_i d\mu_i = 0 \quad \text{or} \quad \sum_i x_i d\mu_i = 0 \tag{7.28}$$

The last of this trio of relations is called the Gibbs-Duhem equation. Its extreme importance to solution thermodynamics is the connection it provides between the nonideal behavior of one component with the deviations from ideality of the other components. This property is especially useful for binary systems, and Equation (7.28) will be applied to A-B solutions later in this chapter.

Equations similar to Equations (7.26) to (7.28) apply to other solution thermodynamic properties: v and $\bar{v}_i$, u and $\bar{u}_i$, and s and $\bar{s}_i$.

7.5 ACTIVITY AND ACTIVITY COEFFICIENT

Although the thermodynamic behavior of species in solution is ultimately tied to their chemical potentials, a connection between this property and the concentration of the component is needed. This connection is made via a quantity called the *activity* of a solution species. The activity is a measure of the thermodynamic "strength" of a component in a solution compared to that of the pure substance; the purer, the stronger. As an example, when alcohol is mixed with water its effectiveness is reduced.*

The link between chemical potential and concentration consists of two parts. In the first part, the chemical potential and the activity of a solution component are related by the definition of the activity a_i:

$$\mu_i = g_i + RT\ln a_i \tag{7.29}$$

The definition has been chosen so that the activity tends to unity for pure i; that is, $\mu_i = g_i$, the molar free energy of pure i. Activity varies monotonically with concentration. Therefore, when component i approaches infinite dilution $a_i \rightarrow 0$ and $\mu_i \rightarrow -\infty$. This inconvenient behavior of the chemical potential at zero concentration is avoided by using the activity in practical thermodynamic calculations involving species in solution. Another reason for the choice of the mathematical form of the relation between μ_i and a_i embodied in Equation (7.29) is that the activity is directly measurable as the ratio of the equilibrium pressure exerted by a component in solution to the vapor pressure of the pure substance. This important connection is discussed in Chapter 8.

The second part of the relation between chemical potential and concentration is the definition of the *activity coefficient* as the ratio of the activity to the mole fraction:

$$\gamma_i = a_i/x_i \tag{7.30}$$

The activity coefficient has the following properties:

- It is unity for a pure substance (i.e., when $x_i = 1$).
- It approaches a constant value as $x_i \rightarrow 0$.
- It can be either greater than or less than unity.
- It is unity for all concentrations if the solution is ideal.
- It is a function of solution composition and temperature.

The reason that $\gamma_A \rightarrow 1$ in an A-B binary solution dilute in component B is because A molecules are surrounded mainly by other A molecules; the interactions are predominantly of the A-A type, so component A behaves as if it were pure. This limiting behavior is called Raoult's law. At the other extreme, the activity coefficient

* Fanciers of single-malt scotches recognize this fact by imbibing it straight, undiluted by water or ice.

of A in a solution dilute in A approaches a constant value characteristic of the A-B intermolecular interactions. This behavior is termed Henry's law. A more detailed description of Raoult's and Henry's laws is presented in Chapter 8.

A useful connection between the activity coefficients of species in a solution is obtained by eliminating a_i between Equations (7.29) and (7.30) and substituting the resulting equation into the Gibbs-Duhem equation, Equation (7.28). This procedure yields:*

$$\sum_i x_i d\ln\gamma_i = 0$$

This equation is particularly useful for two-component (A-B) solutions, where it becomes:

$$x_A d\ln\gamma_A + x_B d\ln\gamma_B = 0 \tag{7.31}$$

Problem 7.4 shows how this equation can be used to assess the validity of formulas for h^{ex}. In an equally important application, the above equation can be integrated to give the Gibbs free energy analog of Equation (7.19) for the enthalpy:

$$\ln\gamma_B = -\int_{\ln\gamma_{A0}}^{\ln\gamma_A} \frac{x_A}{x_B} d\ln\gamma_A' = -\int_0^{x_A} \frac{x_A'}{1-x_A'} \frac{d\ln\gamma_A}{dx_A'} dx_A' \tag{7.32}$$

where γ_{A0} is the limiting value of the activity coefficient of A as x_A approaches zero. In this limit γ_B approaches unity, which accounts for the zero lower limit of the integral of $d\ln\gamma_B$. This extraordinary relation permits the activity coefficient of B to be computed from the measurement of the activity coefficient of A over a composition range starting from pure B. Figure 7.6 shows the graphical implementation of Equation (7.32) using the first equality. The curve intersects the horizontal axis at a value $-\ln\gamma_{A0}$ and asymptotically approaches an ordinate value of infinity as $\ln\gamma_A \to 0$ (i.e., as $x_A \to 1$). The shaded area in this plot is $\ln\gamma_B$ at the solution composition x_A.

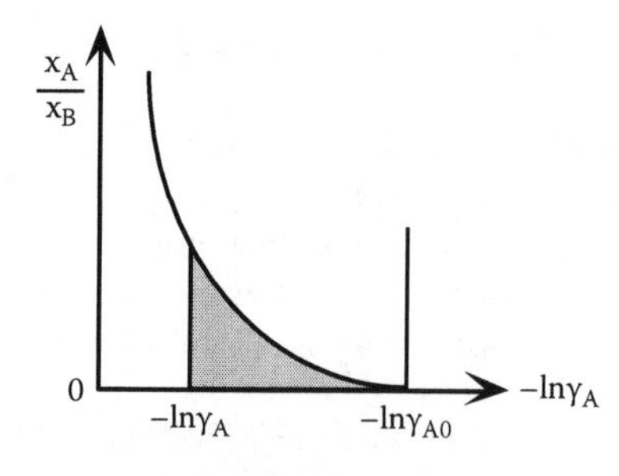

FIGURE 7.6 Graphical determination of γ_B from the dependence of γ_A on x_A.

* This result makes use of $\Sigma x_i d\ln x_i = \Sigma x_i(1/x_i)dx_i = \Sigma dx_i = d\Sigma x_i = 0$.

Problem 7.12 is an example of the use of Equation (7.32). Other exercises dealing with either Equations (7.31) or (7.32) are problems 7.7, 7.8, 7.9, 7.12, 7.13, 7.15, and 7.16.

7.6 EXCESS FREE ENERGY AND THE ENTROPY OF MIXING

As in Equation (7.21) for the enthalpy, the molar free energy of a solution (g) can be written in terms of pure-component contributions (g_A and g_B) and an excess value (g^{ex}). However, an important contribution needs to be added. For a binary solution, the terms contributing to g are:

$$g = x_A g_A + x_B g_B + g^{ex} + \Delta g_{mix} \tag{7.33}$$

g^{ex} contains the effects of solution nonideality. The last term on the right-hand side arises from the entropy of mixing, and is present in ideal as well as nonideal solutions. According to the definition of $g = h - Ts$, the free energies in Equation (7.33) can be expressed in terms of enthalpy and entropy. In particular,

$$\Delta g_{mix} = \Delta h_{mix} - T\Delta s_{mix} = -T\Delta s_{mix} \tag{7.34}$$

because $\Delta h_{mix} = 0$ (all nonideal behavior is included in g^{ex}).

Δs_{mix} for an A-B gas mixture was derived in Section 7.2.2. Here Δs_{mix} is derived for solid solutions by a totally different method. The starting point is Boltzmann's famous equation inscribed on his tombstone (Figure 1.6). For mixing components A and B to form* a mole of solution, it is

$$\Delta s_{mix} = R\ln W_{mix} \tag{7.35}$$

Equation (7.35) represents the difference between the entropy of the solution and the entropy of the pure components. W_{mix} is the number of different arrangements of the atoms on the lattice sites of the crystal. For the pure constituents, $W_{mix} = 1$ because every atom is localized on a site and the atoms are indistinguishable (i.e., they cannot be labeled A_1, A_2 ...).

In order to determine W_{mix} for the solution, the system consists of N lattice sites on which N_A and N_B atoms of A and B, respectively, are distributed at random. There are no defects in the crystal, so $N = N_A + N_B$.

Starting with an empty lattice, A atoms are added one at a time and the number of ways that each can be placed on the unoccupied sites is counted. The first A atom can be placed on any of the N lattice sites; the second A atom has $N - 1$ sites available for placement; the third $N - 2$ available sites, and so on. Since each placement is random, the total number of ways that the entire batch of A atoms can be placed on the available sites is:

$$N(N-1)(N-2)\ldots[N-(N_A-1)]$$

* The relation between the gas constant, Boltzmann's constant and Avogadro's number is $R = kN_{Av}$.

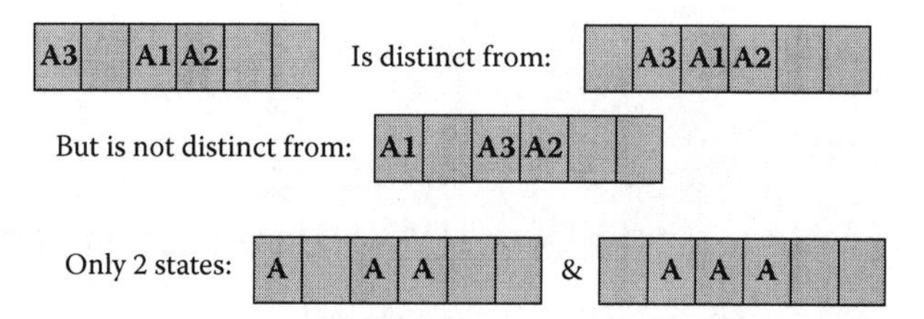

FIGURE 7.7 Indistinguishability of atoms.

However, this method overcounts the total number of ways because it implicitly assumes that the A atoms are distinguishable. For example, the configurations of the first three atoms placed on a six-site crystal are shown in Figure 7.7. The top line shows two configurations of atoms labeled A_1, A_2, and A_3. The atoms occupy a different set of sites in each and so are distinct. The middle line contains a site on which A atoms occupy the same sites as the configuration in the upper left, the only difference being the order. This middle site is not allowed because the atoms cannot be labeled—they are indistinguishable. Three distinguishable atoms can be permuted six (or 3!) ways, but these constitute only one arrangement, namely the one shown at the bottom left of Figure 7.7. Removing the labeling of the atoms on the upper-right-hand configuration gives the single arrangement shown in the bottom right of the figure. To assure indistinguishability, the count arrived at above must be divided by N_A! The correct result for W is:

$$\Delta g_{mix} = \frac{N(N-1)(N-2)...[N-(N_A-1)]}{(N-N_A)!} = \frac{N!}{N_A!\times(N-N_A)!}$$

To compute $\ln W_{mix}$ in Boltzmann's equation, Stirling's approximation, $\ln N! = N\ln N - N$, is used. This converts the above equation to:

$$\ln \Delta g_{mix} = -N_A \ln\left(\frac{N_A}{N}\right) - (N-N_A)\ln\left(\frac{N-N_A}{N}\right)$$

The remaining empty sites after the A atoms have been placed on the lattice just accommodate the N_B atoms of B. There is but one way that the rest of the sites can be filled with B atoms. Setting $N = N_{Av}$ (for a mole of solution), and dividing by N_{Av} (to convert atom number to mole fraction) and substituting the above equation into Equation (7.35) yields Equation (7.10):

$$\Delta s_{mix} = -R(x_A \ln x_A + x_B \ln x_B) \tag{7.10}$$

With this result in hand, substitution of Equations (7.10) and (7.34) into Equation (7.33) gives:

$$g = x_A g_A + x_B g_B + g^{ex} + RT(x_A \ln x_A + x_B \ln x_B) \tag{7.36}$$

Further progress requires expressing g^{ex} in terms of the composition.

7.7 REGULAR SOLUTIONS

This *regular solution* model is the simplest of all theories of the excess free energy, g^{ex}. This solution property includes contributions from the *excess enthalpy* h^{ex} and the excess entropy s^{ex}:

$$g^{ex} = h^{ex} - Ts^{ex} \tag{7.37}$$

7.7.1 EXCESS PROPERTIES

In regular solution theory, the molecules mix randomly as they do in ideal solutions, so that:

$$s^{ex} = 0 \tag{7.38}$$

That is, there is no tendency for either like or unlike molecules to form clusters.

When $s^{ex} = 0$, the excess free energy reduces to the excess enthalpy. The analytical formulation of h^{ex} in terms of composition is restricted by the limiting behavior of the solution enthalpy (h) as the solution approaches pure A and pure B. In these limits, $h \rightarrow h_A$ and $h \rightarrow h_B$, respectively. Examination of Equation (7.21) shows that to satisfy these limits, h^{ex} must be zero at $x_A = 0$ and at $x_B = 0$. The simplest function that obeys these restraints is the symmetric expression:

$$h^{ex} = \Omega x_A x_B \tag{7.39}$$

where Ω is a temperature-independent property of the A-B binary pair called the *interaction energy*.* The form of Equation (7.39) requires that Ω be equal to the difference between the energy of attraction (bond energy) of the A-B pair and the mean of the bond energies of the A-A and B-B interactions (see Gaskell (1981, pp. 360, 366–371)).

Despite its simplicity, Equation (7.39) applies to a remarkably wide variety of solutions, including binary metal alloys such as thallium-tin, binary oxides such as UO_2-Nd_2O_3, and solutions of organic compounds such as benzene-cyclohexane. For the regular solution model to apply, the intermolecular attractions must be of a general nature, not specific such as hydrogen bonding when water is one component. In addition, the two molecules must be approximately the same size to preserve randomness of mixing and thus ensure the validity of Equation (7.38).

Problem 7.2 deals with an extreme case of solution nonideality, illustrated by mixing of sulfuric acid and water. Other problems dealing with regular-solution theory are 7.6, 7.8, 7.11, 7.16 and 7.18.

7.7.2 ACTIVITY COEFFICIENTS

The activity coefficients γ_A and γ_B can be deduced from h^{ex} by applying the free energy analogs of Equations (7.20a) and (7.20b). Replacing h by g and $\bar{h}_i$ by $\bar{g}_i = \mu_i$ in these formulas yields:

---

* Systems that obey Equation (7.38) but not Equation (7.39) are called *nonregular solutions*. However, some restrictions apply to h^{ex} (see Problem 7.4).

$$\mu_A = g + x_B \frac{dg}{dx_A} \qquad \text{and} \qquad \mu_B = g + x_A \frac{dg}{dx_B} \tag{7.40}$$

Substituting Equation (7.36) into the above equations and comparing the resulting formulas for μ_A and μ_B with those obtained by eliminating the activity between Equations (7.29) and (7.30) yields:

$$RT \ln \gamma_A = h^{ex} + x_B \frac{dh^{ex}}{dx_A} \qquad \text{and} \qquad RT \ln \gamma_B = h^{ex} + x_A \frac{dh^{ex}}{dx_B} \tag{7.41}$$

The regular solution simplification, $g^{ex} = h^{ex}$, has been used.* The latter is given by Equation (7.39) and Equation (7.41) reduces to:

$$RT \ln \gamma_A = \Omega x_B^2 \qquad \text{and} \qquad RT \ln \gamma_B = \Omega x_A^2 \tag{7.42}$$

Because the interaction parameter Ω is not a function of temperature, the logarithm of the activity coefficient varies inversely with temperature. The activity coefficients must satisfy the Gibbs-Duhem equation (Equation [7.31]). Not all h_{ex} functions of composition fit this restraint (see Problem 7.4). However, Problem 7.8 demonstrates that the activity coefficients of Equation (7.42) satisfy the Gibbs-Duhem equation.

7.8 CHEMICAL POTENTIALS IN GAS MIXTURES

The analysis in Section 7.2.2 of the entropy change associated with mixing of ideal gases at fixed T and p was based on the absence of an entropy change *if the pure gases are at the partial pressures that they will have in the mixture.* Because the gases are ideal, neither is there an enthalpy change in the mixing process. With both the enthalpy and entropy of each species unaltered, the free energy must also remain constant during this mode of mixing. Because the partial molar free energy of a species in a solution or a gas mixture is the same as its chemical potential, the above argument can be summarized by the equation:

$$\mu_i \text{ (in mixture at } p_i) = g_i \text{ (pure, at } p_i) \tag{7.43}$$

Contrary to condensed phases, the free energy of a pure gas is pressure-dependent. In order to provide a common pressure reference for all pure gases (arbitrarily chosen at 1 atm), g_i in the above equation is expressed in terms of g_i^o, which is the molar free energy of species i at temperature T and 1 atm pressure. The effect of the difference in pressure between p_i and 1 atm is obtained by combining $dg_i/dp_i = v_i$ (for T = constant) with the ideal gas law:

$$\frac{dg_i}{dp_i} = \frac{RT}{p_i}$$

* Proofs of Equations (7.40) to (7.42) are left as an exercise (Problem 7.3).

Integrating this equation from p_i to 1 atm gives the molar free energy of pure *i* at p_i, which, when substituted into Equation (7.3) yields:

$$\mu_i = g_i^o + RT\ln p_i. \tag{7.44}$$

The reference condition denoted by the superscript zero is called a *standard state* of the pure gas. In the above analysis, it has been set at 1 atm, which is convenient when dealing with permanent gases such as air. With this standard state, p_i must be expressed numerically in units of atmospheres.

PROBLEMS

7.1 Two moles of an A-B binary solution with 50 mole% A at 200°C are mixed with three moles of a 20 mole% A solution at 100°C and the resulting mixture is brought to 150°C. The heat capacities of A and B are 36 and 52 J/mole-K, respectively, and the solutions are ideal. Determine the enthalpy and entropy changes for this process and the heat exchanged with the surroundings. (Hint: form the three solutions from pure components at a reference temperature of 100°C.)

7.2 Data obtained by mixing water and pure sulfuric acid are shown below.

	Initial Volumes, cm^3			Temperature, °C	
Solution	H_2O	H_2SO_4	Final Volume	Initial	Final
#1	73	25	92	22	66
#2	48	49	89	22	101

(a) Calculate the molar enthalpy (relative to the initial pure components) for both solutions. The specific heats of sulfuric acid and water are 147.7 and 75.4 J/mole-K, respectively. Assume ideality only for the purpose of calculating the specific heat of the mixture. The density of sulfuric acid is 1.84 g/cm^3.

(b) Plot the data according to the method treated in Problem 7.20 and determine the partial molar quantities. At low acid concentrations, the partial molar volumes of sulfuric acid and water in their solutions are reported to be 45 and 17.5 cm^3/mole, respectively. How does this compare to your result?

(c) Assume that this mixture obeys regular solution theory and determine the interaction parameter Ω.

1. Are the two experiments consistent in the sense that they give the same interaction parameters?
2. Is the sign of Ω physically reasonable?

(d) The plot below shows the partial molar enthalpies for the sulfuric acid/water system. From this graph, calculate the excess enthalpies of solutions #1 and #2 in the table and compare the results with the values of h^{ex} obtained in Part (a).

(e) How would the temperature increase differ if mixing were performed as follows:

1. Water is added to the acid.
2. Acid is added to the water.
3. Acid and water are mixed in proportions equal to their mole fractions in the final solution (i.e., the composition is kept constant as the solution is created).

(f) In order to dispose of the acid solutions, they must first be neutralized with a base such as NaOH.

1. Write the neutralization reaction.
2. Calculate the mass of NaOH needed to neutralize the acid from the tests.
3. A lottery will be held and the person picked will have his or her calculation of Part 2 above checked by placing a drop on his or her arm of the acid neutralized according to the result he or she obtained.

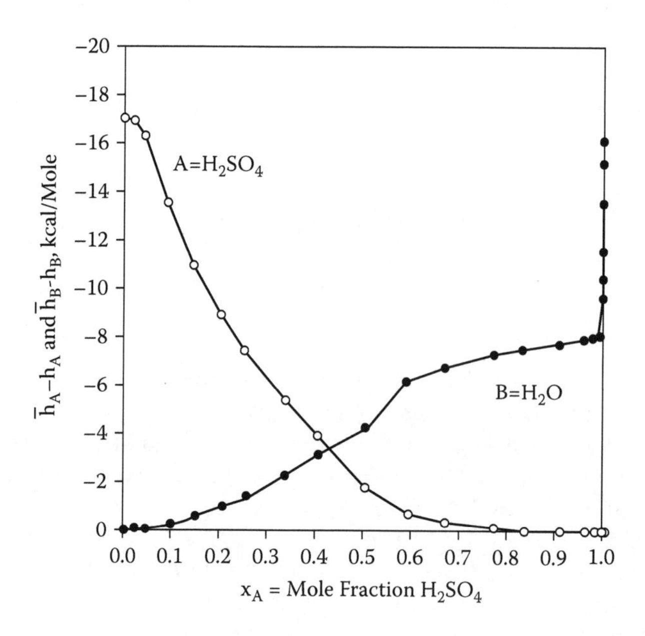

7.3 Complete the following proofs:

(a) Equations (7.40) from Equations (7.27) and (7.26).
(b) Equations (7.41) from Equations (7.33), (7.29) and (7.30).
(c) Equation (7.42) from Equation (7.39).

7.4 The following formula has been proposed for the excess enthalpy of nonideal solutions that do not obey regular solution theory:

$$h^{ex} = CRTx_A^2 x_B$$

(a) What are the expressions for the activity coefficients of A and B?
(b) Do the activity coefficients obey the Gibbs-Duhem relation?
(c) Do the activity coefficients approach physically-acceptable limits as x_A approaches zero and x_A approaches unity?

7.5 Calculate the enthalpy and entropy changes when one mole of *solid* Cr at 1873 K is added to a large quantity of a liquid Fe-Cr alloy ($x_{Fe} = 0.78$) at the same temperature. "Large" means that the composition and solution properties are only slightly changed by addition of the mole of Cr; a Taylor series expansion relating the entropy of mixing of the final composition to that of the initial composition eliminates the need to quantify "large." Assume that the heat capacity difference between solid and liquid Cr is negligible. The heat of fusion of Cr is 21 kJ/mole and its melting point is 2173 K. Assume that Fe-Cr solutions are ideal. The Taylor series expansion of the function $f(x)$ around $x = 0$ is:

$$f(x) = f(0) + (df/dx)_{x=0}\,(x - 0) + \dots$$

7.6 Repeat Problem 7.5 with the following changes:

1. The Cr is initially liquid (rather than solid) at 1873 K.
2. The solution is *regular* rather than ideal, and the activity coefficient of Cr in the liquid alloy is 0.5.

How much heat is given off or absorbed during the mixing process? (Hint: to determine the enthalpy change, first show that $\Delta H = \bar{h}_{Cr} - h_{Cr}$ then use Equation (7.28) to relate this difference to the activity coefficient.)

7.7 What is wrong with the following equation for the activity coefficient of component A in an A-B solution?

$$\ln \gamma_A = Cx_A x_B^2$$

(discuss all potential answers):

(a) The activity coefficient it predicts is less than unity.
(b) It does not obey the Gibbs-Duhem equation.
(c) γ_A has the wrong limit as x_A approaches zero.
(d) γ_A has the wrong limit as x_A approaches unity.

7.8 Prove that the activity coefficients from regular solution theory obey the Gibbs-Duhem equation.

7.9 The activity coefficient of component B in an A-B solution at a fixed temperature is represented by: $\ln\gamma_B = Cx_A^2 - Dx_A^3$. What is the corresponding equation for $\ln\gamma_A$?

7.10 The generic partial molar quantity $\bar{y}_i$ (e.g $\bar{h}_i, \bar{g}_i, \bar{v}_i$, etc) is defined by the following relationship to the total solution property *Y*:

$$dY = \sum \bar{y}_i dn_i \tag{1}$$

and

$$Y = \sum n_i \bar{y}_i \tag{2}$$

where *n* is the number of moles of component *i*. From these equations prove that:

$$dy = \sum \bar{y}_i dx_i \tag{3}$$

where

$$y = Y/\Sigma n_i, \qquad x_i = n_i/\Sigma n_i \tag{4}$$

and Σn_i is the total moles of solution. (Hint: First demonstrate the general form of the Gibbs-Duhem equation):

$$\sum n_i \, d\bar{y}_i = 0 \tag{5}$$

Explain why direct division of (1) by Σn_i does *not* yield (3).

7.11 1 mole of pure A at 400 K is mixed with 3 moles of an A-B solution with 33% B at 600 K. Assume that mixing is adiabatic and that A-B solutions follow regular solution theory with Ω = +30 kJ/mole. The heat capacities of pure components A and B are 30 and 50 J/mole-K, respectively.

(a) What is the temperature of the final solution?
(b) What is the entropy change in the process?

7.12 The activity coefficient of component A of an A-B binary solution is given by:

$$\ln\gamma_A = Cx_B^2(1 + Dx_B)$$

where C and D are constants.

What is the activity coefficient of component B?

7.13 The activity coefficient of component A of an A-B solution is given by:

$$\ln \gamma_A = E x_B (1 - x_A)$$

where E and F are constants. Student no. 1 claims that the equation for $\ln\gamma_B$ can be obtained simply by interchanging the subscripts A and B in the equation for $\ln\gamma_A$. Student no. 2 says that this is wrong, and she can prove it. Show the proof that student no. 2 has in mind.

7.14 Calculate the entropy change when one mole of pure A is added to 4 moles of an equimolar A-B solution at the same temperature. Assume ideal mixing.

7.15 Give two reasons why the following is not an acceptable formula for describing the activity coefficient of component A in a binary A-B mixture.

$$\ln \gamma_A = C x_A^3$$

where C is a constant.

7.16 The activities of Ni in the liquid Fe-Ni alloy at 1600°C are given below.

x_{Ni}	1.0	0.9	0.8	0.7	0.6	0.5	0.4	0.3	0.2	0.1
a_{Ni}	1.0	0.89	0.77	0.62	0.49	0.37	0.28	0.21	0.136	0.067

(a) Over what concentration range does the nickel component obey Henry's law?
(b) Determine the activity coefficient of iron at $x_{Ni} = 0.5$.
(c) Check whether the Fe-Ni alloy obeys regular solution theory.

7.17 One mole of solid A at 1000 K is dissolved adiabatically in 3 moles of liquid B at 1200 K. The resulting solution obeys regular solution theory with an interaction energy of –30 kJ/mole. Find

(a) The final temperature of the solution.
(b) The entropy change of the process.

The melting temperature of A is 1100 K and its heat of fusion is 20 kJ/mole. The heat capacities of both liquid and solid A and of liquid B are all equal to 25 J/mole-K.

7.18 An important process in the production of brass from copper and zinc is the mixing of the two pure liquid metals, each at their respective melting points (steps C3 and Z3 in the flowsheet). The alloying process is adiabatic and conducted at constant pressure. Brass is one of the few liquid-metal alloys that is very nearly ideal. The Cu/Zn mole ratio is 3/2.

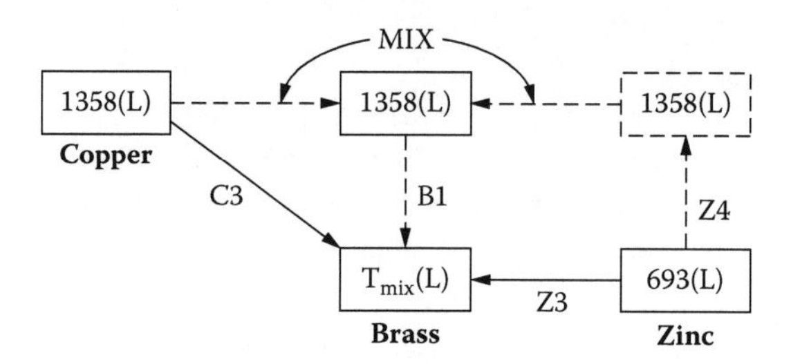

Calculate:

(a) The temperature of the liquid brass after mixing.
(b) The entropy change in this mixing process.

The heat capacities of liquid zinc and copper are 15 and 20 J/mole-K, respectively.

7.19 By considering the dashed line in the diagram of the solution molar property z vs. x_A along with Equations (7.20) with h replaced by z, prove that the intercepts w and q are $\bar{z}_B$ and $\bar{z}_A$, respectively.

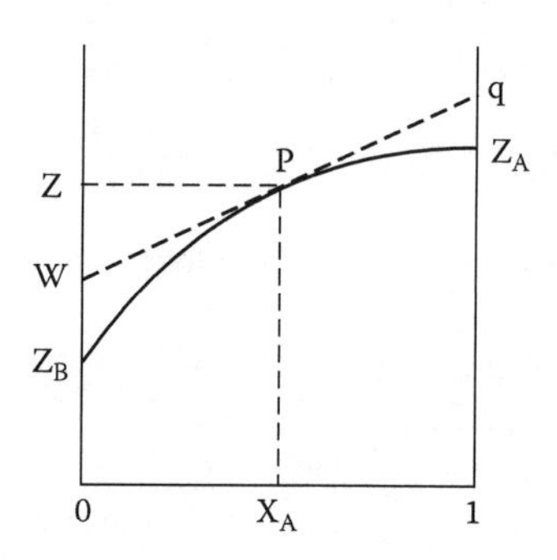

7.20 At a fixed temperature, the molar enthalpy of an A-B solution varies with composition according to:

$$h = a + bx_B + cx_B^2$$

(a) What are the molar enthalpies of pure A and pure B?
(b) What are the partial molar enthalpies of A and B at $x_B = 0.5$?

REFERENCE

Gaskell, D. R. 1981. *Introduction to Metallurgical Thermodynamics,* 2nd ed. New York: McGraw-Hill.

8 Binary Phase Equilibria: Phase Diagrams

8.1 SCOPE

The preceding chapter dealt with the chemical properties of species in single mixture or solution phases. The free energy of a solution or mixture and the chemical potentials of its constituents were quantified. In the present chapter, these properties are applied to determine the phases present and their compositions when a two-component system, or binary system, achieves equilibrium. The two components, denoted by A and B, distribute between two phases labeled I and II. This system, shown in Figure 8.1, is at thermal, mechanical, and chemical equilibrium. The first two of these conditions means that the two phases have the same temperature and pressure. The consequence of the last condition is developed in the following section.

Phases I and II can be solid, liquid or gas in any combination except two gases. Chemical reaction between components A and B is not permitted; the two species retain the same molecular form in the two phases they occupy (with the exception of the dissolution of a diatomic gas in some solids).

Practically important examples of two-component, two-phase equilibria include the following:

Gas-liquid, for example, oxygen dissolution in water. Although the solubility is small, the effect on corrosion of metals is profound.

Gas-solid, for example, dissolution of hydrogen in metals. The special feature of this process is the dissociation of molecular hydrogen into atoms in the metal.

Liquid-liquid, for example, distribution of a solute between two immiscible solvents, such as acetic acid between water and a hydrocarbon. This example contains three components, but because the solvents are immiscible (meaning no mutual solubility), the thermodynamics governing the distribution of the solute is formally identical to the two-component case. Liquid-liquid extraction is a basic unit operation in the chemical and biochemical process industries.

Two condensed phases: solids and liquids are collectively known as condensed phases. Analysis of equilibria between condensed phases usually ignores the gas phase. The condensed phase combinations include liquid-solid and solid-solid. The many crystallographic forms of solids are each regarded as distinct phases, so these equilibria show considerable variety. The subject is known as binary phase diagram representation.

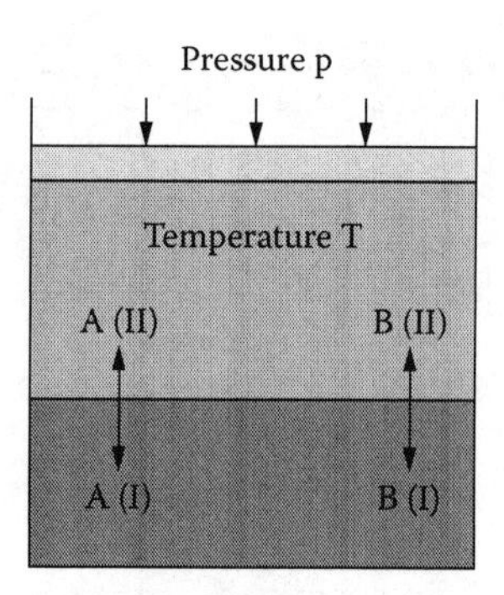

FIGURE 8.1 Components A and B equilibrated in two phases I and II at specified pressure and temperature.

In each of the above examples, the objective is to determine the concentrations of components A and B in the two coexisting phases. In condensed-phase equilibria, identification of the stable phases I and II is also an objective.

8.2 INTERPHASE EQUILIBRIUM

The two-headed arrows in Figure 8.1 indicate chemical equilibrium of components A and B between phases I and II. According to the discussion in Section 1.11.1, the criterion of equilibrium at fixed T and p is the minimization of the total free energy of the contents of the cylinder-piston in Figure 8.1. Since the total free energy is the sum of those of the two phases, this criterion is:

$$dG = dG_I + dG_{II} = 0 \tag{8.1}$$

The free energy of a phase is related to the chemical potentials of its components by Equations (7.27). Using the latter for components A and B in Equation (8.1) gives:

$$\mu_{AI}dn_{AI} + \mu_{BI}dn_{BI} + \mu_{AII}dn_{AII} + \mu_{BII}dn_{BII} = 0$$

where n_{AI} ... n_{BII} are the numbers of moles of each constituent in each phase and μ_{AI} ... μ_{BII} are their chemical potentials.

The change in the state of the system implied by the differentials in Equation (8.1) is the movement of a small quantity of one of the two components from phase I to phase II without altering the other component. Thus, if dn_{AI} moles of A are transferred from I to II, the change in n_{AII} is $dn_{AII} = -dn_{AI}$. Because component B is not moved, $dn_{BI} = dn_{BII} = 0$. Inserting these mole relations into the preceding equation yields the result $\mu_{AI} = \mu_{AII}$. Applying the same argument to component B yields $\mu_{BI} = \mu_{BII}$. In general, for any number of components in the two-phase system, the conditions for chemical equilibrium are:

$$\mu_{iI} = \mu_{iII} \qquad \text{for all i} \tag{8.2}$$

The chemical potentials are seen to be analogous to the thermal and mechanical potentials which provide the equilibrium conditions $T_I = T_{II}$ and $p_I = p_{II}$. Equation (8.2) is the multicomponent generalization of the equilibrium condition for two coexisting phases of a pure substance, namely $g_I = g_{II}$, where g is the molar free energy (Equation [5.2]).

8.3 DISTRIBUTION OF COMPONENTS BETWEEN PHASES

8.3.1 Vapor–Liquid Equilibria

The objective here is to use Equation (8.2) to relate the concentrations of species A and B (mole fractions x_A and x_B) in a condensed phase containing only these two species to their equilibrium partial pressures in the vapor, p_A and p_B. The condensed phase is assumed to contain only A and B (although this is not required). The gas phase may contain an inert diluent such as helium that does not enter the liquid phase. As long as the total pressure is not very large, its value does not affect the distributions of A and B between the two phases. Only the distribution laws for component A are derived; those for component B are determined in the same manner.

From Equation (8.2) the equilibrium condition for component A is $\mu_A(g) = \mu_A(s \text{ or } L)$, where g denotes the gas phase and s and L denote solid and liquid, respectively. The relationships between the chemical potential and the partial pressure in the gas phase and between the chemical potential and the mole fraction in the condensed phase were derived in Sections 7.10 and 7.5, respectively. For the solid or liquid, combination of Equations (7.29) and (7.30) yields:

$$\mu_A(s \text{ or } L) = g_A(s \text{ or } L) + RT\ln(\gamma_A x_A) \tag{8.3}$$

g_A is the molar free energy of pure A and γ_A is the activity coefficient of A in the A-B solution. The chemical potential of A in the gas phase is given by Equation (7.44):

$$\mu_A(g) = g_A^o(g) + RT\ln p_A \tag{8.4}$$

where $g_A^o(g)$ is the molar free energy of pure gaseous A at 1 atm pressure (indicated by the superscript o). The corresponding quantity for the condensed phase, g_A, does not need an indication of 1 atm because it is essentially pressure-insensitive. Both molar free energies are at the same temperature T.

Equating the right-hand sides of Equations (8.3) and (8.4) as required by the equilibrium criterion yields:

$$\frac{p_A}{\gamma_A x_A} = \exp\left(-\frac{g_A^o(g) - g_A(s \text{ or } L)}{RT}\right)$$

Exercises that apply the general partial pressure-composition formulas of Equations (8.5) and (8.6) are part of Problems 8.2, 8.10, 8.15, 8.18, 8.19, and 8.25.

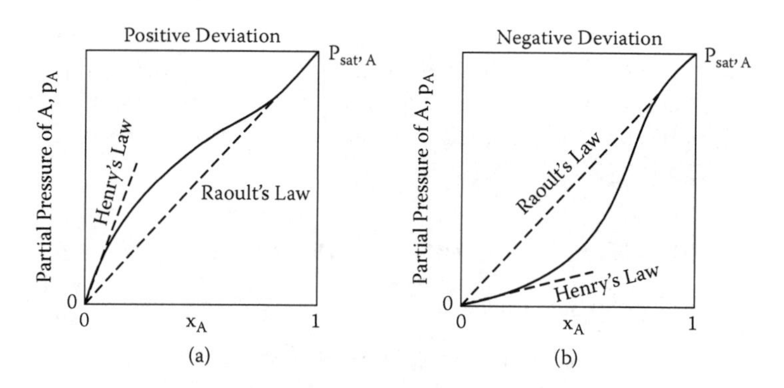

FIGURE 8.2 Equilibrium partial pressures of component A over A-B solutions that exhibit positive and negative deviations from ideality. The temperature is fixed.

For pure component A, the terms on the left-hand side of the above equation are: $\gamma_A = 1$, $x_A = 1$ and $p_A = p_{sat,A}$, the vapor pressure. Therefore, the right-hand side of the above equation is the vapor pressure of pure A, and the formula can be written as:

$$p_A = \gamma_A x_A p_{sat,A} \tag{8.5}$$

A similar equation applies to component B:

$$p_B = \gamma_B x_B p_{sat,B} \tag{8.6}$$

Equations (8.5) and (8.6) are the general relations between the concentrations of a component in the condensed phase and the equilibrium partial pressures in the gas phase. The latter depend on the composition of the condensed phase and on the deviations of the components from ideality, as represented by the activity coefficients. These two quantities are not independent, however; if γ_A is known as a function of composition, γ_B follows from application of the Gibbs-Gibbs-Dihem equation (see Equation [7.31]).

The activity coefficients can be greater or less than unity, depending on the strength of the bonds between A and B molecules compared to the mean of the A-A and B-B bond strengths. The curves in Figure 8.2 illustrate these two cases; activity coefficients greater than unity give positive deviations from ideal behavior and activity coefficients less than unity result in negative deviations. This figure illustrates the type of measurement (i.e., partial pressure as a function of solution composition) that, in conjunction with Equation (8.5), provides experimental values of activity coefficients of components in solution.

8.3.2 Raoult's Law

The line labeled "Raoult's law" represents ideality. For systems that obey this law, the activity coefficients of both A and B are unity over the entire composition range.

The equilibrium partial pressures are lower than the pure-component vapor pressures by factors exactly equal to the corresponding mole fraction in the ideal solution:

$$p_A = x_A p_{sat,A} \qquad p_B = x_B p_{sat,B} \tag{8.7}$$

Such solutions are the exception rather than the rule. For Raoult's law to be followed, the two components must be chemically very similar (e.g., A = benzene and B = toluene—see Problem 8.1). However, even metals as similar as nickel and iron exhibit nonideal behavior when alloyed.

In Figure 8.2, the partial pressure curve joins the Raoult's law line well before x_A reaches one. That is, not only is $\gamma_A \rightarrow 1$ as $x_A \rightarrow 1$, but $d\gamma_A/dx_A \rightarrow 0$ as well.

8.3.3 Henry's Law

At the other limit, as the solution becomes dilute in component A, the partial pressure curve turns into the straight line labeled Henry's law in Figure 8.2. For this functional dependence, Equation (8.5) shows that the activity coefficient must be constant, although not equal to unity. This dilute solution behavior of a constant activity coefficient reduces Equation (8.5) to:

$$p_A = \gamma_A(x_A \rightarrow 0) \times p_{sat,A} \times x_A = k_{HA} x_A \tag{8.8}$$

The product of the activity coefficient and the vapor pressure, k_{HA}, is termed the *Henry's law constant.* Physically, Henry's law behavior in dilute solutions simply reflects the fact that all A molecules are surrounded by B molecules, irrespective of the concentration of A (as long as it is low). In the Henry's law limit for component A, component B obeys Raoult's law.

8.3.4 Liquid–Liquid Equilibrium

Another common phase-distribution system based on Henry's law is the partitioning of a solute species between two immiscible solvents, designated as phases I and II. At equilibrium, the chemical potentials of solute A in the two phases, as given by Equation (8.3), are equal. The pure-A molar free energy is common to both phases, and so cancels, leading to the equilibrium relation:

$$\frac{x_A^{II}}{x_A^{I}} = \frac{\gamma_A^{I}}{\gamma_A^{II}} \tag{8.9}$$

The ratio of the activity coefficients of A in the two solvents is called the *distribution coefficient* of A. As long as A is dilute in both solvents, the activity coefficients are independent of the concentration of A, but in general are not equal.

8.3.5 Gas Dissolution

The second example of the application of Henry's law involves the dissolution of the so-called permanent gases in condensed phases. Examples are helium solution in glass and oxygen dissolution in water. To describe these equilibria, the chemical potentials of the gas (species A) in the two phases are equated. In the gas phase, the chemical potential of A is given by Equation (8.4). In applying Equation (8.3) to A in the condensed phase, however, the assignment of the reference free energy $g_A(s \text{ or l})$ poses a difficulty. It cannot refer to pure A as a solid or liquid because A is gaseous at all temperatures of interest. Consequently, $g_A(s \text{ or l})$ in Equation (8.3) is replaced by the molar free energy of A in its normal gaseous state at 1 atm pressure. That is, $g_A(s \text{ or l})$ is replaced by $g_A^o(g)$. In so doing, the limiting behavior $\gamma_A \rightarrow 1$ as $x_A \rightarrow 1$ is lost, but this is of no practical consequence because high concentrations of the permanent gases in the condensed phases cannot be attained. With this modification in Equation (8.3), equating with Equation (8.4) yields:

$$\frac{x_A}{p_A} = \frac{1}{\gamma_A} = k_{HA} \tag{8.10}$$

where k_{HA} is the Henry's law constant for gas A in the solid or liquid. Note that the definition of k_{HA} is the inverse of that defined in Equation (8.8) for condensable vapors.

The activity coefficient in Equation (8.10) has lost its original meaning as a deviation from solution ideality. γ_A would not equal 1 even if A and B (condensed phase) satisfied the ideality condition (Section 7.3). In the end, all that thermodynamics has been able to elucidate for this case is the proportionality of x_A and p_A.

Table 8.1 lists the Henry's-law constants for atmospheric gases in water. Note that the concentration units are moles per liter (molarity) rather than mole fraction. The former is the customary concentration unit in aqueous solutions.

TABLE 8.1
Henry's-Law Constants for Several Gases in Water at 25°C (Ref. 2)

Gas	Partial Pressure in air, atm	H, atm/(moles/lit)
N_2	0.79	1550
O_2	0.21	730
CO_2	3.5×10^{-4}	63
CO	$\sim 1 \times 10^{-7}$	980

The preceding application of Henry's law does not apply to the very important case of the dissolution of diatomic gases such as H_2, O_2, and N_2 in metals. These gases dissociate into atoms upon entering a metal, and the dissolution process is best treated as a chemical reaction rather than as a physical distribution between phases. Discussion of this process is deferred until the next chapter.

8.4 BINARY PHASE DIAGRAMS—ANALYTICAL CONSTRUCTION

Binary phase diagrams depict the stable condensed phase (or phases) formed by a two-component system as a function of temperature and overall composition. The ordinate of a phase diagram is the temperature and the overall composition is the abscissa.* The phase rule (Equation [1.21]) for a two component system permits $F = 4 - P$ degrees of freedom for a two-component system. Since the diagrams deal only with condensed phases, they are minimally affected by total pressure.** Ignoring the total pressure reduces the number of degrees of freedom by one, thereby allowing $3 - P$ properties to be independently varied. In a single phase ($P = 1$) portion of the phase diagram, two degrees of freedom are permitted. These are the temperature T and the composition, represented by the mole fraction of one of the constituents. Single-phase regions appear as areas in the phase diagram.

In two-phase zones ($P = 2$), only one system property can be specified. Fixing the temperature, for example, determines the compositions of the two coexisting condensed phases. These temperature-composition relationships appear in the phase diagram as lines (or curves) called *phase boundaries*. A three-phase system ($P = 3$) has no degrees of freedom and is represented by a point on the phase diagram.

The distinction between overall compositions and the compositions of individual phases is essential to understanding phase diagrams. For single-phase zones, the two are identical. When two phases coexist, their compositions are different from the overall composition. The latter is the quantity-weighted average of the compositions of the two phases (i.e., the lever rule).

The structure of a phase diagram is determined by the condition of chemical equilibrium. As shown in Section 8.2, this condition can be expressed in one of two ways: either the total free energy of the system is minimized (Equation [8.1]) or the chemical potentials of each component in coexisting phases are equated (Equation [8.2]) . The choice of the manner of expressing equilibrium is a matter of convenience and varies with the particular application. Free-energy minimization is usually used with the graphical method and chemical-potential equality is the method of choice for the analytic approach.

8.4.1 Melting of Ideal Binary Condensed Phases

A pure substance melts at a fixed temperature. A binary solution changes from solid to liquid over a range of temperatures. In the melting range, components A and B in the solid phase are in equilibrium with A and B in the coexisting liquid phase. The equilibrium conditions are:

$$\mu_A(L) = \mu_A(S) \qquad \text{and} \qquad \mu_B(L) = \mu_B(S)$$

---

* The composition of the gas phase in equilibrium with the solid or liquid (as given by Equations [8.5] and [8.6]) is ignored in this representation. To graphically depict the equilibrium partial pressures as functions of temperature and composition would require a third dimension on the plot, which would make the representation unwieldy.

** The total pressure and the equilibrium partial pressures of the components of the condensed phase are independent of each other. For example, an inert gas can be added to the gas phase without affecting the thermodynamics.

The chemical potentials are related to mole fractions by Equation (8.3). If the A-B solutions are ideal in both the solid and liquid phases, the above conditions become:

$$g_A(L) + RT\ln x_{AL} = g_A(S) + RT\ln x_{AS} \tag{8.11a}$$

$$g_B(L) + RT\ln x_{BL} = g_B(S) + RT\ln x_{BS} \tag{8.11b}$$

Because $x_{AL} + x_{BL} = 1$ and $x_{AS} + x_{BS} = 1$, the above equations contain two unknowns. The solution gives the composition of the solid and liquid as functions of temperature, the plot of which is a binary phase diagram. Eliminating the mole fractions of component A gives the following solutions of Equations (8.11):

$$x_{BL} = \frac{1-e^{\alpha}}{e^{\beta}-e^{\alpha}} \quad \text{and} \quad x_{BS} = e^{\beta}\left(\frac{1-e^{\alpha}}{e^{\beta}-e^{\alpha}}\right) \tag{8.12}$$

α and β are the temperature functions:

$$\alpha = \frac{g_A(L)-g_A(S)}{RT} \qquad \beta = \frac{g_B(L)-g_B(S)}{RT}$$

The temperature dependence of the free energy of a pure phase was derived in Section 3.7.4. To a good approximation, the free energy difference between liquid and solid is given by Equation (3.27b):

$$g(L)-g(S) = \left(1-\frac{T}{T_M}\right)\Delta h_M \tag{8.13}$$

where T_M is the melting temperature of the pure substance and Δh_M is its heat of fusion. Combining the above two equations gives α and β as explicit functions of temperature:

$$\alpha = \left(1-\frac{T}{T_{MA}}\right)\frac{\Delta h_{MA}}{RT} \qquad \beta = \left(1-\frac{T}{T_{MB}}\right)\frac{\Delta h_{MB}}{RT} \tag{8.14}$$

Figure 8.3 shows the phase diagram for an ideal binary system calculated from Equations (8.12) using specific values of the melting properties of metals A and B (chosen as U and Zr, see Table 8.2). The upper line (representing T versus x_{BL}) is called the *liquidus*. All points lying above this line are completely liquid. Similarly, all points below the lower curve (the *solidus*, or T versus x_{BS}) are completely solid, which in this case is an ideal solid solution. In the region bounded by the solidus and the liquidus, two phases coexist.

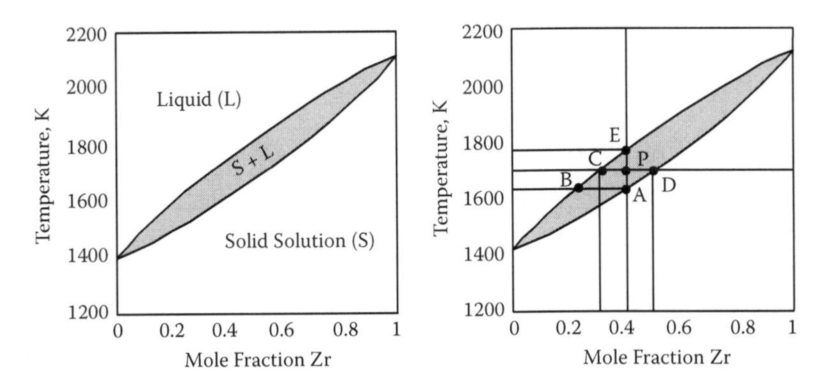

FIGURE 8.3 Phase diagram of the U-Zr binary system with ideal behavior in both liquid and solid.

TABLE 8.2
Melting Properties of Uranium and Zirconium

Component	Element	T_M, K	Δh_M, kJ/mole
A	Uranium	1406	15.5
B	Zirconium	2130	23.0

Example: In the right-hand plot of Figure 8.3, horizontal and vertical lines are superimposed on the phase diagram in order to illustrate important characteristics of the melting process. If the solid solution with a composition $x_{Zr} = 0.4$ is heated, the intersection of the vertical line with the solidus (at point A) shows that the first liquid appears at 1630 K and has a composition $x_{ZrL} = 0.21$ (at point B). As the temperature is increased to 1700 K, the system lies at point P. Here a liquid phase with composition $x_{ZrL} = 0.31$(point C) and a solid phase with $x_{ZrS} = 0.49$ (point D) coexist. The fraction of the mixture present as liquid at point P is obtained from the mole balance, known as the lever rule:*

$$\text{Fraction liquid at point P} = \frac{\overline{PD}}{\overline{CD}} = \frac{x_{ZrS} - x_{Zr}}{x_{ZrS} - x_{ZrL}} = \frac{0.49 - 0.4}{0.49 - 0.31} = 0.50$$

Upon heating from point P, the last solid disappears at $T = 1790$ K (point E on the liquidus). Melting of this binary system at this particular overall composition is spread over a 160 K temperature range.

Problem 8.6 provides additional practice in applying ideal-melting theory to the MnO-FeO binary system. The effect of even slight deviations from ideal solution behavior can result in phase diagrams that are distorted or qualitatively different

* See Section 2.6.2 for application of the lever rule in single-component vapor-liquid systems. A more detailed discussion of the lever rule for binary condensed-phase systems can be found in Section 8.8.

from the diagram shown in Figure 8.3. Problems 8.9, 8.16, and 8.24 explore the nonideality effect on this type of phase diagram using regular solution theory for the solid and liquid phases.

8.4.2 Phase Separation

The single-phase solid and liquid phase regions in Figure 8.3 show no structure because the A-B solutions were assumed to be ideal. However, if the components exhibit positive deviations from ideality (i.e., if the A-B molecular interaction is weaker than the average of the A-A and the B-B interactions), the single-phase solutions separate into two distinct phases, either both liquid or both solid. The system in which phase separation has occurred is termed *partially miscible* because the B-rich phase contains some dissolved A and the A-rich phase contains some dissolved B. Nonideality is assumed (for simplicity) to be represented by the regular solution model (Section 7.7):

$$h^{ex} = \Omega x_A x_B \tag{8.15}$$

where Ω is the interaction energy. If this property is negative, the A-B solution is more stable than the ideal solution, and no phase separation occurs. If $\Omega > 0$, the solution is energetically less stable than an ideal solution. When the destabilizing effect of the excess enthalpy overcomes the stabilizing influence of the entropy of mixing, phase separation occurs.

Labeling the partially miscible phases as I and II, use of Equation (8.3) in Equation (8.2) with $i = A$ yields the following:

$$g_A + RT \ln \gamma_{AI} x_{AI} = g_A + RT \ln \gamma_{AII} x_{AII} \tag{8.16a}$$

The analogous equation for component B is:

$$g_B + RT \ln \gamma_{BI} x_{BI} = g_B + RT \ln \gamma_{BII} x_{BII} \tag{8.16b}$$

The activity coefficients derived from h^{ex} of Equation (8.15) are given by Equation (7.42). Substituting these into Equation (8.16) and replacing x_A by $1 - x_B$ for both phases gives:

$$\Gamma\left(x_{BI}^2 - x_{BII}^2\right) = \ln\left(\frac{1 - x_{BII}}{1 - x_{BI}}\right) \tag{8.17a}$$

$$\Gamma\left[(1 - x_{BI})^2 - (1 - x_{BII})^2\right] = \ln\left(\frac{x_{BII}}{x_{BI}}\right) \tag{8.17b}$$

where

$$\Gamma = \Omega/RT \tag{8.18}$$

is a dimensionless (but temperature-dependent) form of the interaction energy.

Equation (8.17b) can be obtained from Equation (8.17a) by replacing x_{BI} and x_{BII} by $1 - x_{BI}$ and $1 - x_{BII}$, respectively. This mathematical feature implies that the two equations are mirror-image branches of a function that is symmetric about $x_B = 0.5$. Using the mathematical property $x_{BII} = 1 - x_{BI}$ of such a symmetric function, Equations (8.17a) and (8.17b) are seen to be identical, and can be represented by the function:

$$\Gamma = \frac{\ln\left(\dfrac{1-x_B}{x_B}\right)}{1-2x_B} \tag{8.19}$$

This function passes through a minimum at $x_B = 0.5$, at which point, l'Hopital's rule shows that $\Gamma = 2$. No mathematical solutions exist for $\Gamma < 2$, which physically means that the system is a single-phase solution for all compositions. For $\Gamma > 2$, the same value of Γ is obtained for x_B and $1 - x_B$ (i.e., there are two solutions to Equation (8.19)).

To convert Equation (8.19) to the equation for the phase diagram (i.e., T versus x_B), the maximum $\Gamma = 2$ is equivalent to a unique temperature T^* through Equation (8.18):

$$T^* = \Omega/2R \tag{8.20}$$

T^* is called the *critical solution temperature*. It is a property of the A-B binary system that reflects its nonideal behavior, as represented in the regular solution approximation.

With the help of Equations (8.18) and (8.20), Equation (8.19) becomes:

$$\frac{T}{T^*} = \frac{2(1-2x_B)}{\ln\left(\dfrac{1-x_B}{x_B}\right)} \tag{8.21}$$

Equation (8.21) is graphed in Figure 8.4. For $T > T^*$ only a single solution exists. Below T^*, the system spits into two phases. In this range, the composition of phase I (x_{BI}) follows the curve up to $x_B = 0.5$. The remaining portion of the curve represents the compositions x_{BII} of the second phase.

8.5 BINARY PHASE DIAGRAMS BY THE GRAPHICAL METHOD

The cases of melting of two-component ideal solutions and of phase separation in a regular solution described in Section 8.4 were easily treated by analytical methods. However, as the nonideal behavior of the liquid and solid solutions become more complicated (i.e., do not follow regular solution theory), the analytical methods based on Equation (8.2) as the starting point quickly become sufficiently complex to preclude derivation of simple formulas such as Equations (8.12) and (8.21). The graphical method does not have this restriction. Provided only that the free energy versus composition curves can be drawn for each phase, construction of the phase

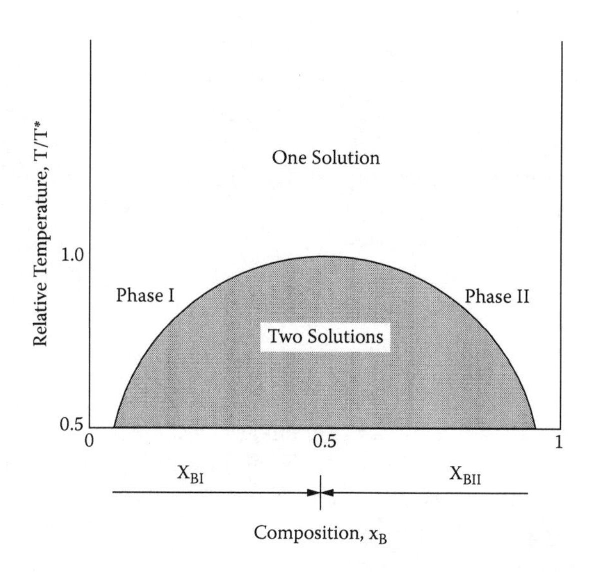

FIGURE 8.4 Phase diagram of a binary system exhibiting phase separation.

diagram is straightforward. Moreover, the graphical method provides a qualitative understanding of the process that would be lost in complex mathematical analysis.

The graphical procedure is based on minimizing the free energy of the system at a fixed temperature (and total pressure). Per mole of solution, the free energy is given by Equation (7.36) with g^{ex} approximated by h^{ex} for a regular solution*:

$$g = g_{comp} + h^{ex} - T\Delta s_{mix} \tag{8.22}$$

where g_{comp} is the sum of the free energies of the pure (unmixed) components:

$$g_{comp} = x_A g_A + x_B g_B \tag{8.23}$$

where x_B represents the overall composition of the two-phase mixture, not the composition of individual phases. Δs_{mix} is the entropy of mixing:

$$\Delta s_{mix} = -R(x_A \ln x_A + x_B \ln x_B) \tag{8.24}$$

The phase diagram depends on the last two terms in Equation (8.22), which represent the change in the free energy of the system when the pure components are mixed. The composition-dependence of g expressed by Equation (8.22) is called a *free energy curve*. These curves are parametric in T.

8.5.1 Common Tangent Rule

The foundation of the graphical method is called the *common tangent rule*. This is the proof that a graphical construction applied to free energy curves provides the

* h^{ex} need not be that of a regular solution; see footnote on p. 201.

link to the phase diagram. The common tangent rule states that: *the compositions of the two coexisting equilibrium phases lie at the points of common tangency of the free energy curves.* The two phases may be both solids, both liquids, or one solid and one liquid. For generality, the two phases are labeled I and II. As usual, the components are A and B.

Equation (7.27) for a binary system is:

$$dg = \mu_A dx_A + \mu_B dx_B$$

Applying this equation to each phase and using $dx_A = -dx_B$ gives:

$$\left(\frac{dg}{dx_B}\right)_I = \mu_{BI} - \mu_{AI} \qquad \text{and} \qquad \left(\frac{dg}{dx_B}\right)_{II} = \mu_{BII} - \mu_{AII}$$

The equilibrium condition of Equation (8.2) requires that $\mu_{AI} = \mu_{AII}$ and $\mu_{BI} = \mu_{BII}$. When these equalities are substituted into either of the above equations, the result is the common tangent condition:

$$\left(\frac{dg}{dx_B}\right)_I = \left(\frac{dg}{dx_B}\right)_{II} \tag{8.25}$$

This equation states that the equilibrium concentrations of two coexisting phases in a binary system are the points on the free energy curves that are tangent to the same straight line.

8.5.2 Phase Separation in Regular Solutions

To aid in visualizing the interplay of these terms, Equation (8.22) is divided by RT and h^{ex} is given by Equation (8.15):

$$\Gamma x_B(1-x_B) + [x_B \ln x_B + (1-x_B)\ln(1-x_B)] = (g - g_{comp})/RT \tag{8.26}$$

where $\Gamma = \Omega/RT$.

Equation (8.26) is plotted in Figure 8.5, which shows how the excess enthalpy and the entropy of mixing combine to produce mixing free energy curves. Γ is varied either by changing the interaction energy Ω or by changing the temperature. The development of the curves with two minima for $\Gamma > 2$ is clearly shown. In this range, the positive (repulsive) interaction energy overcomes the negative contribution to the free energy provided by the entropy of mixing (second term) and the solution becomes unstable. The dashed lines represent common tangents to these minima and the points of tangency are the compositions of the two partially miscible phases.

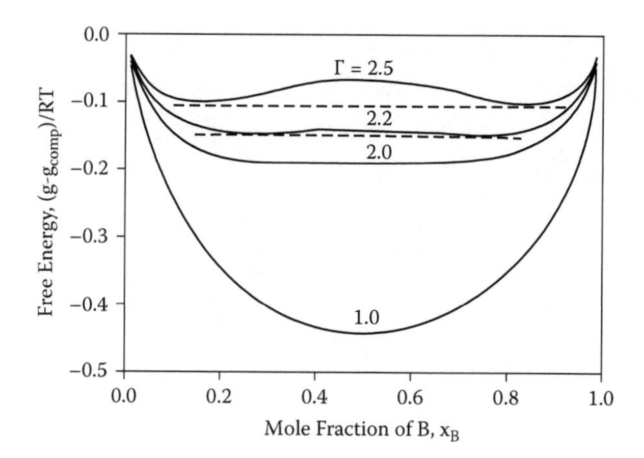

FIGURE 8.5 Graphical solution of the phase diagram for partially-miscible solution formation.

This is a cumbersome procedure for this particularly simple system, which is easily treated analytically (Section 8.4.2). However, as shown in the next section, graphical construction of phase diagrams from free energy versus composition curves becomes the method of choice for complicated systems.

Example: Show (analytically) that the compositions of the two phases at the off-center minima in Figure 8.5 are identical to those obtained by the analytical method for the value of T^* corresponding to $\Gamma = 3$ applied to the graphical method.

Analytical method: Eliminating Ω between Equations (8.18) and (8.20) gives:

$$\frac{T}{T^*} = \frac{2}{\Gamma} = \frac{2}{3}$$

Solving Equation (8.21) (numerically) for this value of the relative temperature gives $x_{BI} = 0.07$ and a second solution at $x_{BII} = 0.93$.

Graphical method: The common-tangent points $\Gamma \geq 2$ in Figure 8.5 are found by setting the derivative of Equation (8.26) equal to zero:

$$\Gamma \frac{d}{dx_B}\left[(1-x_B)x_B\right] + \frac{d}{dx_B}\left[(1-x_B)\ln(1-x_B) + x_B \ln x_B\right] = \frac{d}{dx_B}\left[\frac{g - g_{comp}}{RT}\right] = 0$$

For $\Gamma = 3$, this equation yields the pair of solutions $x_{BI} = 0.07$, $x_{BII} = 0.93$. The exact correspondence of these phase compositions with those obtained by the analytical method is expected because both methods are based on the same model.

Problem 8.3 offers an additional exercise in analyzing phase separation in a binary regular solution.

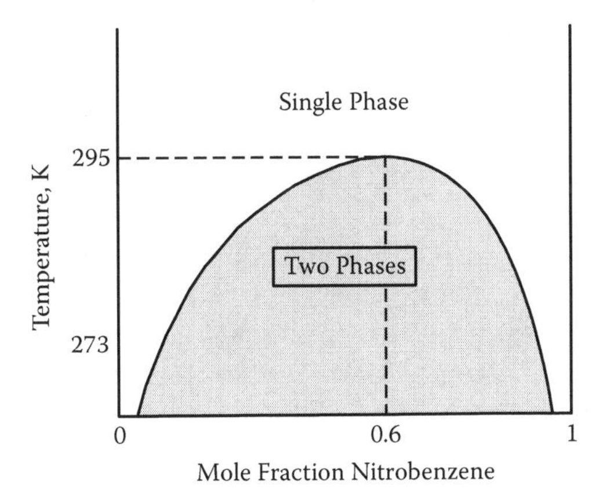

FIGURE 8.6 Phase diagram for nitrobenzene and *n*-hexane. This phase diagram can be used to find the excess enthalpy-composition relation for this binary system (Problem 8.26).

8.5.2.1 Real Systems Exhibiting Phase Separation

The symmetry of the two-phase boundary in Figure 8.4 arises from the use of regular solution theory to account for nonideality. Deviations from this model are common. Figure 8.6 shows the phase diagram of the binary liquid system consisting of *n*-hexane and nitrobenzene. The critical solution temperature of 295 K is achieved at a mole fraction of nitrobenzene of 0.6. The symmetry of the curve in Figure 8.4 is lost because h^{ex} cannot be represented by the single-parameter regular-solution formula given by Equation (8.15).

Real systems may exhibit the characteristic features of both ideal (or near-ideal) melting (Figure 8.3) and phase separation (Figure 8.4) in the same phase diagram. Figure 8.7 shows the phase diagram of the ZrO_2–ThO_2 system. Although technically three-component systems, oxides can generally be represented as pseudo-binary systems consisting of the two very stable compounds.

8.5.3 Melting/Solidification of an Ideal Two-Component System

The melting characteristics of a binary system that is ideal in both liquid and solid states were derived analytically in Section 8.4.1. Here, the same analysis is performed graphically.

All graphical determinations of binary phase diagrams begin with the free energy versus composition curves for all possible phases in the system. For ideal systems, these curves are given by Equations (8.22) to (8.24) with $h^{ex} = 0$ in both solid and liquid phases:

$$g(L) = x_A g_A(L) + x_B g_B(L) + RT(x_A \ln x_A + x_B \ln x_B) \tag{8.27a}$$

$$g(S) = x_A g_A(S) + x_B g_B(S) + RT(x_A \ln x_A + x_B \ln x_B) \tag{8.27b}$$

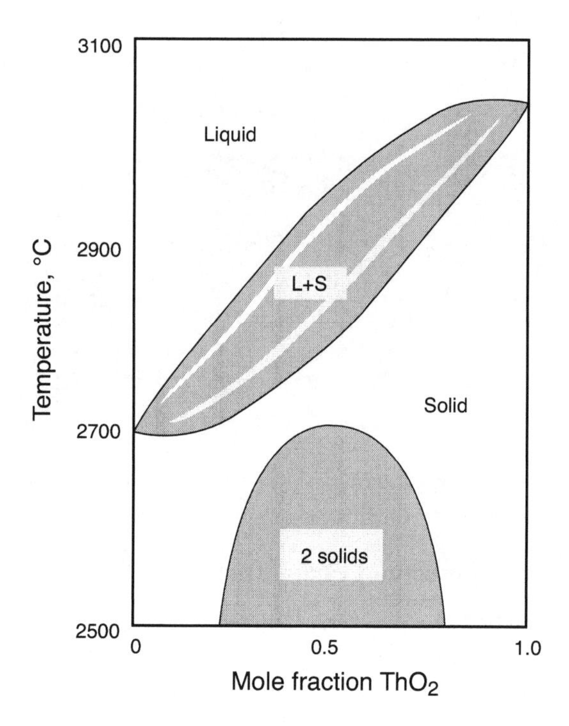

FIGURE 8.7 The thoria–zirconia phase diagram.

In order to construct the free energy–composition curves using these equations, the four pure-component molar free energies must be specified. Equation (8.13) relates the molar free energies of the pure liquids and pure solids:

$$g_A(L) = g_A(S) + \left(1 - \frac{T}{T_{MA}}\right)\Delta h_{MA}; \qquad g_B(L) = g_B(S) + \left(1 - \frac{T}{T_{MB}}\right)\Delta h_{MB} \quad (8.28)$$

As discussed in Section 1.6, the free energy (in common with u and h) has no absolute value. Therefore, two molar free energies, say $g_A(S)$ and $g_B(S)$, can be specified arbitrarily. Of course, this choice affects the shapes and positions of the free energy curves, but the compositions at the common tangency points are unaffected.* For convenience, $g_A(S) = 0$ and $g_B(S) = 0$ are choosen.

To illustrate the common tangent construction, a temperature of 1500 K is chosen. The melting properties of the two components are those of uranium (component A) and zirconium (component B) and are given in Table 8.1. Equation (8.28) gives $g_A(L) = -1.04$ kJ/mole. The minus sign indicates that liquid U is more stable than the solid at a temperature above the melting point. For Zr, Equation (8.28) yields $g_B(L) = 6.80$ kJ/mole.

* This can be proved by substituting Equation (8.27) into (8.26) and then into Equation (8.25). The resulting common-tangent equation does not contain $g_A(S)$ and $g_B(S)$.

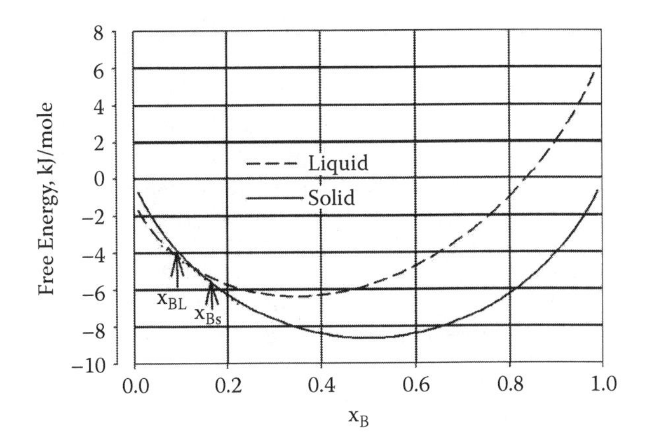

FIGURE 8.8 Free-energy curves for ideal solid and liquid solutions.

Plots of Equations (8.27a) and (8.27b) are shown in Figure 8.8. The common tangency points are $x_{BL} = 0.10$ on the liquid curve and $x_{BS} = 0.17$ on the solid curve. For $x_B < x_{BL}$ the free energy of the liquid is lower than that of the solid, so the system is a single-phase liquid. Similarly, for $x_B > x_{BS}$, $g(L) > g(S)$ and the system is a solid solution. At this temperature, the two-phase region occupies the overall composition interval $x_{BL} < x_B < x_{BS}$. The same results are obtained from the analytical method of Section 8.4.1.

Problems 8.11 and 8.17 use free energy plots and the common-tangent rule to deduce more complex types of phase diagrams than the ideal systems treated above. The following section shows how this method generates a particularly common phase diagram.

8.6 A EUTECTIC PHASE DIAGRAM

The binary systems treated in the preceding sections were either ideal (melting-solidification) or deviated positively from ideality according to regular solution theory (phase separation). These simple types are rarely found in real binary systems. First, there may be more than one solid phase, each with a distinct crystal structure, just as there are in pure substances (see Section 5.6). Second, the liquid phase and the solid phase(s) are generally nonideal. The extent of deviation from ideality is usually different in each phase, and may not be adequately represented by regular solution theory. Very negative deviations from ideal solution behavior, indicative of strong A-B interaction, often lead to the formation of distinct compounds that appear in the phase diagram (e.g., AB_2, A_2B, AB). The increased physical complexity of such systems renders analytic calculation of the phase diagram amenable only to analysis by computer codes.

Irrespective of the complexity of the nonideal behavior of the phases involved, the phase diagram can always be constructed if the free energy versus composition curves for each phase can be drawn. The link between the two graphical representations is the common-tangent rule. Because of the wide variations in the shapes of free-energy curves, the types of phase diagrams deduced from them reaches zoological proportions.

In this section, a common variety called the *eutectic phase diagram** is developed by the graphical method.

The prototypical eutectic system consists of one liquid and two solid phases, labeled α and β. The α phase has the crystal structure of pure solid A and the β phase that of pure B. The two structures are usually different, as opposed to the phase-separation system, in which solids A and B have the same structure (see Figure 8.7).

8.6.1 Free Energy–Composition Curves

In the free energy formula of Equation (7.36), nonideality is expressed by the general form g^{ex}, the excess free energy. The simplifications used in the prior analyses of ideal melting and phase separation, namely neglecting s^{ex} and confining h^{ex} to the regular solution model, are not valid for most binary systems. In order to construct phase diagrams by the common-tangent technique, more elaborate solution models are needed to relate free energy to composition for all likely phases. Figure 8.9 shows plots of g versus x_B for the three phases at six temperatures, with T_6 the highest and T_1 the lowest. In the six graphs, the curves for each phase keep approximately the same shape but shift relative to each other. Examination of the six plots shows that the liquid phase curve shifts upward as the temperature is reduced more rapidly than do the curves for the two solid phases. This feature arises from the thermodynamic relation of Equation (6.16):

$$\left(\frac{\partial g}{\partial T}\right)_p = -s$$

Because the entropy of a liquid is always larger than that of a solid, g_L moves up with decreasing temperature faster than g_α or g_β. The relative movement with T of each of the phases and the shape of their g versus x_B curves determines the phase diagram.

Consider first the composition-dependence of the two solid phases. The free energies of the α and β solids at first decrease from the pure-component value then increase rapidly as the other component is added. The initial decreases are due chiefly to stabilization of the dilute solutions by the entropy of mixing. The shape of the curve is also affected by the composition-dependence of the nonideality term g^{ex}.

The subsequent rise in free energy of the two solid phases is due for the most part to crystal structure effects. The intercept of the α curve with the left-hand axis (e.g., point a in the T_6 plot) represents the free energy of pure A in the α crystal structure. If the α phase curve were extended to intersect the right-hand axis, the free energy here would represent that of pure B in the α crystal structure. That this intercept is higher than the intercept of the β curve with the right-hand axis simply reflects the fact that pure B is more stable in the β crystal structure than it is in the structure of the α phase.

* "Eutectic" is the Greek word for low-melting, indicating a melting temperature lower than that of either of the pure components.

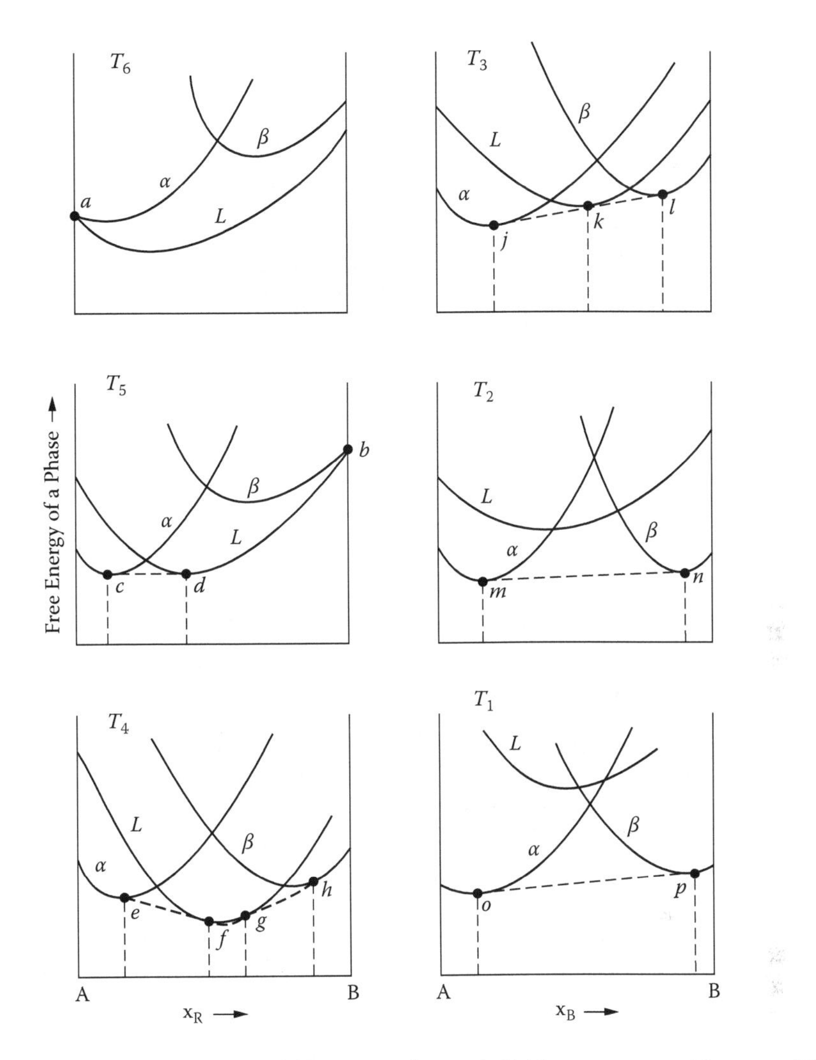

FIGURE 8.9 Free energy–composition curves for an A-B binary system with two solid phases (α and β) and a liquid phase. (From Gaskell, D. R. 1981. *Introduction to Metallurgical Thermodynamics*, 2nd ed. New York: McGraw-Hill. With permission.)

8.6.2 The Phase Diagram

The phase diagram deduced from the free energy plots in Figure 8.9 is shown in Figure 8.10. The mapping of the former to the latter is explained below at each temperature, which is reduced in steps from T_6 to T_1.

Temperature T_6 is the melting point of pure A, and is the first temperature at which a solid phase appears from the liquid. At this point, the free energy–composition plot shows that $g_L = g_\alpha$. For the entire composition range, the liquid is the stable phase because its free energy is lower than that of either of the two solid phases. Transferring this information from the T_6 plot in Figure 8.9 to the T_6 isotherm in the

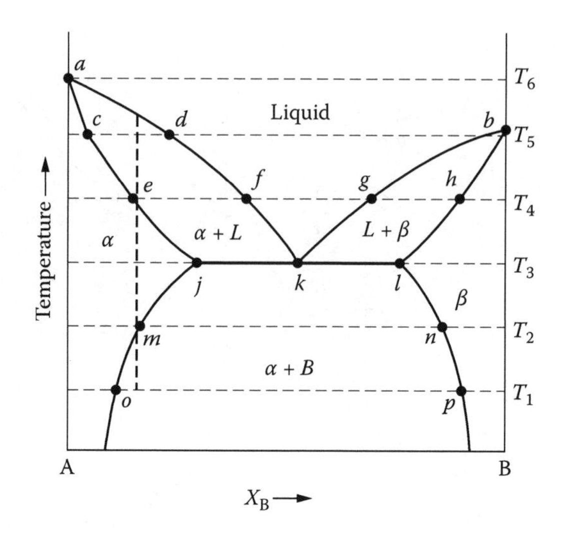

FIGURE 8.10 Phase diagram derived from Figure 8.9. (After Gaskell, 1981.)

phase diagram of Figure 8.10 shows a liquid-to-solid α transition at point a. Only liquid is present over the composition range $0 < x_B \leq 1$.

At temperature T_5, point b is the melting temperature of pure B in the β crystal structure. In addition, the α and L curves are joined by a common tangent at points c and d. At all compositions to the left of c, solid solutions of B in A with the α crystal structure exist. Between c and d, the system's lowest free energy lies along the common tangent. In this interval, liquid of composition at point d and solid α of composition c coexist. For compositions larger than that at d, liquid is the lowest free energy phase. Translating points b, c, and d to the T_5 isotherm in the phase diagram shows the phases indicated by the free energy curves.

At temperature T_4, the liquid free energy curve has risen relative to the solid curves to the extent that two common tangents can be drawn, one to the α curve and the other to the β curve. This implies that both α + L and β + L two-phase zones are present in the phase diagram at this temperature, as shown in Figure 8.10.

Temperature T_3 is unique because a single common tangent links the three free energy curves in Figure 8.9. When transferred to the phase diagram, the isotherm at T_3 shows that three phases coexist at equilibrium: α and β solid solutions with compositions at points j and l, respectively, and a liquid with composition at point k. This point is called the *eutectic point*, meaning the lowest-temperature in the phase diagram at which the system is all liquid.

At the two temperatures T_2 and T_1 common tangents join the two solid phases. In Figure 8.10, A-rich solid solutions with crystal structure α are stable up to mole fractions of B corresponding to points m and o. Further addition of B to the system results in precipitation of a B-rich β phase at the corresponding compositions n and p. The two-branch curve omj/acej is called the *terminal solubility* of B in α-A. Similarly, the counterpart on the B-rich side of the phase diagram is the terminal solubility of A in β-B.

8.6.3 Heat Up Behavior

The solid-to-liquid transformation in the eutectic system is more complex than the melting process in the ideal-solution phase diagram discussed in connection with Figure 8.3. In Figure 8.10, suppose the system starts at temperature T_1 with a mole fraction of B a bit greater than the terminal solubility at o. The initial state is in the $\alpha + \beta$ two-phase region. Upon increasing the temperature, holding the overall composition constant, the β phase disappears when the terminal solubility curve is reached at point m. Further temperature increase moves the system through the single-phase α region without phase changes until point e is reached. Here the first liquid with composition at point f appears. Additional heat up moves the system through the two-phase α + L region, with the composition of the α phase decreasing from e to c and the liquid phase composition moving from f to d. Complete liquefaction occurs when the liquidus adfk is reached. For a two-phase mixture with overall B mole fraction corresponding to point k, heat up from T_1 would result in complete melting at the eutectic temperature T_3, but nothing else.

8.6.4 A Degenerate Eutectic System: Gold/Silicon

The Au–Si eutectic phase diagram is characterized by negligible solubility of A (gold) in B and of B (silicon) in A. The single-phase α and β regions in Figure 8.10 disappear and points j and l are displaced to their respective temperature axes. Figure 8.11 shows the gold–silicon phase diagram, which is a prototypical degenerate eutectic system. As shown in the following example, information regarding the nonideality of the liquid phase can be garnered from analysis of the liquidus lines in the phase diagram.

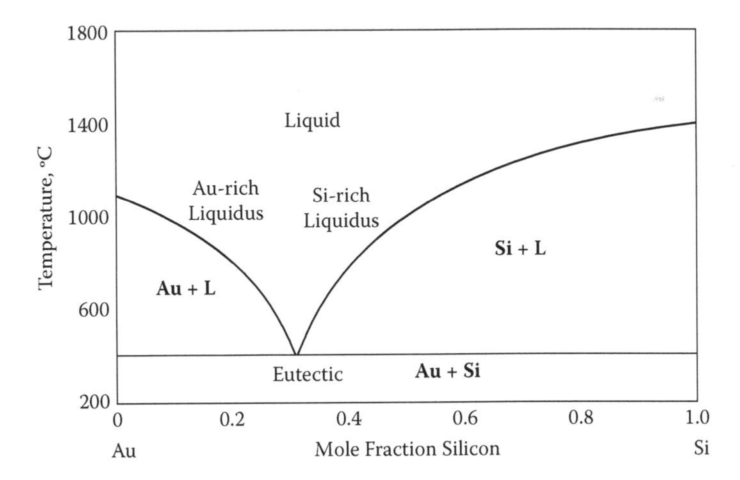

FIGURE 8.11 The gold-silicon phase diagram.

Example: Calculate the activity coefficients of the two components in the liquid phase along the two liquidus curves in Figure 8.11. Is regular solution theory obeyed? Show that the data are consistent with the Zeroth law of thermodynamics.

In the Au + L two-phase region, pure solid gold is in equilibrium with a liquid phase with compositions along the Au-rich liquidus. Equating the chemical potentials of these two phases as expressed by Equation (8.3) gives:

$$g_{Au,S} = g_{Au,L} + RT\ln(\gamma_{Au,L} x_{Au,L}) \tag{8.28}$$

where $x_{Au,L}$ is the mole fraction of gold along the Au-rich liquidus. Using the terminology of Equations (8.13) and (8.14), Equation (8.28) can be solved for $\gamma_{Au,L}$:

$$\gamma_{Au,L} = e^{-\alpha} / x_{Au,L} \tag{8.29}$$

where

$$\alpha = \left(1 - \frac{T}{T_{M,Au}}\right)\frac{\Delta h_{M,Au}}{RT} = \left(1 - \frac{T}{1336}\right)\frac{1540}{T} \tag{8.30}$$

Similarly, the activity coefficient of silicon in the Si-rich liquidus is:

$$\gamma_{Si,L} = e^{-\beta} / x_{Si,L} \tag{8.31}$$

where

$$\beta = \left(1 - \frac{T}{T_{M,Si}}\right)\frac{\Delta h_{M,Si}}{RT} = \left(1 - \frac{T}{1677}\right)\frac{6110}{T} \tag{8.32}$$

Tables 8.3a and 8.3b list the activity coefficients along the two liquidus curves calculated from the above equations. The activity coefficients approach unity as the mole fraction of the component approaches unity, which is expected behavior. At lower concentrations, the activity coefficients are less than unity, indicating attractive Au-Si molecular interactions in the liquid phase. This behavior contrasts with the strong repulsive Au-Si interactions in the solid phase, as indicated by the complete immiscibility of the two components.

TABLE 8.3A
Activity Coefficients of Gold along the Au-Rich Liquidus in the Au–Si System

T, °C	a	$x_{Au,L}$	$\gamma_{Au,L}$	$RT\ln\gamma_{Au,L}/(1 - x_{Au,L})^2$
1000	0.057	0.93	1.00	—
800	0.283	0.78	0.97	–700
600	0.611	0.73	0.74	–3600
370[a]	1.242	0.69	0.42	–5800

[a] Eutectic temperature.

TABLE 8.3B
Activity Coefficients of Silicon Along the Si-Rich Liquidus in the Au–Si System

T, °C	b	$x_{Si,L}$	$\gamma_{Si,L}$	$RT\ln\gamma_{Si,L}/(1 - x_{Si,L})^2$
1300	0.241	0.83	0.95	–2800
1200	0.505	0.70	0.86	–2500
1000	1.156	0.54	0.58	–3300
800	2.051	0.44	0.30	–4100
600	3.355	0.37	0.094	–5200
370[a]	5.859	0.31	0.009	–6300

[a] Eutectic temperature.

The last column in the tables constitutes a test of whether the system obeys regular solution theory. If it did, according to Equation (7.42), the combination of variables in the last column should be a constant. However, the data show substantial variation of this group of parameters, indicating that nonideality in liquid solutions of gold and silicon cannot be described by regular solution theory.

8.6.4.1 "Proof" of the Zeroth Law

The zeroth law of thermodynamics states that two systems (phases) each in equilibrium with a third are in equilibrium with each other. This dictum is illustrated in Figure 8.12 for the Au-Si system.

The compositions along the Si-rich liquidus in Figure 8.11 are in equilibrium with pure solid silicon. The zeroth law requires that the two phases must be in equilibrium with the same vapor.

Using Equation (8.5), the analytic expression of this 3-way equilibrium is:

$$\begin{array}{ccccc} \text{vapor} & = & \text{liquid} & = & \text{solid} \\ p_{Si} & = & \gamma_{Si,L} x_{Si,L} p_{sat,Si(L)} & = & p_{sat,Si(S)} \end{array} \tag{8.33}$$

which must hold along the entire Si-rich liquidus. An alternate form of Equation (8.33) is:

$$\gamma_{Si,L} x_{Si,L} = {p_{sat,Si(s)}}/{p_{sat,Si(L)}} \tag{8.33a}$$

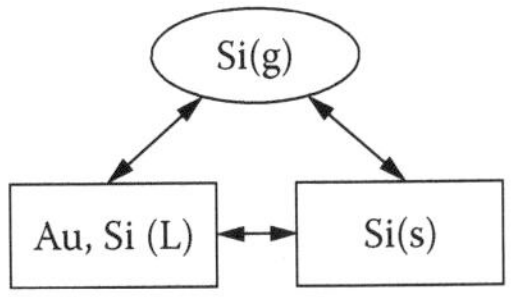

Si(g)

Au, Si (L)

Si(s)

FIGURE 8.12 Three-way equilibrium in the Au–Si system.

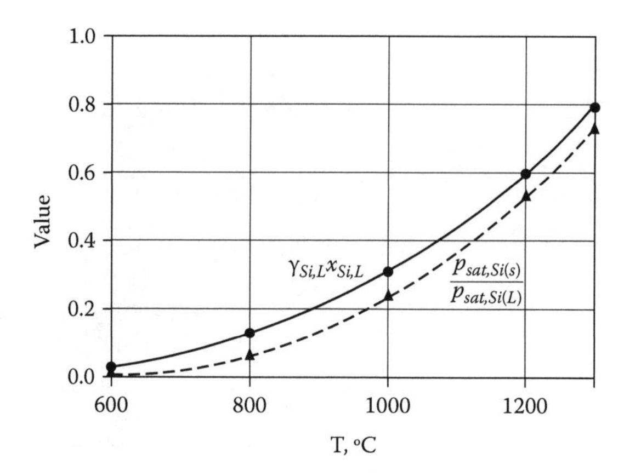

FIGURE 8.13 Comparison of Si–Au phase diagram data with vapor pressure data on pure Si.

The left-hand side of this equation is strictly phase diagram information (Table 8.2b). The right-hand side involves data on the vapor pressures of the solid and liquid phases of pure silicon.* The two sides of Equation (8.33a), which come from independent sources, are plotted in Figure 8.13. Agreement between the two curves is satisfactory, which validates the quality of the data more than it does the Zeroth law (which does need to be proven by experimental data).

The Ag–Pb binary system exhibits features similar to those shown in Figure 8.11. This system is the subject of problem 8.5.

8.6.5 A Complex Phase Diagram: Iron–Uranium

Eutectic features often appear in parts of more complex phase diagrams, as shown in the iron–uranium diagram in Figure 8.14. If the diagram is divided into three parts at 1/3 and 6/7 mole fraction uranium, the result is two simple eutectic diagrams and a more complex diagram (on the right).

The Fe-rich side resembles Figure 8.11, with two added features. The first is the number of phases of the pure components. The left-hand ordinate of Figure 8.14 makes provision for three crystallographic modifications of iron: the α phase (designated as αFe) is stable up to 912°C; γFe exists in the range $912 \leq T \leq 1394$°C; the δFe phase is the equilibrium form from 1394°C to the melting point. The αFe/γFe and the γFe/δFe transitions are indicated by horizontal lines at 912 and 1394°C. Uranium is completely insoluble in all three crystal forms of iron. The eutectic point is at 17 mole percent uranium and 1080°C.

The other feature that distinguishes the Fe-rich portion of the iron–uranium phase diagram from the gold–silicon diagram of Figure 8.11 is the right-hand border at $x_U = 0.333$. In the Au–Si system, the eutectic is bounded on the right by pure Si. The left-hand portion of the Fe–U diagram, on the other hand, has an *intermetallic compound*, Fe_2U, as the right-hand border. Intermetallic compounds form when

* Because the temperatures in Table 8.2 are below the melting temperature of Si, vapor pressures of the liquid are obtained by extrapolation from values above the melting temperature.

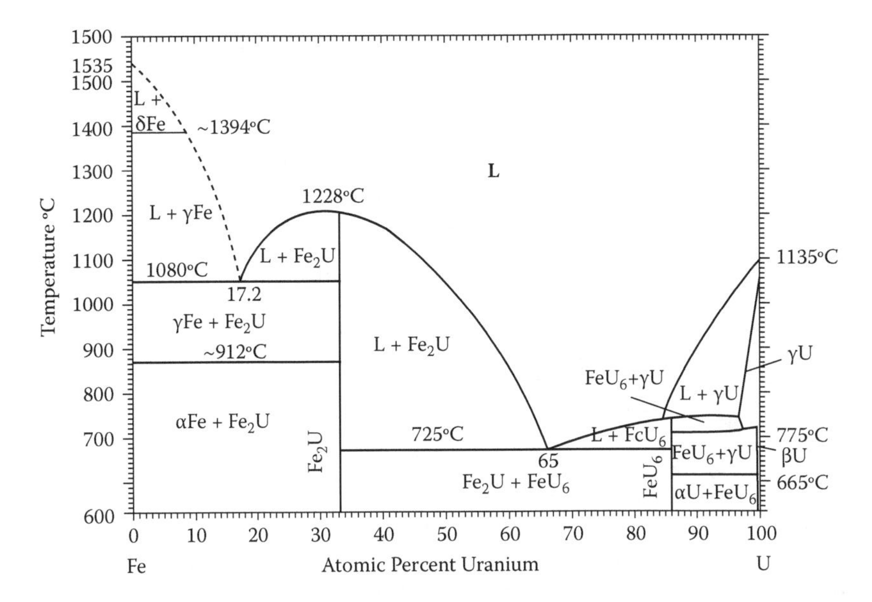

FIGURE 8.14 The iron-uranium phase diagram.

integer atom ratios of the two components form a crystallographic structure of high stability. These compounds often have high melting points, as does Fe_2U, thus enabling it to serve as the second "wall" of the left-hand field. The Fe-rich portion of the Fe-U phase diagram consists solely of two-phase regions (except for the liquid). The two-phase regions contain various combinations of the three phases of pure iron and the liquid or Fe_2U.

The diagram in Figure 8.14 contains another eutectic point at 85.7 mole percent U and 725°C. This eutectic region is bounded on the left by Fe_2U and on the right by another intermetallic compound, FeU_6. It has exactly the same structure as the Au–Si phase diagram of Figure 8.11.

Pure uranium exhibits three crystal forms: αU stable up to 660°C, βU from 660°C to 776°C, and γU from 776°C to the melting point. Contrary to the total insolubility of U in solid Fe, solid uranium dissolves small amounts of iron in each of the three crystal structures. The three zones labeled αU, βU, and γU in Figure 8.14 are single-phase zones resembling the β region in Figure 8.10. The three U-rich solutions appear in adjacent two-phase regions along with the liquid or with FeU_6. These regions are analogous to the two-phase zones in the Fe-rich portion of the diagram.

Additional practice in identifying the species and phases present in regions of phase diagrams is provided in Problems 8.7, 8.12, 8.13, 8.14 and 8.20. Problem 8.21 shows how to describe the change in the relative amounts and the compositions of phases that appear as a binary system is cooled from the liquid state.

8.6.6 Metal–Nonmetal: The Fe/O Phase Diagram

In addition to the metal–metal phase diagrams discussed to this point, the other large class of phase diagrams are the metal–nonmetal systems. In particular, the most

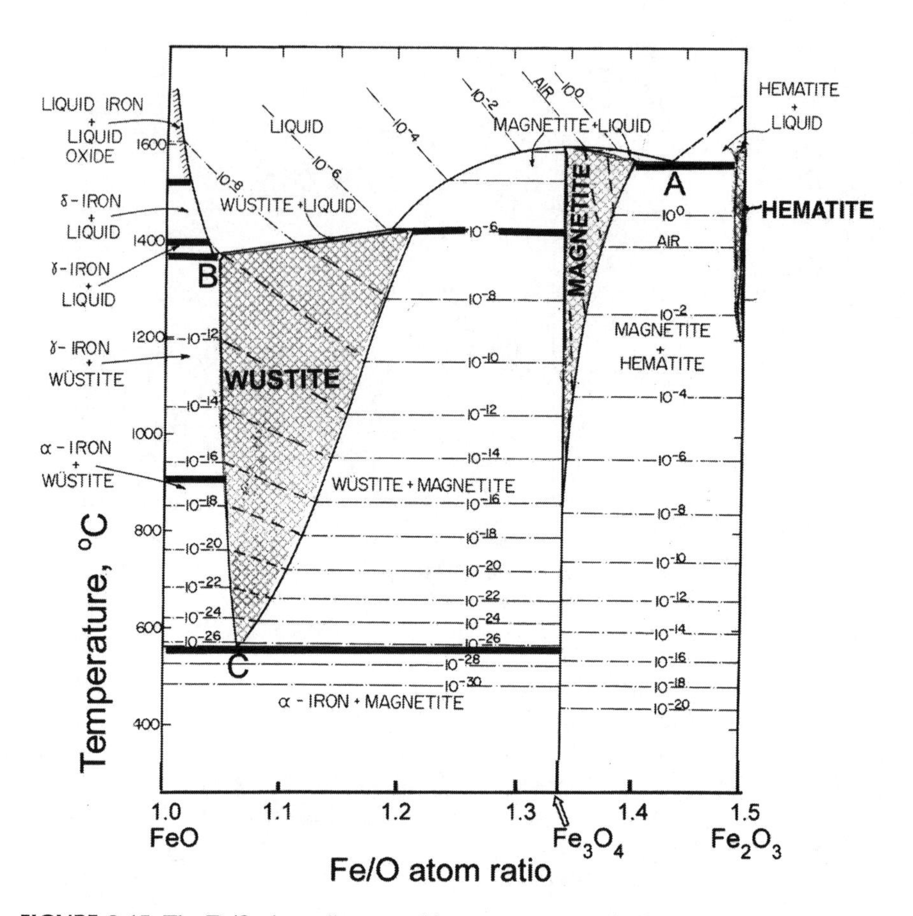

FIGURE 8.15 The Fe/O phase diagram with oxygen pressure isobars.

important nonmetal is oxygen. Of all the metal/oxygen systems, the Fe/O phase diagram, shown in Figure 8.15, has received the closest study. On the diagram, the concentration units are the O/Fe atom ratio rather than the mole fraction. The diagram begins at O/Fe = 1 ends at O/Fe = 1.5. The latter corresponds to Fe_2O_3, with the highest oxidation state of the metal 3+ (ferric). The other valence of iron is 2+ (ferrous). The diagram does not continue to O/Fe = 0 because there is no change from the two-phase regions listed on the left-hand side. Heavy horizontal lines in the diagram separate two-phase regions. The dashed lines represent isobars of the oxygen pressure in equilibrium with each point in the diagram. Single-phase regions are shaded.

Other features of the diagram include:

- The 2+ valence state of iron should result in the compound FeO. However, what actually exists is a broad (wide range of O/Fe) single-phase oxide called wustite, which is FeO deficient in Fe^{2+}. This oxide is symbolized by "FeO," even though it is not stoichiometric FeO (i.e., it does not have an oxygen-to-iron ratio of unity). Alternatively, wustite is written as $Fe_{1-x}O$, x indicating the degree of hyperstoichiometry (meaning O/Fe > 1). To maintain electrical neutrality, some of the iron must be in the 3+ state.

- The single congruently-melting (meaning that the solid and the liquid into which it transforms have the same O/Fe ratio) compound magnetite, Fe_3O_4. Since ions cannot have a fractional valence, this compound is a mixture of 2/3 Fe^{3+} and 1/3 Fe^{2+}. The average valence is 8/3 which, multiplied by 3, balances the 8 negative charges from the four O^{2-} ions in the compound.
- The single-phase magnetite region of the phase diagram consists of the iron-deficient compound $Fe_{3-y}O_4$. This O/Fe ratio requires that the fraction of Fe^{3+} be greater than the value of 2/3 in the stoichiometric compound Fe_3O_4.
- The highest oxide, hematite, has a very narrow single-phase region above 1200°C. This oxide is oxygen-deficient, with the formula Fe_2O_{3-x}.
- Two liquids exist above 1535°C: metallic iron and a liquid oxide.
- The diagram shows two eutectics,
  - an oxygen-rich one at point A at the confluence of the magnetite and hematite solids and a liquid oxide
  - another at point B where the two solids are wustite and γ-Fe and the liquid is again an oxide.
- Point C is a eutectic-like invariant point involving α-Fe, magnetite and wustite.

8.6.7 Freezing Point Depression: Salt in Water

As discussed in Section 1.1, the scale of the first mercury thermometer was established by the temperature of a mixture of salt (NaCl) and water (both liquid and solid). The phase diagram of the NaCl/H_2O system is shown in Figure 8.16.

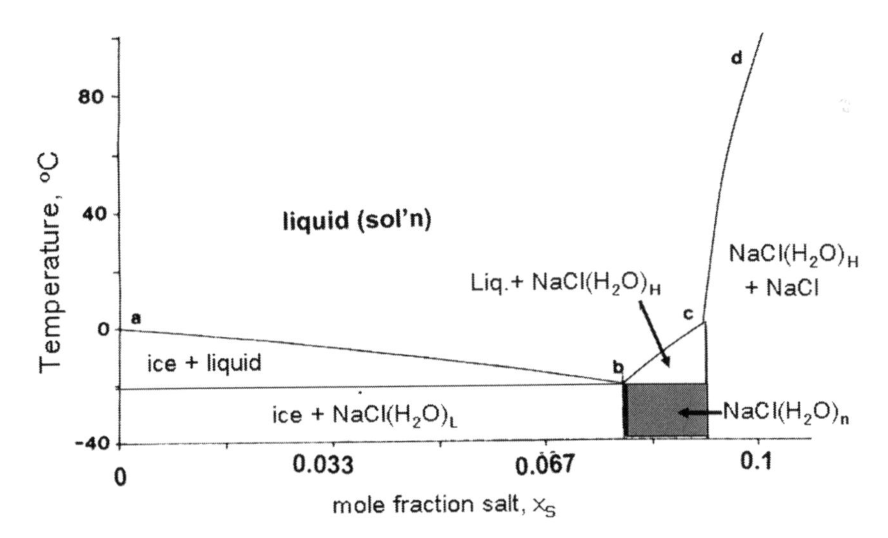

FIGURE 8.16 The NaCl/H_2O phase diagram.

The upper portion of the diagram is a large single-phase region entitled *liquid* (*sol'n*); the common name is *brine*. At the left-hand abscissa ($x_S = 0$) is pure ice and the right-hand abscissa ($x_S = 1$, not shown) is pure salt. The shaded zone in Figure 8.16 is the only single-phase solid region in the diagram. It is a compound $NaCl(H_2O)_n$, with $n^{-1} \approx x_S$, the mole fraction of salt. The upper and lower boundaries of this phase are $x_S \approx 0.092$ and $x_S \approx 0.078$, respectively. These phase boundaries are labeled $NaCl(H_2O)_H$ and $NaCl(H_2O)_L$, so that the range of the water-to-salt mole ratio in the single-phase region is $L \leq n \leq H$.

The freezing point of the salt-water solution decreases with salt concentration along the curve **ab**, which is given by:

$$T(°C) = -150x_S - 1.1\times10^3 x_S^2 \tag{8.34}$$

As the temperature of the salt solution in the concentration range $0 \rightarrow 0.078$ is reduced, pure solid water (ice) appears when the temperature reaches the value given by Equation (8.34). The maximum depression occurs at point **b**, where the salt mole fraction is $x_L = 0.078$ and $T = -18°C$, or 0°F. Point **a** represents the freezing point of pure water, which is 32°F. These two reference temperatures formed the basis of the Farenheit temperature scale, named after its eponymous author (see Section 1.1.1).

An approximation to Equation (8.34) is obtained by equating the chemical potentials of the two phases in equilibrium along the curve **ab**:

$$g_I = g_W + RT\ln(\gamma_W x_W) \tag{8.35}$$

where g_I and g_W are the molar free energies of pure ice and pure liquid water, respectively. $x_W = 1 - x_S$ is the mole fraction of water and γ_W is the activity coefficient of water in the salt solution. The difference in the free energies of the liquid and ice phases of water is given by Equation (3.27b):

$$g_W - g_I = \Delta h_M(1 - T/T_M) = -RT\ln[\gamma_W(1 - x_S)] \tag{8.36}$$

Here $\Delta h_M = 6008$ J/mole is the enthalpy of melting of ice and $T_M = 273$ K is the freezing point of pure water.

Example: What is the activity coefficient of water in an aqueous solution containing 120 g NaCl per liter?

Converting 120 g to 2.05 moles of NaCl and 1000 g water to 55.6 moles gives $x_S = 0.036$ and Equation (8.34) gives T = – 6.7°C, or 266 K. Solving Equation (8.36) yields $\gamma_W = 0.968$.

Determination of the activity coefficient of salt in the brine would require use of Equation (7.32), with $B = S$ and $A = W$. This involves calculating γ_W as illustrated above for a sufficient number of salt concentrations to permit the integral on the right side of Equation (7.32) to be accurately evaluated.

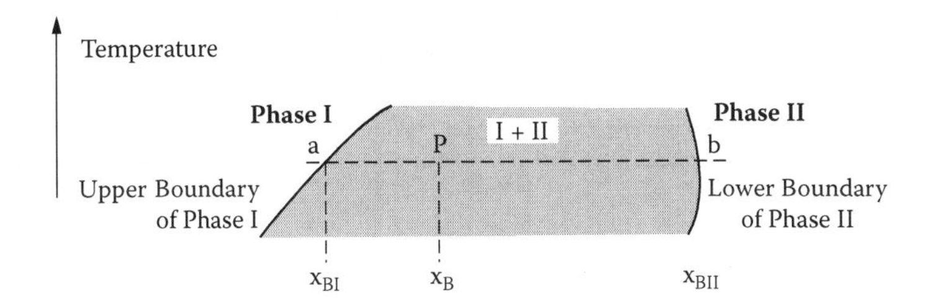

FIGURE 8.17 Illustration of the lever rule.

8.7 THE LEVER RULE

At temperature-composition points within a single-phase region, the abscissa of the phase diagram is the actual composition. In two-phase regions, on the other hand, the abscissa gives the overall composition of the two coexisting phases. Figure 8.17 shows a two-phase region (shaded area) bounded on the left by phase I and on the right by phase II. This diagram represents any of the two-phase regions with single-phase neighbors in the phase diagrams depicted in this chapter. At point P in the two-phase region of Figure 8.17, the overall composition is x_B. The compositions of the two phases present are located at the intersections of the horizontal line through P with the upper and lower phase boundaries of the adjacent single phases—that is, at points a and b. The lengths of the line segments a-P and P-b give the relative quantities (in moles) of phases I and II in the mixture. This can be shown by material balances. If the mixture contains n_I moles of phase I of composition x_{BI} and n_{II} moles of phase II of composition x_{BII}, the total number of moles is $N = n_I + n_{II}$ and the total moles of component B in the mixture is:

$$x_B N = x_{BI} n_I + x_{BII} n_{II}$$

These two equations are solved for the fraction of phase I in the mixture:

$$\textit{fraction phase}\ \mathrm{I} = \frac{n_I}{N} = \frac{x_{BII} - x_B}{x_{BII} - x_{BI}} = \frac{\overline{Pb}}{\overline{ab}} \tag{8.37}$$

This formula is known as the lever rule. It applies to any two-phase region in a phase diagram but has no meaning if applied to a single-phase zone.

PROBLEMS

8.1 Benzene and toluene form nearly ideal solutions. At 20°C the vapor pressures of pure benzene and toluene are 74 Torr and 22 Torr, respectively. A 1:1 solution is prepared, and the external pressure is varied such that it boils at 20°C. Assuming isothermal boiling, calculate:

(a) The total pressure when the solution first begins to boil.

(b) The composition of the vapor at the onset of boiling.
(c) The total pressure when only a few drops of liquid remain.

8.2 The vapor pressures of dilute solutions of HCl in liquid $GeCl_4$ (germanium tetrachloride, molecular weight 214) are given in the following table for $T = 300$ K.

x_{HCl}	p (kPa)
0.005	32.0
0.012	76.9
0.019	121.8

Assuming that the vapor pressure is due entirely to hydrochloric acid,

(a) Show that the HCl component obeys Henry's law in this concentration range; calculate the Henry's law constant.
(b) Predict the vapor pressure of HCl above the solution if the concentration is 0.10 mole HCl/kg $GeCl_4$.

8.3 A solution of 35 mole% aniline and 65 mole% hexane is cooled from a high temperature. Phase separation occurs at 56.5°C. Aniline and hexane obey regular solution theory.

(a) What is the interaction energy Ω?
(b) What is the critical solution temperature?
(c) The vapor pressure of pure aniline and pure hexane at 70°C are 0.011 atm and 1.11 atm, respectively. What is the total pressure of a 50 mole% mixture at this temperature?

8.4 Derive the equation for the elevation of the normal boiling point of water in which a nonvolatile solute is present at mole fraction x. Assume that Raoult's law applies to the water solvent.

(a) Method 1: Consider the partial pressure of water over the solution to be a function of temperature and mole fraction of the solute, or $p_w(T,x)$. Take the differential of p_w and set it equal to zero (because the normal boiling point is defined as the condition that $p_w = 1$ atm). Use Equation (5.13) for the vapor pressure of water and where needed, assume $x << 1$.
(b) Method 2: Use the condition of the equality of the chemical potentials of water in the vapor and in the solution. In evaluating the chemical potentials, the water pressure over the solution is 1 atm for all x. The temperature is $T = T° + \Delta T$, where $T°$ is the normal boiling point of pure water and ΔT is the boiling point elevation. Taylor series approximations may be used because $\Delta T/T° << 1$ and $x << 1$.

(Hint: You will need to use the analog of Equation (3.27b) applied to vapor-liquid equilibrium.)

(c) What is ΔT for a solute mole fraction of 0.04?

8.5 Pure solid Ag is in equilibrium with an ideal liquid Ag-Pb solution. The phase diagram for the system resembles the left-hand portion of Figure 8.11 with Au replaced by Ag.

(a) Starting from the criterion of chemical equilibrium (i.e., equality of chemical potentials), derive the relation for the solubility of Ag in liquid Pb at a temperature T. The melting point of Ag is T_{MAg} and the enthalpy of fusion of pure Ag is Δh_{MAg}.

(b) Alternatively, the equation developed in (a) can be viewed as giving the "melting point" of a solid Ag-Pb solution of specified mole fraction of Ag. The melting point means the temperature at which the last solid disappears on heating the alloy. Pure silver melts at 962°C and has an enthalpy of fusion of 11.3 kJ/mole. What is the melting point of an alloy with a mole fraction of silver of 0.945, assuming it to be ideal?

(c) The observed melting point of the Ag-Pb solution in (b) is 930°C. Use this information to determine the activity coefficient of Ag in the liquid in which its mole fraction is 0.945.

(d) Sketch the Ag-rich portion of the phase Ag–Pb diagram and show the Ag-rich liquidus curves that correspond to the ideal solution assumption and to the nonideal case.

8.6 FeO and MnO form ideal liquid and solid solutions. The melting properties of the pure oxides are:

Component	Δh_M (kJ /mole)	T_M (K)
FeO	31.0	1643
MnO	54.4	2148

(a) At what temperature does an equimolar solid solution first melt? What is the composition of the first-formed liquid?

(b) At what temperature is melting complete?

8.7 In the phase diagram shown below, a congruently melting compound of A and B is formed at $x_B = 0.667$.

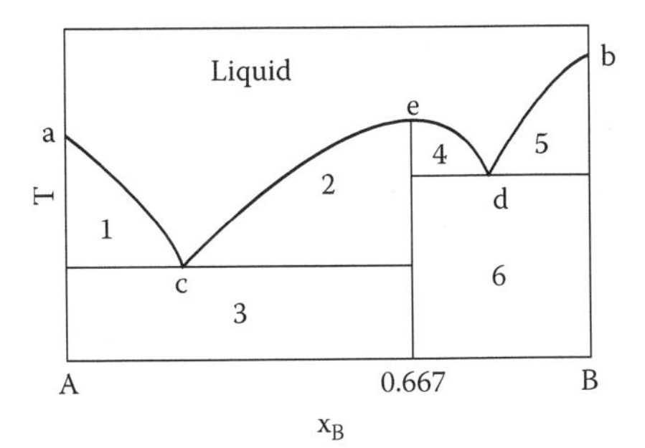

(a) What is the formula of this compound?
(b) Identify the phases present in the numbered areas of the phase diagram.
(c) Identify the points a through e.
(d) What are the solubilities of A in B and of B in A?

8.8 Cesium and rubidium form complete solutions in both the solid and liquid phases. In the phase diagram shown below, a minimum in the liquidus-solidus occurs at 282 K where the mole fractions in both liquid and solid are 0.5.

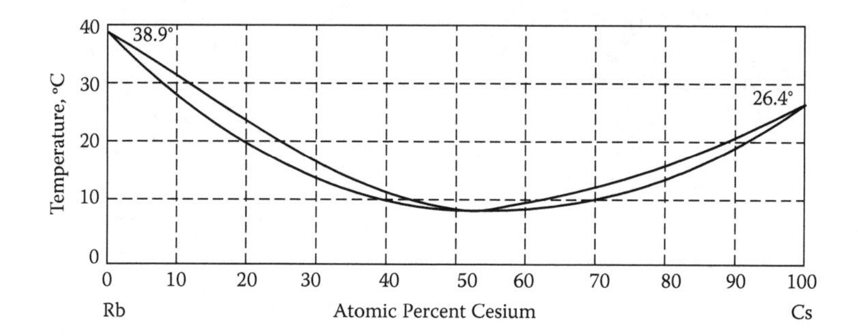

(a) Assuming the liquid solution to behave ideally, calculate the activity coefficients of Cs and Rb in the solid phase.
(b) Does the solid phase obey regular solution theory?

Element	Δh_M (kJ /mole)	T_M (K)
A = Rb	2.2	312
B = Cs	2.1	301.4

8.9 The points on the liquidus and solidus curves for completely miscible components A and B are shown below:

T (°C)	60	63	70	80	90	100
x_{BL}	0	0.06	0.19	0.42	0.65	1
x_{BS}	0	0.18	0.42	0.70	0.88	1

(a) Draw the phase diagram of this binary system.
(b) If a solid solution with $x_B = 0.5$ were just melted, what is the composition of the liquid in equilibrium with it?
(c) Assuming regular solution behavior, calculate the interaction energies of the liquid and solid phases using the phase compositions at 80°C. The fusion enthalpies of A and B are 3 kJ/mole and 4 kJ/mole, respectively. (Hint: modify the analysis of Section 8.5 to account for nonideality.)

8.10 The partial pressures of species A in equilibrium with A-B alloys at 1000 K are given below (pressures are in atmospheres). The vapor pressure of pure B is 5×10^{-5} atm at this temperature.

x_A	1.0	0.9	0.8	0.7	0.6	0.5	0.4	0.3	0.2
$p_A \times 10^6$	5.0	4.4	3.8	2.9	1.8	1.1	0.8	0.6	0.4

(a) Tabulate the activity coefficient of A as a function of composition.
(b) Over what range of composition is Henry's law obeyed by A?
(c) What is the equation for the equilibrium partial pressure of B in the composition range in which component A obeys Henry's law?

8.11 Shown below are the free energy versus composition plots for an A-B binary system at four temperatures. The ordinate is the free energy and the abscissa is the mole fraction of component B.

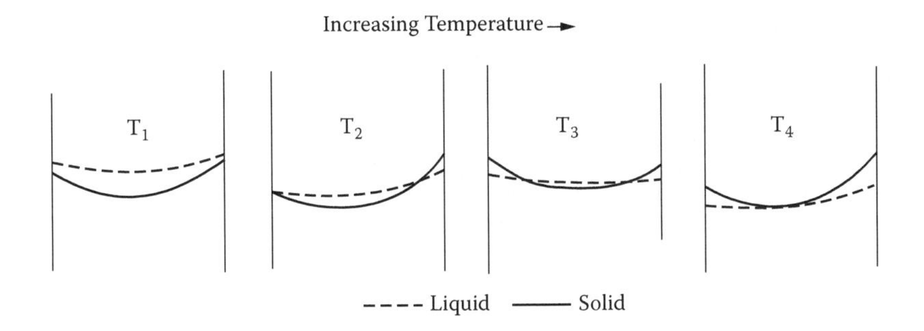

(a) On these plots, label the melting temperatures of the pure components. Draw common tangent lines on the plots and indicate with letters the points of common tangency on the solid and liquid curves.
(b) From the identifications made in Part (a), sketch the phase diagram derived from the free energy plots.

8.12 The silver–praseodymium phase diagram is shown below.

(a) Identify the 12 numbered regions.
(b) What is the partial pressure of silver in equilibrium with the liquid at point P on the diagram? The vapor pressure of pure solid silver at 900°C is 4×10^{-8} atm and its heat of fusion is 2.7 kJ/mole. The enthalpy of sublimation is 89 kJ/mole. (Hint: note the similarity to Figure 8.11. You must calculate the vapor pressure of pure liquid silver at 900°C in order to use Equation [8.5].)
(c) Assuming regular solution theory to apply, what is the partial pressure of silver in equilibrium with the liquid at point Q?

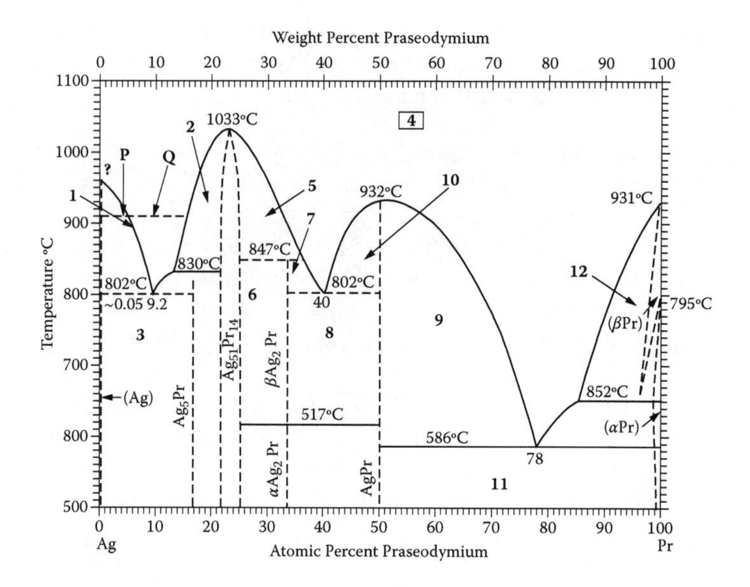

8.13 Identify the phases present in the numbered areas of the following phase diagram:

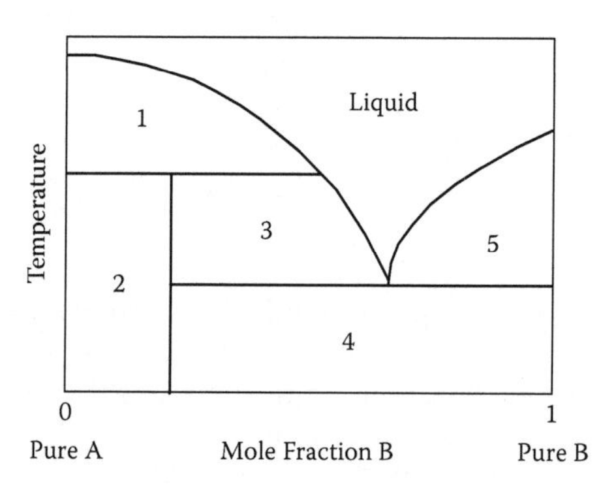

8.14 The A-B phase diagram shown below is typical of two substances that show partial miscibility in the solid state. Identify the numbered phases, the point P and the curve RS.

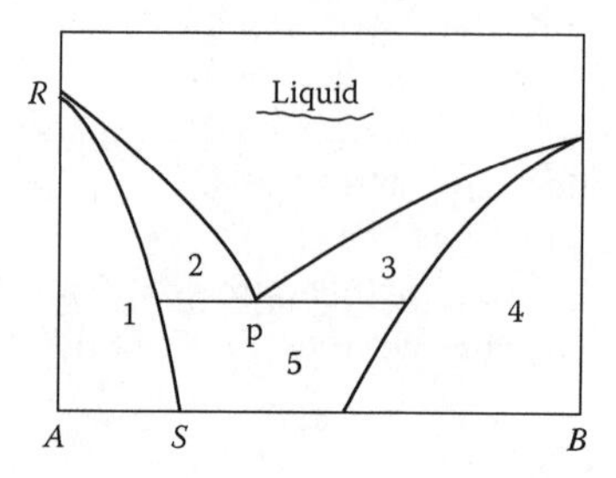

8.15 The Pb–Sn system exhibits regular solution behavior with an interaction energy $\Omega = -0.7$ kcal/mole.

(a) If the ratio of the vapor pressure of pure lead to that of pure tin at 200°C is 2.5, what is the ratio of the partial pressures of lead and tin at the same temperature in equilibrium with a solution containing 60 mole% lead?
(b) At what temperature does the solution separate into two immiscible phases?
(c) What is the activity coefficient of lead in a solution with $x_{Pb} = 0.3$ at 500°C?

8.16 Consider a mixture of A and B with 50 mole% B whose temperature is increased from room temperature. The phase diagram of this system is shown below.

(a) What is the composition of the liquid phase that first forms?
(b) At 70°C what is the mole ratio of solid to liquid?
(c) What are the activity coefficients of A and B in the liquid and solid phases at 80°C? Assume regular solution behavior, with $\Delta H_{MA} = 10$ kJ/mole and $\Delta H_{MB} = 12$ kJ/mole.

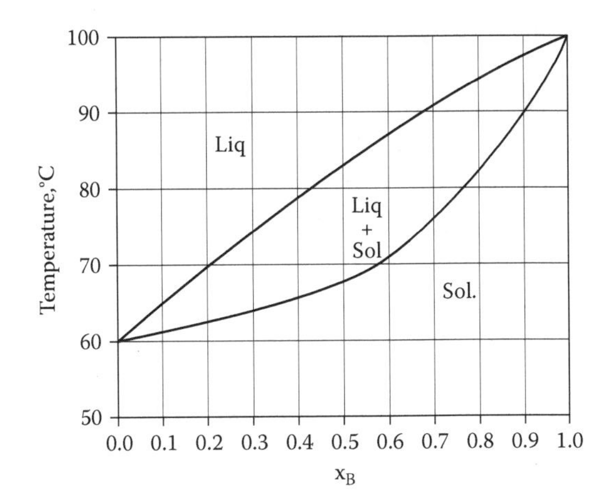

8.17 Sketch the phase diagram corresponding to the free energy curves shown below. Note that in this case, $C_S > C_L$ and $T_o > T_1 > T_2$. (Hint: refer to Problem 8.11.)

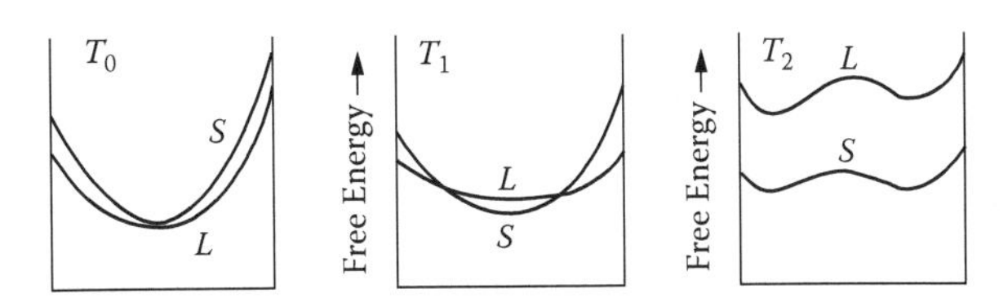

8.18 What is the partial pressure of water in equilibrium with an aqueous solution containing 30 mole% alcohol? The temperature is 25°C and alcohol and water form a regular solution with an interaction energy of $\Omega = -20$ kJ/mole.

8.19 The following table gives the mole fractions of component A in a binary solution and in the equilibrium gas phase and their total vapor pressure ($p = p_A + p_B$). Complete the next three columns of the table to include:

(a) In the fourth column, enter the partial pressures of component A.
(b) In the fifth column, tabulate the ratio p_A/x_{AL}. From these entries, deduce the Henry's law constant for component A.
(c) In the sixth column, calculate and enter the activity coefficient of A in the liquid.

x_{AL}	x_{Ag}	p, kPa	p_A, kPa	p_A/x_{AL}	γ_{AL}
0	0	36.066			
0.0898	0.0410	34.121			
0.2476	0.1154	30.900			
0.3577	0.1762	28.626			
0.5194	0.2772	25.239			
0.6036	0.3393	23.402			
0.7188	0.4450	20.693			
0.8019	0.5435	18.592			
0.9105	0.7284	15.496			
1	1	12.295			

8.20 This phase diagram depicts a solution of A and B.

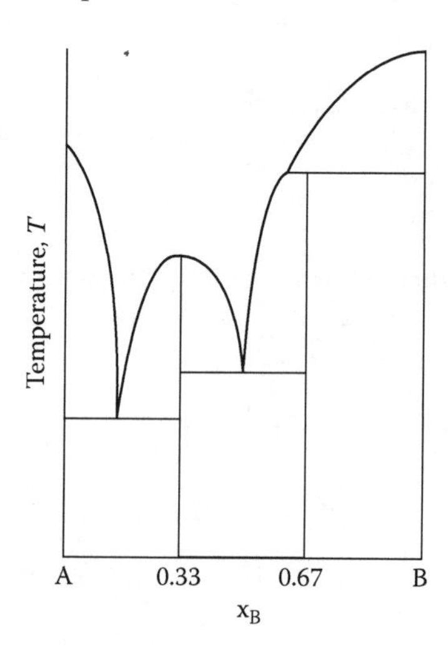

(a) Label each region in the diagram with a letter identifier.
(b) Identify the species present in the different regions in terms of their A and /or B constituents. (If there are compounds, give their formulae.)
(c) State whether each component in a region is solid, liquid, or gas.

8.21 The phase diagram below shows two constant-composition paths, labeled (1) at $x_{B(1)} = 0.54$ and (2) at $x_{B(2)} = 0.70$, along which the binary A-B liquid is cooled from high temperature.

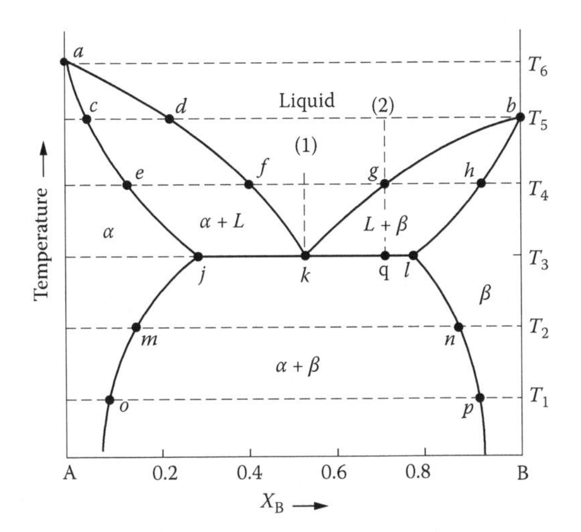

(a) When cooling along path (1), what phases appear when the temperature reaches T_3?
(b) When cooling along path (2), determine:

1. The phase that appears when the temperature reaches T_4.
2. The compositions and proportions of the coexisting phases at a temperature half-way between T_3 and T_4 (numerical values required).
3. The phases that coexist when the temperature reaches T_3.

8.22 The phase diagram of the U–O system from U to UO_2 is shown below. The composition axis is expressed as the oxygen-to-uranium ratio, O/U. When O/U < 2, the single-phase oxide is designated as UO_{2-x}, where O/U = 2-x.

(a) Identify the phases present in each of the numbered regions of the diagram.
(b) What phases are present at point P? How many degrees of freedom are there at this point?
(c) At point x in the diagram:

1. Identify the phases present.
2. If there is more than one phase, determine their relative amounts and compositions.

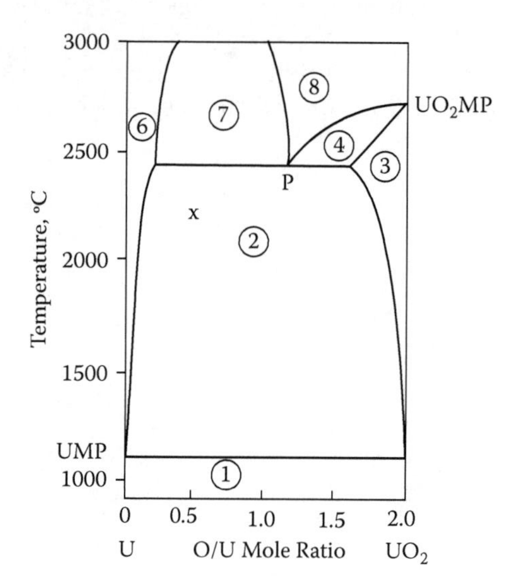

8.23 The phase diagram below represents an A-B system in which the liquid solution is ideal but the solid is not. Calculate the activity coefficient of A on the solidus curve at 880°C. The enthalpy of melting of pure A is 25 kJ/mole.

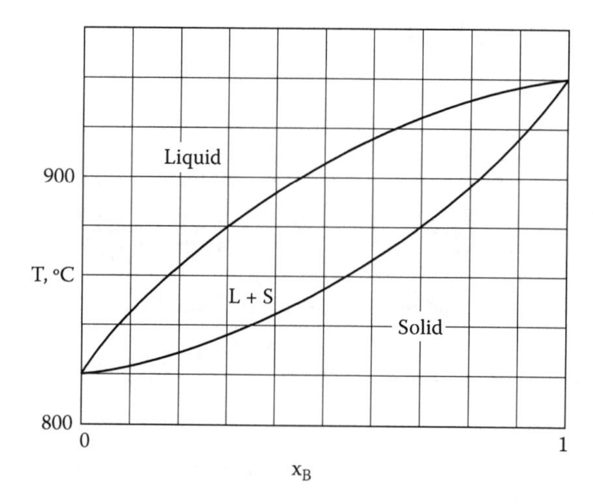

8.24 The vapor in equilibrium with a solid A-B alloy has the same composition as the solid. For $x_{BS} = 0.1$, the total vapor pressure ($p_A + p_B$) is 0.04 atm. For the pure components at the same temperature, $p_{satA} = p_{satB} = 0.03$ atm.

(a) If the solid solution were ideal, what should be the total vapor pressure over the alloy?

(b) Assuming that the solid solution obeys regular solution theory, derive the equation from which the interaction energy, $\Gamma = \Omega/RT$ can be calculated. Use the observed total vapor pressure at $x_{BS} = 0.1$ and solve the equation for Γ.

8.25 (a) Apply the phase-separation analysis of Section 8.4.2 with the excess enthalpy given by:

$$h^{ex} = \Omega x_A x_B \left(1 - \alpha x_B\right)$$

instead of Equation (8.15). α is a constant.

(b) Determine Ω and α for the n-hexane/nitrobenzene system by fitting the two-phase dome in Figure 8.6 to the results of part (a).

REFERENCE

Gaskell, D. R. 1981. *Introduction to Metallurgical Thermodynamics*, 2nd ed., New York: McGraw-Hill.

9 Chemical Thermodynamics

9.1 CHEMICAL REACTIONS

In all of the thermodynamic processes and properties that have been treated in the previous eight chapters, the substances involved retained their molecular identities. When atoms switch molecular units, a chemical reaction has taken place. This chapter presents the methods for determining the extent of reaction, or the equilibrium composition.

9.1.1 CATEGORIES OF REACTIONS: STOICHIOMETRY

Chemical reactions can be classified according to the number of phases involved. A reaction that is confined to a single phase is termed *homogeneous*. An example of this type of reaction is the combustion of methane:

$$CH_4(g) + 2O_2(g) = CO_2(g) + 2H_2O(g) \tag{9.1}$$

Another example of a homogeneous reaction is the transfer of electrons between ions in aqueous solution, as in the reduction of tetravalent plutonium by ferrous ions:

$$Pu^{4+}(aq) + Fe^{2+}(aq) = Pu^{3+}(aq) + Fe^{3+}(aq) \tag{9.2}$$

The medium in which the reaction occurs is designated in parentheses: g for the gas phase; aq for an aqueous solution; L for a pure, nonaqueous liquid (e.g., a molten metal); soln for a nonaqueous solution; and s for a solid phase. Discussion of aqueous ionic chemical equilibrium is deferred until Chapter 10.

Reactions that involve species present in more than one phase are termed *heterogeneous*. For example, in the oxidation of solid or liquid metal M to form the solid oxide MO_2, each participant is present as a pure phase:

$$M(s \text{ or } L) + O_2(g) = MO_2(s) \tag{9.3}$$

If one or more of the participants is dissolved in an inert diluent, the oxidation reaction is written as:

$$M(soln) + O_2(g) = MO_2(s) \tag{9.4}$$

The purpose of the present chapter is to apply thermodynamic theory to characterize the state of equilibrium of chemical reactions. At equilibrium, the concentrations of the species involved are unchanging and the properties of the chemical reaction lead to a relation between these concentrations. This relation is known as the *law of mass action.*

Rather than deal with each reaction as a specific case, a generalized reaction in which molecular species A and B interact to form new molecular entities C and D is analyzed:

$$aA + bB = cC + dD \tag{9.5}$$

The coefficients a,…,d are the *stoichiometric numbers* or *balancing numbers* that serve to conserve elements on the two sides of the reaction. They are usually chosen so that one of them is unity.

9.1.2 Equilibrium

The equal sign in Equation (9.5) signifies that the equilibrium state has been achieved. By convention, the molecular species on the left-hand side of the reaction are called *reactants* and those on the right-hand side are termed *products.* At equilibrium, there is no fundamental distinction between reactants and products; Equation (9.5) could just as well have been written with C and D on the left and A and B on the right. As long as the element ratios are the same, the equilibrium composition does not depend on the initial state. For example, an initial state composed of $\underline{c}$ moles of C and $\underline{d}$ moles of D produces the same equilibrium mixture as an initial state having $\underline{a}$ moles of A and $\underline{b}$ moles of B. This is the analog of mechanical equilibrium—a rock dropped from the north rim of the Grand Canyon ends up in the Colorado River just as surely as one dropped from the south rim.

Chemical reactions release or absorb heat. In the nineteenth century, it was thought that the equilibrium composition was the one that released the maximum amount of heat. As a corollary, reactions that absorbed heat were not supposed to occur. Chemical equilibrium was identified with minimum system enthalpy. The obvious shortcomings of this theory were rectified when it was recognized that the property that is minimized at equilibrium is the free energy, not the enthalpy.

Chemical equilibrium is universally analyzed by holding temperature and pressure constant. These conditions are chosen because they represent most practical situations in which chemical reactions occur. With these constraints, the condition of equilibrium is the minimum of the free energy (Equation [1.20a]).

The physical reason for minimizing G rather than H is that G includes the effects of entropy changes during the reaction. Since $G = H - TS$, a reduction in H is a direct contribution to a reduction of G. In this sense, the old theory is partially correct. However, as the reaction in Equation (9.5) proceeds from the initial reactants A and B towards the side that has the lowest enthalpy (assumed to be the products C and D), the entropy increases because the randomness of the mixture is greater when all four species are present (entropy of mixing). However, the internal entropies of the individual species need to be taken into account. The difference of entropies of the product species C and D and the reactant species A and B needs to be added to the mixing entropy to determine how S changes with the extent of reaction.

Subtracting the TS term from H results in a minimum G at a composition between complete conversion and no reaction. This buffering effect of the mixing entropy on the extent of a chemical reaction is particularly important in homogeneous reactions, where all species occupy a single phase. At the other extreme, in heterogeneous reactions in which each phase is a separate, pure species (as in Equation [9.3]), there is no entropy of mixing and complete conversion or total lack of reaction is not only possible, but must occur.

Example: The entropy of mixing effect can be illustrated by the gas phase reaction in which an initial mixture of two moles of H_2 and one mole of O_2 partially combine to form H_2O gas. The enthalpy change accompanying complete conversion of the initial H_2 and O_2 to H_2O is labeled ΔH^o (the significance of the superscript will be explained in the next section). Because heat is released in this reaction, ΔH^o is negative. If f denotes the fraction of the reaction completed, the enthalpy relative to the initial state is $H = f\Delta H^\circ$.

Ignoring for simplicity the entropy carried by the individual species (i.e., in translation, rotation, etc.), the entropy of mixing at some intermediate state of the reaction is given by application of Equation (7.10) extended for a three-component gas:

$$S = -R\left(n_{H_2} \ln x_{H_2} + n_{O_2} \ln x_{O_2} + n_{H_2O} \ln x_{H_2O}\right)$$

where n and x represent the mole numbers and mole fractions, respectively, at some intermediate state of partial conversion. Based on the initial 2:1 mole ratio of H_2 and O_2, the mole numbers at partial reaction expressed in terms of the fraction reacted f are $n_{H_2} = 2(1-f)$, $n_{O_2} = 1-f$, and $n_{H_2O} = 2f$. Adding these three gives $3 - f$ total moles. The mole fractions can be computed from these mole numbers. With H and S expressed in terms of the fraction reacted, the free energy of the mixture is:

$$\frac{G}{RT} = \frac{H}{RT} - \frac{S}{R} = \left(\frac{\Delta H^o}{RT}\right) f + 2(1-f)\ln\left(\frac{2(1-f)}{3-f}\right) + (1-f)\ln\left(\frac{1-f}{3-f}\right) + 2f\ln\left(\frac{2f}{3-f}\right)$$

Division of the equation by RT renders all terms dimensionless. The terms expressing H, TS, and G obtained from the above equation are plotted in Figure 9.1 for $\Delta H^o/RT = -4$ (this value is not the correct one for the reaction being considered, but it shows the interplay of the H and TS terms more clearly).

For this choice of reaction ($O_2 + 2H_2 = 2H_2O$), the initial reactant ratio ($H_2/O_2 = 2$) and the enthalpy change upon complete reaction ($\Delta H^o/RT = -4$), the equilibrium occurs at $f \sim 0.72$. Irrespective of the values chosen for these two parameters, the fraction reaction is greater than zero but less than one.

The method of calculating equilibrium concentrations used in this example is too cumbersome for practical applications. The approach based on the law of mass action described in Section 9.5 is preferable. This method requires knowledge of two basic properties of a reaction: the enthalpy change ΔH^o and the entropy change ΔS^o. The latter was (for simplicity) set equal to zero in the above example.

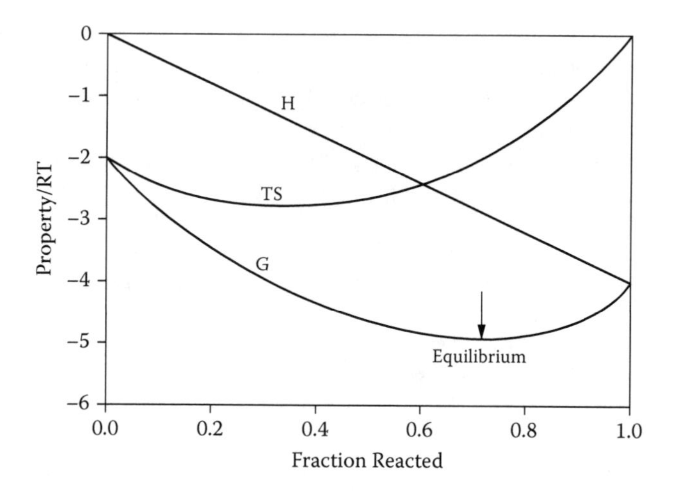

FIGURE 9.1 Contributions to the free energy of a reacting gas mixture from the enthalpy change of the reaction and the entropy of mixing.

9.2 ENTHALPY CHANGE OF A REACTION

ΔH^o is the enthalpy change when reactants are completely converted to products at a fixed temperature and at a reference, or standard, pressure. For the purpose of defining ΔH^o, the reactants are unmixed and pure, as are the products. With reference to Equation (9.5), if $\underline{a}$ moles of A and $\underline{b}$ moles of B are converted entirely to $\underline{c}$ moles of C and $\underline{d}$ moles of D, the enthalpy change is:

$$\Delta H^o = H^o(products) - H^o(\text{reactants}) = (dh_D^o + ch_C^o) - (ah_A^o + bh_B^o) \tag{9.6}$$

where h_i^o is the molar enthalpy of pure species i. The superscript ^{o} indicates that the enthalpies of the pure species are evaluated at the reference pressure, which is arbitrarily set at 1 atm (actually 1 bar, which is 1.01 atm). A pure substance at this pressure is said to be in its *standard state*. As a practical matter, the effect of pressure on ΔH^o is small; the enthalpies of ideal gases are pressure-independent (Section 2.4) and those of solids and liquids are only weakly pressure-dependent. However, the entropy of an ideal gas is highly pressure-dependent, and adherence to the standard pressure convention is essential for proper specification of ΔS^o.

Because the enthalpy change is equal to the heat exchanged with the surroundings in a constant pressure process, ΔH^o is also called the *heat of reaction*. If the reaction releases heat (ΔH^o negative), it is called *exothermic*. If heat is absorbed (ΔH^o positive), the reaction is *endothermic*. The reactions given by Equations (9.1) and (9.4) are exothermic. Dissociation reactions such as $N_2(g) = 2N(g)$ are endothermic, since energy (or enthalpy) is required to break the N-N bond.

Contrary to the effect of pressure, the molar enthalpy of a pure substance is strongly temperature-dependent. However, because ΔH^o is the difference between the molar enthalpies of product and reactants, it varies less with temperature. Because enthalpy is not a property with an absolute value, a reference temperature must be chosen for the calculation of the individual h^o values of the participants in the reaction. However, ΔH^o, being the difference in molar enthalpies, does not depend on the choice of the reference temperature.

At the reference temperature and pressure, each substance taking part in the reaction is assumed to exist in its normal state. Thus, O_2 is gaseous, water is liquid, and aluminum is solid. The molar enthalpy at temperatures T (but still at 1 atm) is given by Equations (3.19) and (3.20) for solids and liquids, respectively, and for gases, by:

$$h^o_{i(g)}(T) = h^o_{i,ref} + C_{Pi}(T - T_{ref}) \tag{9.7}$$

where C_{Pi} is the heat capacity at constant pressure.

ΔH^o for the reaction given by Equation (9.5) is obtained by substituting $h^o_A, \ldots, h^o_D$ into Equation (9.6):

$$\Delta H^o = \Delta H^o_{ref} + \Delta C_P(T - T_{ref}) \tag{9.8}$$

where:

$$\Delta C_P = (dC_{PD} + cC_{PC}) - (aC_{PA} + bC_{PB}) \tag{9.9}$$

The superscript ^{o} denotes 1 atm pressure for the pure species and the subscript *ref* indicates values at an arbitrarily-chosen reference temperature. In Equation (9.7)–(9.9), specific heats are temperature-independent.

Example: The oxidation of metal M according to reaction (9.3) takes place at a temperature T above the melting point of M but below the melting point of MO_2. Using T_{ref} = 298 K and neglecting the difference between the heat capacities of solid and liquid M, the enthalpy change of the reaction is:

$$\Delta H^o = h^o_{MO_2,298} + C_{PMO_2}(T - 298) - \left[h^o_{O_2,298} + C_{PO_2}(T - 298)\right]$$

$$- \left[h^o_{M,S298} + C_{PM}(T - 298) + \Delta h_{M,M}\right]$$

$$= \Delta H^o_{298} + \Delta C_P(T - 298) + \Delta h_{M,M}$$

$\Delta h_{M,M}$ is the enthalpy increase of M on melting.

The second term on the right of this equation is usually small compared to the first and third terms. In this case, ΔH^o is independent of temperature.

Example: Alternate method of calculating ΔH^o for reaction (9.3) when M = Zr.

Values of h^o_i for reactants and product at two temperatures are shown below (in kcal/mole).

T, K	ZrO_2	Zr	O_2
298	–262.3	0	0
1500	–239.9	9.6	9.6

The convention for this example is $h^o_{i,ref} = 0$ for all elements in their normal states at 298 K and 1 atm pressure. For this reaction, Zr and ZrO_2 are solids and O_2 is a gas. Also, $\Delta H^o_{298} = -262.3$ kcal/mole at 1500 K, the enthalpy change of the reaction is:

$$\Delta H^o(1500) = h^o_{ZrO_2}(1500) - h^o_{Zr}(1500) - h^o_{O_2}(1500) = -239.9 - 9.6 - 9.6$$

$$= -259.1 \text{ kcal/mole}$$

The enthalpy change of the reaction is ~ 3 kcal/mole higher at 1500 K than at 298 K. All things considered, though, this change is insignificant.

The effect of ΔC_P in Equation (9.8) is analyzed for gas-phase equilibrium calculations in Problem 9.24.

9.3 ENTROPY CHANGE OF REACTION: THE ENTROPY RULE OF THUMB

The entropy change due of reaction (9.5) that converts pure, unmixed reactants to pure, unmixed products, all at the standard pressure of 1 atm, is given by

$$\Delta S^o = S^o(products) - S^o(\text{reactants}) = (ds^o_D + cs^o_C) - (as^o_A + bs^o_B) \qquad (9.10)$$

Just as for the molar enthalpy, the molar entropies in Equation (9.10) can be referenced to the values for the normal physical state of the substances at T_{ref}. s^o can be obtained from Equations (3.22) and (3.23) for solids and liquids (see Figure 3.26), and for gases, from

$$s^o_i = s^o_{i,ref} + C_{Pi}\ln(T / T_{ref}) \qquad (9.11)$$

Equation (9.10) becomes

$$\Delta S^o = \Delta S^o_{T_{ref}} + \Delta C_P \ln(T / T_{ref}) \qquad (9.12)$$

Although Equation (9.11) applies in principle to gases as well as to condensed phases, the absolute entropy of an ideal gas can be calculated from statistical thermodynamics; if $T_{ref} = 0$ K.

The molar entropy of an ideal gas consists of contributions from translation (i.e., motion of the molecules as a whole) and from internal modes of motion, such as rotation and vibration. The translational contribution is much larger than the others, which means that all gas species have approximately the same entropy. The translational entropy of an ideal gas at the standard pressure of 1 atm is given by:

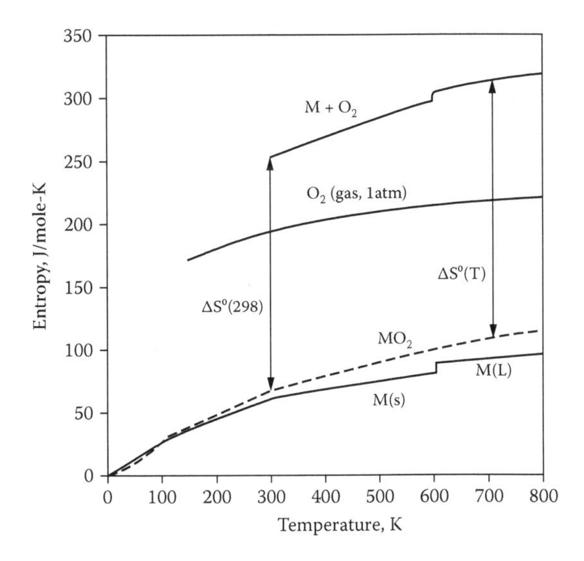

FIGURE 9.2 Standard entropy change of the reaction $M + O_2 = MO_2$.

$$s^o = R[\ln(M^{3/2}T^{5/2}) - 1.97] \tag{9.13}$$

where M is the molecular weight of the gas.

For solids, the third law is $s = 0$ and $T = 0$ K, so an absolute entropy can be calculated from:

$$s^o = \int_0^T \frac{C_P(T')}{T'} dT'$$

The molar entropies of the participants in the reaction $M + O_2 = MO_2$ are shown in Figure 9.2, which is a specialized version of Figure 3.6. Note that the entropies of the solid forms of M and MO_2 vanish at the absolute zero in temperature, as required by the third law. Equation (9.11) does not apply at very low temperatures, where C_P becomes strongly temperature dependent. The discontinuity in the curve for the metal M at approximately 600 K is due to melting.

Figure 9.2 shows that the entropy change of the reaction is dominated by the molar entropy of gaseous oxygen. This is so because the chaotic motion of molecules in the gaseous state is a much more highly disordered state than the regularity of a solid crystal, or even of a liquid. In addition, the entropies of the two condensed phases are very nearly equal, and so cancel in the reaction entropy change:

$$\Delta S^o = s^o_{MO_2} - s^o_M - s^o_{O_2} \approx -s^o_{O_2} \tag{9.14}$$

This result illustrates the very useful *entropy rule of thumb*:

The approximate magnitude and the sign of ΔS^o for any nonaqueous reaction can be estimated by ignoring the condensed phases and noting the number of moles of gaseous species in the product and reactants. Each of these species can be assigned an entropy of 200 J/mole-K.

Example: For O_2 at 300 K, Equation (9.13) gives $s^o = 152$ J/mole-K. The value obtained from the O_2 curve in Figure 9.1 is approximately 190 J/mole-K. The discrepancy is due to the contributions of the internal modes of motion in the oxygen molecule, which are not accounted for in Equation (9.13).

200 J/mole-K is a reasonable approximation to the total entropy of a gaseous species. Equation (9.13) shows that s^o varies slowly with M and T. For the $M + O_2 = MO_2$ reaction, the rule of thumb gives $\Delta S^o = -200$ J/mole-K. Figure 9.1 shows that ΔS^o varies from −180 J/mole-K at 298 K to −200 J/mole-K at 700 K.

Example: For reaction (9.1), the rule of thumb yields $\Delta S^o \sim 0$ because both sides of the reaction have the same number of moles of gas species. The exact value for this reaction is $\Delta S^o \sim -4$ J/mole-K.

Example: The table below is the entropy analog of the enthalpy table in the example at the end of the last section. In this table, the reference temperature is 0 K, so the entropies listed are absolute values.

	s^o, cal/mole-K		
T, K	ZrO_2	Zr	O_2
298	12.0	9.3	49.0
1500	40.2	21.4	61.6

The entropy change of reaction (9.3) for M = Zr is:

$$\Delta S^o(298) = s^o_{ZrO_2}(298) - s^o_{Zr}(298) - s^o_{O_2}(298) = 50 - 39 - 205 = -194 \text{ J/mole-K}$$

$$\Delta S^o(1500) = s^o_{ZrO_2}(1500) - s^o_{Zr}(1500) - s^o_{O_2}(1500) = 168 - 90 - 258 = -180 \text{ J/mole-K}$$

At both temperatures, the entropy rule of thumb ($\Delta S^o > -200$ J/mole-K for the Zr oxidation reaction) is within 10% of the exact valus.

Problems 9.15, 9.23, and 9.24 illustrate the usefulness of the entropy rule of thumb in analyzing chemical equilibria.

9.4 CRITERION OF CHEMICAL EQUILIBRIUM

As in any system constrained to constant temperature and pressure, the equilibrium of a chemical reaction is attained when the free energy is a minimum. Specifically, this means that $dG = 0$, where the differential of G is with respect to the composition of the mixture. In order to convert this criterion to an equation relating the equilibrium concentrations of the reactants and products, the chemical potentials are the essential intermediaries. At equilibrium, Equation (7.27) provides the equation:

$$dG = \sum_i \mu_i dn_i = 0 \tag{7.27}$$

where n_i is the number of moles of species i and the summation includes all reactants and product species. The equation applies to an equilibrium involving multiple phases. If more than one reaction occurs, Equation (7.27) applies to each.

In order to further develop Equation (7.27) into a usable form, it is applied to generic reaction (9.5), yielding:

$$\mu_A dn_A + \mu_B dn_B + \mu_C dn_C + \mu_D dn_D = 0$$

The changes in the mole numbers are related to each other by the stoichiometric coefficients in Equation (9.5). For example, for every mole of A consumed, b/a moles of B disappear and c/a and d/a moles of C and D, respectively, are produced. These stoichiometric restraints yield the following:

$$dn_B = \frac{b}{a} dn_A \qquad dn_C = -\frac{c}{a} dn_A \qquad dn_D = -\frac{d}{a} dn_A$$

The equilibrium condition for the generic reaction becomes:

$$a\mu_A + b\mu_B = c\mu_C + d\mu_D \tag{9.15}$$

The general version of the above equation is:

$$\sum_{reactants} \nu_i \mu_i = \sum_{products} \nu_i \mu_i \tag{9.16}$$

where ν_i are the balancing numbers in the chemical reaction.

There remains only to express the chemical potentials in terms of the concentrations, which will then lead to the law of mass action.

9.5 GAS-PHASE REACTION EQUILIBRIA

Suppose generic reaction (9.5) consists exclusively of gaseous species. Except for the small fraction of the intermolecular collisions that result in changing the elements in the molecules A and B into the products C and D, the components can be treated as ideal gases inhabiting the reaction vessel at partial pressures p_A, ... p_D. Consequently, the chemical potentials in Equation (9.15) are related to the partial pressures by Equation (7.44):

$$\mu_i = g_i^o + RT \ln p_i \tag{9.17}$$

where g_i^o is the molar free energy of pure gas i at 1 atm pressure. The partial pressures must be expressed in atmospheres. Substituting Equation (9.17) into (9.15) for i = A, ... D yields:

$$RT \ln\left(\frac{p_C^c p_D^d}{p_A^a p_B^b}\right) = -(cg_C^o + dg_D^o - ag_A^o - bg_B^o) = -\Delta G^o \tag{9.18}$$

The free energy change of the reaction, ΔG^o, is the analog of the enthalpy and entropy changes of the reaction introduced in Sections 9.2 and 9.3. These properties of the generic reaction are related by the definition of the free energy:

$$\Delta G^o = \Delta H^o - T\Delta S^o \tag{9.19}$$

Equation (9.18) can be expressed in the alternate fashion:

$$K_p = \frac{p_C^c p_D^d}{p_A^a p_B^b} = \exp\left(-\frac{\Delta G^o}{RT}\right) = \exp\left(\frac{\Delta S^o}{R}\right)\exp\left(-\frac{\Delta H^o}{RT}\right) \tag{9.20}$$

K_P is the equilibrium constant for the generic reaction (9.5). The subscript P on K_p indicates that the concentration units are the partial pressures of the reacting species. The first equality in Equation (9.20) is a form of the law of mass action. This "law" relates the partial pressures in the equilibrium reacting gas to a constant (K_P) that is a function of temperature only. The last equality in Equation (9.20) shows that the basic properties of the reaction are ΔH^o and ΔS^o.

The preceding analysis of the generic reaction of Equation (9.5) can be generalized for any gas-phase reaction. The criterion of equilibrium, Equation (9.16), when combined with the chemical potentials of Equation (9.17) yields:

$$K_P = \frac{\prod_{products} p_i^{\nu_i}}{\prod_{reactants} p_i^{\nu_i}} = \exp\left(-\frac{\Delta G^o}{RT}\right) = \exp\left(\frac{\Delta S^o}{R}\right)\exp\left(-\frac{\Delta H^o}{RT}\right) \tag{9.21}$$

where

$$\Delta H^o = \sum_{products} \nu_i h_i^o - \sum_{reactants} \nu_i h_i^o \tag{9.22}$$

$$\Delta S^o = \sum_{products} \nu_i s_i^o - \sum_{reactants} \nu_i s_i^o \tag{9.23}$$

Example: Check on the calculation of the equilibrium conversion in the reaction: $O_2 + 2H_2 = 2H_2O$ that produced Figure 9.1 using the method introduced in this section. Since the internal entropy difference between the product water and the reactant gases was ignored, $\Delta G^o = \Delta H^o = -4RT$. The equilibrium constant is:

$$K_P = \exp(-\Delta G^o/RT) = \exp(-\Delta H^o/RT) = e^4.$$

The mass action law is:

$$K_P = \frac{p_{H_2O}^2}{p_{O_2}p_{H_2}^2} = \frac{x_{H_2O}^2}{x_{O_2}x_{H_2}^2} \tag{9.21a}$$

The second equality originates from Dalton's rule (Equation [7.3]) and the total pressure of 1 atm. Because the initial H_2/O_2 ratio is 2, the stoichiometry of the reaction $O_2 + 2H_2 = 2H_2O$ gives the additional condition $x_{H_2} / x_{O_2} = 2$. Finally, the sum of the mole fractions is unity, $x_{H_2} + x_{O_2} + x_{H_2O} = 1$. With a little algebra, these three equations are combined to give the solution:

$$\frac{4}{27}K_P = \frac{4e^4}{27} = 8.1 = \frac{x_{H_2O}^2}{(1 - x_{H_2O})^3}$$

Numerical solution of this equation gives $x_{H_2O} = 0.632$. In terms of the fraction reacted, f, the water mole fraction is: $x_{H_2O} = 2f/(3-f)$, or $f = 3x_{H_2O}/(2 + x_{H_2O}) = 0.72$. This is the same as the value of f at the minimum G in Figure 9.1.

9.5.1 Effect of Pressure on Gas-Phase Chemical Equilibria

Instead of partial pressures, mixture compositions are often more conveniently expressed in terms of the mole fractions of the species present, as in the previous example. The following formulation illustrates how total pressure affects the equilibrium composition. Using Dalton's rule (Equation [7.3]), $p_i = x_i p$, where x_i is the mole fraction of species i and p is the total pressure, Equation (9.21) becomes:

$$K = p^{\left(\sum_{reactants} \nu_i - \sum_{products} \nu_i\right)} \times K_P = \frac{\prod_{products} x_i^{\nu_i}}{\prod_{reactants} x_i^{\nu_i}} \tag{9.24}$$

Although K_P is a function of temperature only, the equilibrium constant in terms of mole fractions, K, is also total-pressure dependent. The pressure term multiplying K_P demonstrates Le Chatelier's principle: increasing the total pressure in a gas phase reaction favors the side of the reaction with the fewest number of moles. With respect to Equation (9.24), the pressure effect is determined by the exponent of p; if the sum of the stoichiometric coefficients of the reactants is greater than the sum of the coefficients of the reaction products, increasing pressure increases K, which drives the equilibrium composition to the reaction-product side.

Example: The equilibrium composition of the reaction of Equation (9.1) is independent of pressure because the sum of the balancing numbers is the same on both sides of the equation. According to Equation (9.24), $K = K_P$.

The total pressure effect on the gas-phase reaction: $O_2 + 2H_2 = 2H_2O$ is $p^{3-2} = p$, or $K = pK_P$.

9.5.2 Effect of Temperature on Gas-Phase Chemical Equilibria

Assume that the reaction entropy and enthalpy changes (ΔS^o and ΔH^o) appearing in the last term in Equation (9.21) are independent of temperature. Taking the logarithms of the first and last terms, and differentiating with respect to temperature yields:

$$\frac{d \ln K_P}{d(1/T)} = -\frac{\Delta H^o}{R} \tag{9.25}$$

This useful relation is known as the *Van't Hoff equation.* It can be shown to be valid even if ΔH^o and ΔS^o are temperature dependent (see Problem 9.17). However, the discussions in Sections 9.2 and 9.3 suggested that these reaction properties vary little with temperature, and in most cases can be treated as constants. According to Equation (9.25), a plot of experimental measurements in the form of $\ln K_P$ versus $1/T$ should be a straight line with a slope equal to $-\Delta H^o/R$.

Equation (9.25) is also the source of the useful rules shown in the plots in Figure 9.3.

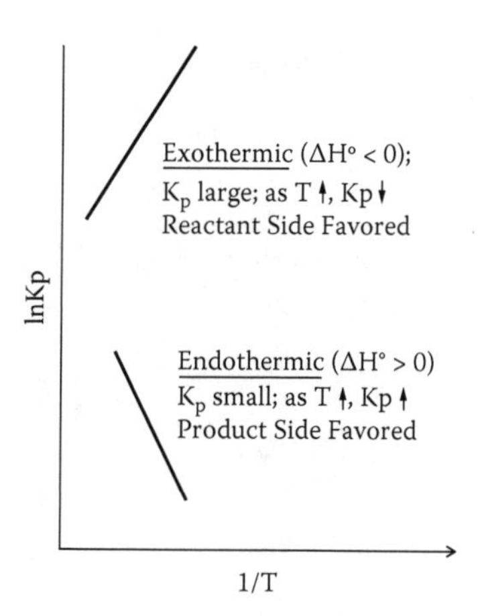

FIGURE 9.3 Effect of temperature on the equilibrium constant.

9.6 SOLVING FOR THE EQUILIBRIUM COMPOSITION

The law of mass action for a particular reaction (Equation [9.24]) is but a single equation with more than one variable. As an example, Equation (9.21a) contains three unknown mole fractions. There are two principal methods for incorporating the conservation equations into the analysis: the "element conservation" method and the "reaction progress variable" method. Both of these methods require the following input information:

- The temperature and total pressure
- ΔH^o and ΔS^o of the reaction

This information fixes K_P by Equation (9.20) and K by Equation (9.24). Thereafter, the two methods differ.

9.6.1 The Element-Conservation Method

Element conservation is expressed as ratios of the number of moles of one element to the number of moles of another element. For a reaction involving N molecular species, $N - 2$ of these mole ratios are required. This method is best explained by applying it to the methane combustion reaction of Equation (9.1). For this reaction, the mass-action law is:

$$K = \frac{x_{CH_4} x_{O_2}^2}{x_{CO_2} x_{H_2O}^2} \tag{A}$$

To solve for the four mole fractions in this equation requires three other equations. The first is the summation of the mole fractions:

$$x_{CH_4} + x_{O_2} + x_{CO_2} + x_{H_2O} = 1 \tag{B}$$

A four-species reaction such as Equation (9.1) must involve three elements. This reaction interchanges C, H, and O among four molecular entities. Element conservation requires that two ratios of the number of moles of the elements, irrespective of their molecular states, be fixed and independent of the extent of reaction. Any two ratios of the three elements can be selected. Choosing the C/H and C/O as specified ratios yields the equations:

$$\left(\frac{C}{H}\right) = \frac{x_{CH_4} + x_{CO_2}}{4x_{CH_4} + 2x_{H_2O}} \qquad \left(\frac{C}{O}\right) = \frac{x_{CH_4} + x_{CO_2}}{2x_{O_2} + x_{H_2O}} \tag{C}$$

The coefficients of the mole fractions in the above ratios are the number of element moles per mole of a molecular species (e.g., 4H in CH_4, 2O in O_2).

The initial molecular forms in the mixture do not affect the equilibrium composition as long as the element ratios are preserved. For example, the element ratios (C/H) = (C/O) = 1/4 apply to an initial (nonequilibrium) mixture of 1 mole of CH_4 and 2 moles of O_2. Alternatively, the same element ratios can be obtained from an initial mixture of 1 mole of CO_2 and 2 moles of H_2O. The same equilibrium composition would result from either of these initial states.

The general approach is to first solve Equations (B) and (C) for three of the mole fractions in terms of the fourth, then substitute these results into Equation (A).

Example: Given $K = 2 \times 10^{10}$; (C/H) = (C/O) = $\frac{1}{4}$

From the (C/H) equation in (C), $x_{H_2O} = 2x_{CO_2}$; from the (C/O) equation, $x_{O_2} = 2x_{CH_4}$

Substituting these into Equation (B) gives $x_{CH_4} = \frac{1}{3} - x_{CO_2}$, so that $x_{O_2} = \frac{2}{3} - 2x_{CO_2}$.

With three mole fractions expressed in terms of x_{CO_2}, Equation (A) becomes:

$$K = \frac{27x_{CO_2}^3}{(1-3x_{CO_2})^3} \qquad \text{or} \qquad x_{CO_2} = \tfrac{1}{3}\left(1 + K^{-\frac{1}{3}}\right)^{-1}$$

Substituting the given value of K into the above yields the CO_2 mole fraction, from which the remaining concentrations are obtained from the previous equations. The results are:

$$x_{CO_2} = 0.3332 \qquad x_{H_2O} = 0.6664 \qquad x_{CH_4} = 1.2 \times 10^{-4} \qquad x_{O_2} = 2.5 \times 10^{-4}$$

As expected from the very large equilibrium constant, reaction (9.1) is driven nearly completely to the right.

This solution is relatively easily solved because:

1. The initial element ratios are stoichiometric, meaning that both of the reactants would disappear if the reaction went to completion.
2. The final equation did not require numerical solution.

9.6.2 The Reaction Progress Variable Method

In this method, the problem is reduced to a single unknown at the outset, rather than at the end as in the element-ratio method described above. The alternative method is illustrated with the gas-phase reaction:

$$2CO(g) + O_2(g) = 2CO_2(g) \tag{9.26}$$

Example: The temperature is 2000 K and the total pressure is $\frac{1}{4}$ atm. The initial mixture is stoichiometric which means consisting of the ratio of reactants that would produce pure product if completely reacted.

The standard enthalpy and entropy changes of this reaction are $\Delta H^o = -564$ kJ/mole and $\Delta S^o = -174$ J/mole-K (note that the entropy rule of thumb given at the end of Section 9.5 is satisfied for this reaction). At 2000 K, the equilibrium constant from Equation (9.20) is:

$$K_P = \exp\left(\frac{-174}{8.314}\right)\exp\left(-\frac{-564 \times 10^3}{8.314 \times 2000}\right) = 4.4 \times 10^5$$

The mass action law in terms of mole fractions is:

$$K = pK_P = \frac{x_{CO_2}^2}{x_{CO}^2 x_{O_2}} = 1.1 \times 10^5 \quad (9.27)$$

The initial gas consists of 1 mole of CO and 0.5 moles of O_2. A table is set up that describes the equilibrium composition in terms of a single variable called the *reaction progress variable*, denoted by ξ. Exact definition of this quantity is somewhat arbitrary; in the present example, the natural choice is the number of moles of the CO_2 product present in the equilibrium mixture. Table 9.1 shows the entries in the table that lead to a single equation to solve.

TABLE 9.1
Reaction Progress Variable Table for the CO Oxidation Reaction

Species	CO	O_2	CO_2	Total
Initial moles	1	0.5	0	1.5
Moles at equilibrium	$1 - \xi$	$0.5 - 0.5\xi$	ξ	$1.5 - 0.5\xi$
Equilibrium mole fractions	$\frac{1-\xi}{1.5-0.5\xi}$	$\frac{0.5(1-\xi)}{1.5-0.5\xi}$	$\frac{\xi}{1.5-0.5\xi}$	1

Substituting the mole fractions in the last row of the table into the mass action law of Equation (9.27) yields the equation:

$$\frac{\xi^2(3-\xi)}{(1-\xi)^3} = 0.25 \times 4.4 \times 10^5 = 1.1 \times 10^5$$

Solving this equation numerically (trial-and-error) yields $\xi = 0.974$ and from the equations in Table 9.1, the mole fractions:

$$x_{CO} = 0.026 \qquad x_{O_2} = 0.013 \qquad x_{CO_2} = 0.961$$

The reaction goes to 96% of completion because the equilibrium constant K is large.

This method can also be applied to a reaction that occurs in a constant-volume vessel rather than at constant pressure by expressing p in terms of ξ for use in Equation (9.27) (see Problem 9.25).

9.7 REACTIONS IN A FLOWING GAS

Flowing a mixture of CO_2 and CO or H_2O and H_2 through a furnace at a temperature high enough for reaction (9.26) to come to equilibrium is a common method of controlling the O_2 partial pressure for a variety of high-temperature experiments

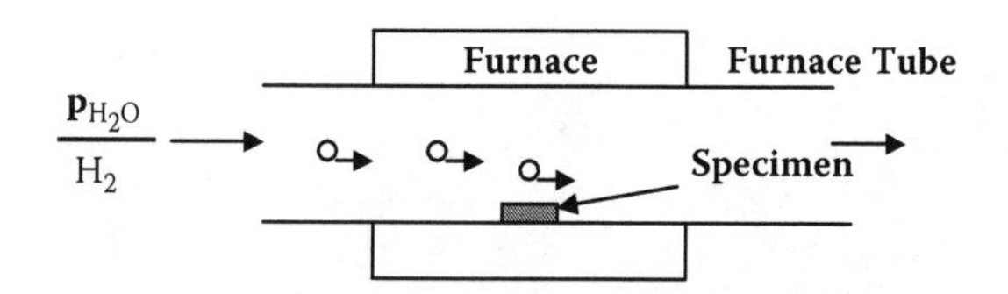

FIGURE 9.4 Gas-flow furnace for generating very small oxygen pressures.

(Figure 9.4). The method is useful when the required oxygen pressures are too small to be reliably maintained in an He/O_2 gas mixture. Typical applications where the oxygen pressure is critical are corrosion testing and control of the oxygen contents of ceramics.

The closed-system analysis can be applied to this open system by considering the gas to consist of small packets (circles in Figure 9.4) in which equilibrium is attained.

Example: The CO_2/CO ratio in the gas entering a furnace at 2000 K is 2 and no O_2 is present. Because the equilibrium concentration of oxygen in the hot zone is quite small, a good approximation is to set the CO_2/CO ratio in the equilibrium gas equal to 2 as well. Using Equation (9.27), this yields the result:

$$p_{O_2} = \frac{(p_{CO_2} / p_{CO})^2}{K_P} = \frac{2^2}{4.4 \times 10^5} = 9.1 \times 10^{-6} \text{ atm} \tag{9.27a}$$

9.8 SIMULTANEOUS GAS-PHASE REACTIONS

When more than one chemical reaction achieves equilibrium, a mass-action law must be written for each. The analytical tool of choice is the reaction-progress-variable method.

Example: Dissociation of water vapor.

At low temperatures, water vapor (steam) does not dissociate, and chemical reactions can be neglected. At high temperature (> ~ 1500 K), dissociation into H_2 and O_2 becomes significant; at very high temperatures, as might occur in the exhaust of a rocket, the hydroxyl radical, OH, appears. The equilibrium reactions governing the concentrations of the four species are:

$$2H_2O(g) = 2H_2(g) + O_2(g) \tag{9.28a}$$

$$2H_2O(g) = H_2(g) + 2OH(g) \tag{9.28b}$$

Assuming the total pressure to be 1 atm, the mass-action expressions for the above equilibria are:

$$K_a = \frac{x_{H_2}^2 x_{O_2}}{x_{H_2O}^2} \qquad K_b = \frac{x_{H_2} x_{OH}^2}{x_{H_2O}^2} \tag{9.29}$$

Two reaction-progress variables are needed. These are chosen as:

$$\xi = O_2 \text{ produced by reaction (9.28a)}$$

$$\psi = H_2 \text{ produced by reaction (9.28b)}$$

TABLE 9.2
Reaction Progress Variable Table for the Water Decomposition Reactions

Species	H_2O	H_2	O_2	OH	Total
Initial moles	1	0	0	0	1
Moles at equilibrium	$1 - 2\xi - 2\psi$	$2\xi + \psi$	ξ	2ψ	$1 + \xi + \psi$
Equilibrium mole fractions	$\frac{1-2\xi-2\psi}{1+\xi+\psi}$	$\frac{2\xi+\psi}{1+\xi+\psi}$	$\frac{\xi}{1+\xi+\psi}$	$\frac{2\psi}{1+\xi+\psi}$	1

Other choices of reaction progress variables are possible, but the calculated equilibrium composition would be unaffected. Assuming that the initial gas is pure water vapor, the setup is shown in Table 9.2. From the definitions above and the balancing numbers in the equilibrium reactions, the moles of OH produced are twice those of H_2 produced. Substituting the equilibrium mole fractions from the table into Equations (9.29) yields:

$$K_a = \frac{(2\xi+\psi)^2\xi}{(1-2\xi-2\psi)^2(1+\xi+\psi)} \tag{9.30a}$$

$$K_b = \frac{(2\xi+\psi)\ (2\psi)^2}{(1-2\xi-2\psi)^2(1+\xi+\psi)} \tag{9.30b}$$

At 3000 K, the equilibrium constants are $K_a = 2.1 \times 10^{-3}$ and $K_b = 2.8 \times 10^{-3}$. Using these values in Equations (9.30a) and (9.30b) and solving simultaneously (using numerical methods) yields $\xi = 0.0540$ and $\psi = 0.0543$. Substituting these values into the last row of Table 9.2 gives the equilibrium composition of the gas:

$$x_{H_2O} = 0.707 \qquad x_{H_2} = 0.146 \qquad x_{O_2} = 0.047 \qquad x_{OH} = 0.098$$

Twenty-nine percent of the steam has decomposed.

Gas phase equilibrium calculations are the subject of Problems 9.1, 9.3, 9.10, 9.11, 9.18, and 9.24. For exothermic reactions, the heat released can influence the equilibrium composition (Problem 9.1). Endothermic reactions such as dissociation of a diatomic gas at high temperature provide an energy-storage mechanism that affects its heat capacity (Problem 9.10).

9.9 REACTIONS BETWEEN GASES AND PURE CONDENSED PHASES

The class of reactions in which an element reacts with a diatomic gas to form a compound is both of practical importance and amenable to simple thermodynamic analysis.

$$M(s) + X_2(g) = MX_2(s) \tag{9.31}$$

Reactions in this category include oxidation, nitriding, and hydriding of metals and halogenation of the electronic material silicon, usually by Cl_2 or F_2. The system is schematized in Figure 9.5.

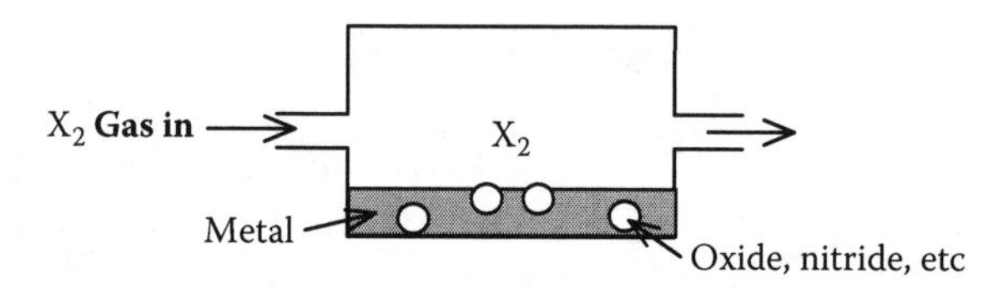

FIGURE 9.5 Metal M and compound MX_2 in environment of X_2 gas.

The simplicity of the thermodynamics stems from the immiscibility of M and MX_2; both are in their pure states, so the chemical potentials are equal to their molar free energies. Consider oxidation of a metal to form a dioxide according to reaction (9.3). The chemical potential of oxygen gas is dependent on its partial pressure, and is described by Equation (9.17). With these simplifications, the equilibrium criterion, Equation (9.16), becomes:

$$g_M^o + \left(g_{O_2}^o + RT \ln p_{O_2}\right) = g_{MO_2}^o \tag{9.32}$$

Rearranging this equation into a more convenient form gives:

$$p_{O_2} = e^{\Delta G^o/RT} = e^{-\Delta S^o/R} e^{\Delta H^o/RT} \tag{9.33a}$$

or

$$RT \ln p_{O_2} = \Delta G^o = \Delta H^o - T\Delta S^o \tag{9.33b}$$

where

$$\Delta G^o = g_{MO_2}^o - g_{O_2}^o - g_M^o \tag{9.34}$$

9.9.1 Implications of the Phase Rule

The metal oxidation system must satisfy the phase rule, Equation (1.21a). Reaction (9.31) involves two components: the elements M and O; or alternatively three species, M, MO_2, and O_2 less one for the equilibrium reaction involving all three. There are three phases: the solids M and MO_2, and O_2 gas. The total pressure is not a degree of freedom because it is equal to the oxygen pressure, which cannot be adjusted independent of temperature. For this system, the phase rule reduces to:

$$\mathbf{F = C + 2 - P} \quad = 2 + 2 - 3 = 1 \tag{9.35}$$

The single degree of freedom is temperature, which determines the oxygen pressure.

Digression 1. The total pressure can be made an additional independent variable by adding an inert gas to the oxygen. However, the inert gas becomes an additional component, so the phase rule for this case is:

$$F = 3 + 2 - 1 = 2$$

The two degrees of freedom are temperature and total pressure. However, total pressure has little effect on the properties of the two solids and none on the O_2 pressure, so can be ignored as an additional degree of freedom for this system.

Digression 2. The oxygen gas could be totally eliminated by completely enclosing the two solids in a tight-fitting, impervious boundary at an external pressure greater than the oxygen pressure. The resulting mixture of two immiscible solid phases would still possess a virtual O_2 pressure even though no gas phase is present. This interpretation of the O_2 pressure exerted by the M + MO_2 mixture (often termed a couple) is the reason that $RT\ln p_{O_2}$ in Equation (9.33b) is called the *oxygen potential* of the M + MO_2 couple.

As the temperature is varied, the oxygen pressure, or equivalently, the oxygen potential, follows by either of Equations (9.33). Thus p_{O_2} is to be interpreted as a property of the M + MO_2 couple, analogous to the vapor pressure of a pure substance.

9.9.2 Stability Diagrams

Equations (9.33a) and (9.33b) are plotted in Figure 9.6. These plots are called *stability diagrams* because the lines separate regions in which only one of the two phases is present. The line represents the p_{O_2}-T combinations where both the metal and its oxide coexist. The oxide-metal stability diagram is similar to the p-T phase diagram of a single substance such as water, where lines separate existence regions of solid, liquid, and vapor phases (see Figures 5.1 and 5.3).

The zones above and below the lines in Figure 9.6 represent regions in which Equation (9.31) is not in equilibrium. If a gas phase containing an O_2 pressure above the line is imposed on an initial M + MO_2 mixture, all M will be oxidized and only MO_2 will remain. Conversely, $p_{O_2} - T$ points below the lines represent conditions where only M is stable.

Note that the slope of the line in Figure 9.2b is positive, which, from Equation (9.33b), implies that ΔS^o is negative. This is consistent with the rule of thumb for

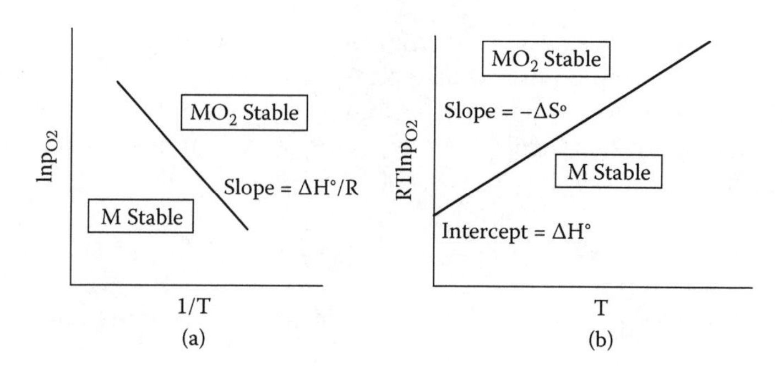

FIGURE 9.6 Two forms of the stability diagram for the M + MO_2 couple.

ΔS^o discussed in Section 9.3; the reactant side of the metal oxidation equation has one gaseous molecule and the product side none. Removal of the gas by the reaction results in a decrease in system entropy. Thus $\Delta S^o < 0$, making the slope of the line described by Equation (9.33b) positive.

9.9.3 Oxygen Isobars on a Phase Diagram

Phase diagrams are a convenient vehicle for displaying the equilibrium oxygen pressures generated by a metal and its oxides. In elements with multiple oxidation states and/or crystal structures, the equilibrium may not involve only the metal and an oxide, as in the MO_2/M couple discussed in the previous section. In particular, two-phase regions separating two different oxides are represented by reactions of the following type:

$$\mathbf{w}M_mO_n + \mathbf{z}O_2 = M_aO_b$$

The stoichiometric coefficients **w** and **z** are determined by balances on the elements M and O:

$$\text{for M:} \qquad \mathbf{w}m = a, \qquad \text{or} \qquad \mathbf{w} = a/m$$

$$\text{for O:} \qquad \mathbf{w}n + 2\mathbf{z} = b, \qquad \text{or} \qquad \mathbf{z} = \tfrac{1}{2}\left[b\text{-}an/m\right]$$

The integers m, n, a and b characterize the two oxides. The equilibrium oxygen pressure is dependent on the free energy of formation of the above reaction:

$$p_{O_2} = \exp(\Delta G^o/zRT)$$

p_{O_2} is actually a surface in the third dimension of the temperature-composition plane; it is a more complex version of the equation of state surface of water (Figure 2.8). As in the case with the EOS of water, projection of the p_{O_2} surface onto the T, O/M plane permits semiquantitative use of the information contained in the surface.

Figure 8.15 displays the p_{O_2} projection onto the Fe–O phase diagram. This diagram is the two-component analog of the familiar diagrams for a single-component substance such as water (Figure 2.9). In single-phase regions, p_{O_2} is a function of T and O/Fe; in the two-phase zones, p_{O_2} is a function of temperature only (the phase rule: add a phase, lose a degree of freedom).

In the four two-phase zones in Figure 8.15 the oxygen pressure is controlled by the following equilibria:

In the iron-rich region:

$$3Fe(\alpha) + 2O_2 = Fe_3O_4 \text{ (magnetite)}$$

$$(1\text{-}x)Fe(\alpha \text{ or } \gamma) + {}^1/_2\, O_2 = Fe_{1\text{-}x}O \text{ (wustite)}$$

In the oxygen-rich region:

$$\frac{3}{1-x} Fe_{1-x}O\text{(wustite)} + \frac{1-4x}{2(1-x)} O_2 = Fe_3O_4\text{(magnetite)}$$

$$\frac{2}{3-y} Fe_{3-y}O_4\text{ (magnetite)} + \frac{1-3y}{2(3-y)} O_2 = Fe_2O_3\text{(hematite)}$$

The stoichiometric coefficients in the above reactions were determined by the method described at the beginning of this section.

9.9.4 Reactive Gas in Contact with a Reactive Metal

The values of the O_2 pressure required for coexistence of M and MO_2 are usually quite small, because, except for the noble metals, oxides are much more stable than the elemental metals. Reaction (9.3) releases substantial heat, so ΔH^o is large and negative. This term dominates ΔG^o, which is also large and negative. For instance, if $\Delta G^o = -200$ kJ/mole at 1000 K, Equation (9.33a) gives $p_{O_2} = 3.6 \times 10^{-11}$ atm. From practical considerations, such a low pressure of O_2 is difficult to produce and control in a process or an experiment. However, oxygen pressures in this range can be reliably established by exploiting the equilibria of gas mixtures that exchange O_2 in a reaction. Gas-phase equilibria of this type include $2CO + O_2 = 2CO_2$ and $2H_2 + O_2 = 2H_2O$. Mixtures with a preset CO_2/CO ratio or H_2O/H_2 ratio generate oxygen pressures in the desired range (Section 9.7). How this O_2 pressure affects the oxidation of a metal is explained below.

By "reactive gas" is meant a mixture of two gases that fix the partial pressure of a third species by an equilibrium reaction. The CO/CO_2 combination that establishes an oxygen partial pressure by reaction (9.26) is an example of a reactive gas.

A "reactive metal" in the present discussion is one that, together with one of its oxides, fixes an equilibrium partial pressure of oxygen by reaction (9.3).

When the reactive gas contacts the reactive metal, the equilibrium can be expressed in one of two ways.

The first way is by the overall reaction:

$$M(s) + 2CO_2(g) = MO_2(s) + 2CO(g) \tag{9.36}$$

The law of mass action for reaction (9.36) is:

$$K_A = \left(p_{CO} / p_{CO_2}\right)^2 = \exp(-\Delta G_A^o / RT) \tag{9.37}$$

The implication of this equation is that only a unique CO/CO_2 ratio at each temperature permits M and MO_2 to coexist.

The second way of expressing the equilibrium of the gas-solid reaction is to equate the oxygen pressure generated by each:

$$\left(p_{O_2}\right)_{CO/CO_2} = \left(p_{O_2}\right)_{MO_2/M} \tag{9.38}$$

That the equilibrium expressions of Equations (9.37) and (9.38) are equivalent is demonstrated as follows. Assume that the mixed CO/CO_2 gas flows over the solid in a furnace, as in Figure 9.5. The equilibrium oxygen pressure in the gas phase is given by Equation (9.27a) (for 2000 K), or in general, by:

$$\left(p_{O_2}\right)_{CO/CO_2} = (p_{CO_2} / p_{CO})^2 \exp\left(\Delta G_{CO/CO_2}^o / RT\right) \tag{9.39}$$

The equilibrium O_2 pressure generated by the MO_2/M couple is a function of temperature only:

$$\left(p_{O_2}\right)_{MO_2/M} = \exp\left(\Delta G_{MO_2/M}^o / RT\right) \tag{9.40}$$

Equating (9.39) and (9.40) yields:

$$\left(p_{CO}/p_{CO_2}\right) = \exp\left[(\Delta G_{CO/CO_2}^o - \Delta G_{MO_2/M}^o) / RT\right] \tag{9.41}$$

Comparing Equations (9.37) and (9.41) shows that:

$$\Delta G_A^o = 2\left(\Delta G_{MO_2/M}^o - \Delta G_{CO/CO_2}^o\right) \tag{9.42}$$

Reaction (9.36) is simply reaction (9.3) minus reaction (9.26). This algebraic relation of the overall reaction to its component reactions also applies to the free-energy changes. This equivalence demonstrates that the standard approach of Equation (9.37) is no different from equating oxygen partial pressures, as in Equation (9.38).

Example: Powdered nickel metal is contacted with a flowing mixture of CO_2 and CO at 1 atm total pressure in a furnace at 2000 K. The quantity of metal is limited, but because of continual flow, the quantity of the gas mixture is unlimited. Therefore, the oxygen pressure established in the gas phase is imposed on the metal, and determines whether or not it oxidizes.

(a) At what CO_2/CO ratio do both Ni and NiO coexist?

The O_2 pressure established by the gas mixture is given by:

$$\left(p_{O_2}\right)_{gas} = \frac{(p_{CO_2} / p_{CO})^2}{K_P}$$

with $K_P = 4.4 \times 10^5$.

The nickel/nickel oxide equilibrium reaction is: $2Ni + O_2 = 2NiO$, for which $\Delta G^o = -46$ kJ/mole at 2000 K. The condition for coexisting Ni and NiO is an oxygen pressure of:

$$\left(p_{O_2}\right)_{solid} = \exp\left(\frac{\Delta G^o}{RT}\right) = \exp\left(\frac{-46\times 10^3}{8.314\times 2000}\right) = 6.3\times 10^{-2} \text{ atm}$$

The mixed solid and the mixed gas are in equilibrium when $(p_{O_2})_{gas} = (p_{O_2})_{solid}$. From the above equations, this condition yields the required ratio of CO_2 to CO in the gas:

$$p_{CO_2} / p_{CO} = \sqrt{K_P\left(p_{O_2}\right)_{solid}} = \sqrt{\left(4.4\times 10^5\right)\left(6.3\times 10^{-2}\right)} = 166$$

(b) What is the solid phase if the CO_2/CO ratio is 1?

For this ratio in the gas, the gas-phase law of mass action gives:

$$\left(p_{O_2}\right)_{gas} = \frac{1}{K_P} = \frac{1}{4.4\times 10^5} = 2.3\times 10^{-6} \text{ atm}$$

Since $(p_{O_2})_{gas} < (p_{O_2})_{solid}$, the gas removes oxygen from the solid and only Ni remains.

A number of problems involving reactions between gases and pure solids are provided at the end of this chapter. Most deal with metal-metal oxide couples in which the oxygen pressure is controlled by CO_2/CO mixtures (Problems 9.5, 9.6, and 9.20). In some applications, more than one solid oxide must be considered (Problem 9.6). In other systems, one or more metal oxides are gaseous (Problems 9.15 and 9.23).

In this class of reactions, the solid phases need not be metals and metal oxides for the theory to apply. In Problem 9.16, sulfur replaces oxygen as the element combining with the metal. In Problem 9.2, the solid is pure carbon and the gas phase is a mixture of hydrogen and methane. This problem is another example of coupling of a gas-phase reaction with a gas-solid reaction. It is analogous to the simultaneous gas-phase equilibrium involving CO and CO_2 and the gas-solid equilibrium between oxygen and a metal-metal oxide couple.

9.10 REACTIONS INVOLVING SOLUTIONS

The only difference between chemical reactions in which liquid or solid reactants and/or products are in solution and the pure-component reactions treated in Section 9.9 is the modification of the chemical potential of the components in solution. The chemical potential for dissolved species is given by Equation (7.29):

$$\mu_i = g_i^o + RT \ln a_i \tag{9.43}$$

If species i were pure, its activity would be unity, and the condition $\mu_i = g_i^o$ used in Section 9.9 would be recovered. For most solutions, g_i^o in Equation (9.43) is the pure-substance value. This assignment assumes that the physical state of the species in solution and in the pure state are the same. For example, an aqueous alcohol solution and pure alcohol are both liquids. On the other hand, a gas such as hydrogen dissociates upon entering a metal, and the H atoms are part of a solid lattice. Pure solid hydrogen would be an inappropriate standard or reference state. This complexity is taken up in the following section.

9.10.1 Solution of a Reactant Species in an Inert Solvent

With these preliminaries, the metal oxidation example used in the preceding section is modified by dissolving reactant metal M in another metal P (not phosphorus) that is inert and does not react with oxygen. To account for the presence of M in this alloy, the equilibrium reaction reads:

$$M(sol'n) + O_2 = MO_2(s) \tag{9.44}$$

Using Equation (9.43) for the chemical potential of M, Equation (9.32) is replaced by:

$$\left(g_M^o + RT \ln a_M\right) + \left(g_{O_2}^o + RT \ln p_{O_2}\right) = g_{MO_2}^o \tag{9.45}$$

a_M is related to the mole fraction of M in the alloy and its activity coefficient according to Equation (7.30). However, for simplicity of notation, the activity is retained. Solving the above equation for the oxygen pressure gives the solution equivalent of Equation (9.33a):

$$a_M p_{O_2} = e^{\Delta G^o / RT} \tag{9.46a}$$

or, in terms of the oxygen potential:

$$RT \ln p_{O_2} = \Delta G^o - RT \ln a_M \tag{9.46b}$$

In these equations, the meaning of ΔG^o is the same as given by Equation (9.34).

Since the activity of M in the alloy must be less than unity, Equation (9.46) shows that the oxygen pressure, or the oxygen potential, is increased by dilution of M with the inert species P.

Although the solvent metal P is inert chemically, it constitutes an additional component as far as the phase rule is concerned. Applying Equation (9.35) with $P = 3$ (as before) but changing C to 3 (M, P and O) gives $F = 3 + 2 - 3 = 2$. The two degrees of freedom are temperature and composition, the latter being the mole fraction of M in the alloy. The stability diagram in Figure 9.7 for oxidation of M in the alloy consists of a family of lines lying above but parallel to the line for pure M. Each line represents a different activity (i.e., mole fraction) of M in the M-P alloy.

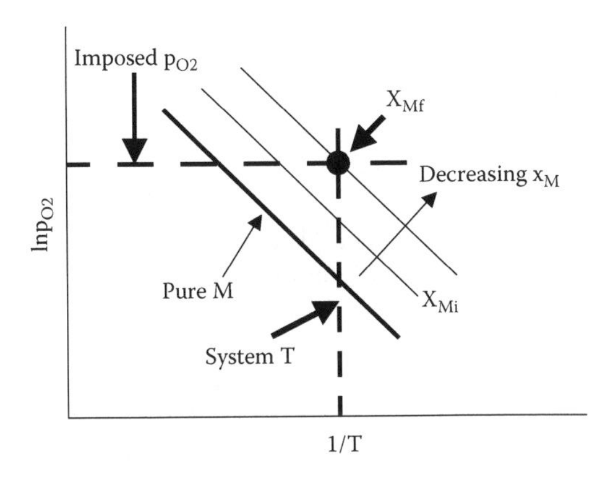

FIGURE 9.7 Stability diagram for reactive metal M in an alloy with inert metal P.

The lowest line is unalloyed M. As the mole fraction of M in the alloy decreases, the lines shift upward but remain parallel. As an example of the use of the diagram, consider an alloy with mole fraction of M equal to x_{Mi}. If an oxygen pressure is imposed on the M-P alloy and the system temperature is fixed, the equilibrium mole fraction of M is reduced to x_{Mf}. Some of the M in the initial alloy becomes oxidized and converted to MO_2 until the required reduction in the concentration of M in solution is achieved.

9.10.2 Reactions in Solution with Two Reactive Species

If the alloying component P is chemically active and forms solutions with M in both the reactant and product phases, the equilibria involved are:

$$M(melt) + O_2(g) = MO_2(slag) \tag{9.47a}$$

$$P(melt) + O_2(g) = PO_2(slag) \tag{9.47b}$$

To distinguish the two solutions, the metal alloy is termed the *melt* and the mixed oxide is called the *slag*. At the temperature of the reaction, the standard free energy changes of the two reactions are ΔG_M^o and ΔG_P^o. Assuming for simplicity that both solutions are ideal, the activities can be replaced by mole fractions and the mass action laws for reactions (9.47a) and (9.47b) are:

$$K_M = e^{-\Delta G_M^o/RT} = \frac{x_M^{slag}}{x_M^{melt} p_{O_2}} \qquad K_P = e^{-\Delta G_P^o/RT} = \frac{x_P^{slag}}{x_P^{melt} p_{O_2}} \tag{9.48}$$

The solution method starts by dividing the above equations to eliminate the oxygen pressure, yielding:

$$\frac{K_M}{K_P} = \frac{x_M^{slag} x_P^{melt}}{x_P^{slag} x_M^{melt}} \tag{9.49}$$

Allowing for chemical reactivity of both species has not altered the number of components ($C = 3$) or the number of phases ($P = 2$), so there are still two degrees of freedom. The first is the temperature and the second is the composition of either one of the two liquid phases. If one is known, the other follows from Equation (9.49), and so is not an independently-variable quantity.

To complete the mathematical solution for the compositions of the two phases, the relative amounts of M, P, and O need to be specified. These mole relationships are most readily followed by combining Equations (9.47a) and (9.47b) into the single reaction:

$$\text{M(melt)} + \text{PO}_2\text{(slag)} = \text{MO}_2\text{(slag)} + \text{P(melt)} \tag{9.50}$$

Suppose the fixed M/P mole ratio is unity and O/P = 2. This is equivalent to charging one mole each of M and PO_2 to the system and not permitting exchange of oxygen with the surroundings. The reaction progress variable ξ is defined as the number of moles of MO_2 (slag) and P(melt) at equilibrium. The moles of M remaining in the melt and the moles of PO_2 in the slag are both equal to $1 - \xi$. By this choice of the initial charge, the total number of moles in each phase remains constant at unity as the reaction moves towards equilibrium. Thus, the number of moles in each phase is equal to the mole fraction, and Equation (9.49) becomes:

$$\frac{K_M}{K_P} = \frac{\xi \times \xi}{(1-\xi)\times(1-\xi)} = \left(\frac{\xi}{1-\xi}\right)^2 \tag{9.51}$$

from which ξ can be determined for specified values of K_M and K_P. With ξ known, the mole fractions in each phase are fixed. The virtual oxygen pressure of the system follows from either of Equations (9.48).

Contrary to the single-metal reactant case, the two-metal reaction system does not exhibit stability diagrams such as those shown in Figure 9.7, or the modified version for nonreactive component P. An imposed oxygen pressure different from the intrinsic oxygen pressure simply dictates the relative amounts of slag and melt according to Equations (9.48).

Example: The temperature is 1000 K, $\Delta G_M^o = -200$ kJ/mole and $\Delta G_P^o = -250$ kJ/mole. The initial charge is 1 mole each of M and PO_2 (or any combination of the metals and their oxides with element mole ratios M/P = 1 and O/P = 2). In Equation (9.48), the equilibrium constants are $K_M = 2.8 \times 10^{10}$ and $K_P = 1.15 \times 10^{13}$. Using these values in Equation (9.51) and solving for the reaction-progress variable at equilibrium gives $\xi = 0.047$. The mole fractions in the melt and slag at equilibrium are:

$$x_M^{slag} = x_P^{melt} = 0.041 \qquad x_P^{slag} = x_M^{melt} = 0.953$$

Using these compositions in either of Equations (9.48) gives $p_{O_2} = 1.7 \times 10^{-12}$ atm. This value is about a factor of 20 lower than the oxygen pressure in equilibrium with the pure M/pure MO_2 couple. The reason is the presence of element P, which has a lower free energy of formation of its oxide than does element M. Element P binds oxygen more strongly than element M, thereby lowering the equilibrium, or intrinsic oxygen pressure, of the system. The stronger oxygen-gettering ability of element P forces 95% of M to remain as a metal while 95% of P forms the oxide and moves into the slag phase.

There is no unique oxygen pressure at which the system is all metal or all oxide, as there is in the M/MO_2 couple. If the imposed oxygen pressure is greater than 1.7×10^{-12} atm, more of the metal oxidizes and the slag becomes somewhat richer in MO_2 than in the case above where the restriction was an oxygen-to-total metal mole ratio of unity.

Problems 9.7 to 9.9 explore several variations of the above melt-slag example. Problem 9.4 treats the industrial process for production of uranium metal by a reaction similar to Equations (9.47) except that fluorides rather than oxides are involved.

9.11 THERMOCHEMICAL DATABASES

The reaction analyses in the previous sections required equilibrium constants K that in turn were determined by the free energy changes ΔG^o. Whether the reaction involves gaseous species, condensed phases, or both, the connection between K and ΔG^o is:

$$K_P = \exp(-\Delta G^o/RT) \tag{9.52}$$

ΔG^o is independent of total pressure because it refers to reactants and products in pure, standard states at 1 atm. However, ΔG^o is a function of temperature, mainly via the relation:

$$\Delta G^o = \Delta H^o - T\Delta S^o \tag{9.53}$$

The linear temperature dependence in this equation accounts for the major portion of the variation of reaction free energy change with T. The reaction enthalpy change ΔH^o and the reaction entropy change ΔS^o are much less temperature sensitive, and in many cases can be taken as constant properties of the reaction (see Sections 9.2 and 9.3).

9.11.1 Standard Free Energy of Formation

Even though the thermochemical database need contain only ΔG^o (or, equivalently, ΔH^o and ΔS^o), the number of reactions that would have to be included in such a compilation would be intractably large. The key to reducing data requirements to manageable size is to provide the standard free energy changes of forming the individual molecular species from their constituent elements. Particular reactions are constructed from these so-called *formation reactions.*

For molecular compounds containing two or more elements, the basic information is the free energy change for reactions by which the compound is created from its constituent elements, the latter in their normal state at the particular temperature. These reaction free energy changes are called *standard free energies of formation* of the compound. For example, the methane combustion reaction (9.1) involves one elemental compound (O_2) and three molecular compounds (CH_4, CO_2, and H_2O). The formation reactions for the latter and their associated free energy changes are:

$$C(s) + 2H_2(g) = CH_4(g) \qquad \Delta G^o_{1a} \tag{9.1a}$$

$$C(s) + O_2(g) = CO_2(g) \qquad \Delta G^o_{1b} \tag{9.1b}$$

$$H_2(g) + \tfrac{1}{2}O_2(g) = H_2O(g) \qquad \Delta G^o_{1c} \tag{9.1c}$$

No formation reaction is needed for the O_2 reactant because this species, although molecular in form, is in its normal elemental state. The above formation reactions contain species not included in Equation (9.1), namely C(s) and $H_2(g)$. However, when the above reactions are combined algebraically to produce reaction (9.1) by the formula:

$$(9.1) = 2x\ (9.1c) + \ (9.1b) - \ (9.1a)$$

the extraneous species cancel out. Similarly, the free energies of formation are combined to produce the free energy change of reaction (9.1):

$$\Delta G^o_1 = 2\Delta G^o_{1c} + \Delta G^o_{1b} - \Delta G^o_{1a} \tag{9.54}$$

In this fashion, a universal chemical thermodynamic database can be constructed from formation free energies of those compounds that have been investigated experimentally.

9.11.2 Graphical Representation: Ellingham Diagrams

The most compact display of free energies of formation is a plot of ΔG^o versus T. According to Equation (9.53), if ΔH^o and ΔS^o are independent of temperature, such a plot is a straight line with a slope equal to $-\Delta S^o$ and an intercept (at $T = 0$ K) of ΔH^o. Such lines are shown in Figure 9.6b for formation of a metal oxide from the elemental constituents. A family of these lines for a number of metals with a common nonmetal (e.g., oxygen, chlorine) is called an *Ellingham diagram.*

In order to permit comparison of different metal-metal oxide couples in the diagram, the formation reactions are written on a per-mole-O_2 basis, rather than for one mole of the oxide. The general formula of a metal oxide is M_aO_b, where the integers a and b depend on the oxidation state (i.e., the valence) of the metal ion in the oxide. For all oxides, the valence of the oxygen ion is –2. The values of a and b must satisfy electrical neutrality of the crystal. Starting from the lowest-valence metal, the oxide formation reactions included in the Ellingham disgram are:

Monovalent (M = Ag, Cs,...)	$4M + O_2 = 2M_2O$
Divalent (M = Mg, Ba,...)	$2M + O_2 = 2MO$
Trivalent (Al, La,...)	$4/3M + O_2 = 2/3\ M_2O_3$
Tetravalent (Zr, U, ...)	$M + O_2 = MO_2$
Pentavalent (Nb, V, ...)	$4/5M + O_2 = 2/5M_2O_5$
Hexavalent (Mo, U, ...)	$2/3M + O_2 = 2/3MO_3$

The tetravalent metal reaction was treated in detail in Section 9.9. The general reaction is:

$$\frac{2a}{b} M + O_2 = \frac{2}{b} M_aO_b \tag{9.55}$$

Since the metal and its oxide are pure in formation reactions, the analysis of Section 9.6 applies; the free energy of formation of the oxide is identical to the oxygen potential of the M/M_aO_b couple:

$$RT \ln\left(p_{O_2}\right)_{M/M_aO_n} = \Delta G^o_{M/M_aO_b} \tag{9.56}$$

Figure 9.8 provides a direct visual assessment of the relative stabilities of the oxides of the metals. The metal-oxide couple with the lowest oxygen pressure is the most stable. According to Equation (9.56), this condition is equivalent to assigning oxide stability to the metal whose oxide has the most negative ΔG^o.

Example: Mixing calcium metal with uranium dioxide at any temperature results in complete reduction of UO_2 by the reaction:

$$UO_2(s) + 2Ca(s) \Rightarrow 2CaO(s) + U(s\ \text{or}\ L)$$

The arrow separating reactants and products in place of an equal sign indicates that the reaction is not an equilibrium reaction. With three components and four phases, the phase rule allows this system no degrees of freedom. The only condition that would permit equilibrium of the above reaction is $\Delta G^o_{U/UO_2} = \Delta G^o_{Ca/CaO}$. However, Figure 9.8 shows that the Ca/CaO line does not intersect the U/UO_2 line. Hence Ca and UO_2 react until one of the two is consumed.

Problems 9.12 and 9.21 deal with the conditions for equilibrium in other systems containing two different metals and their oxides.

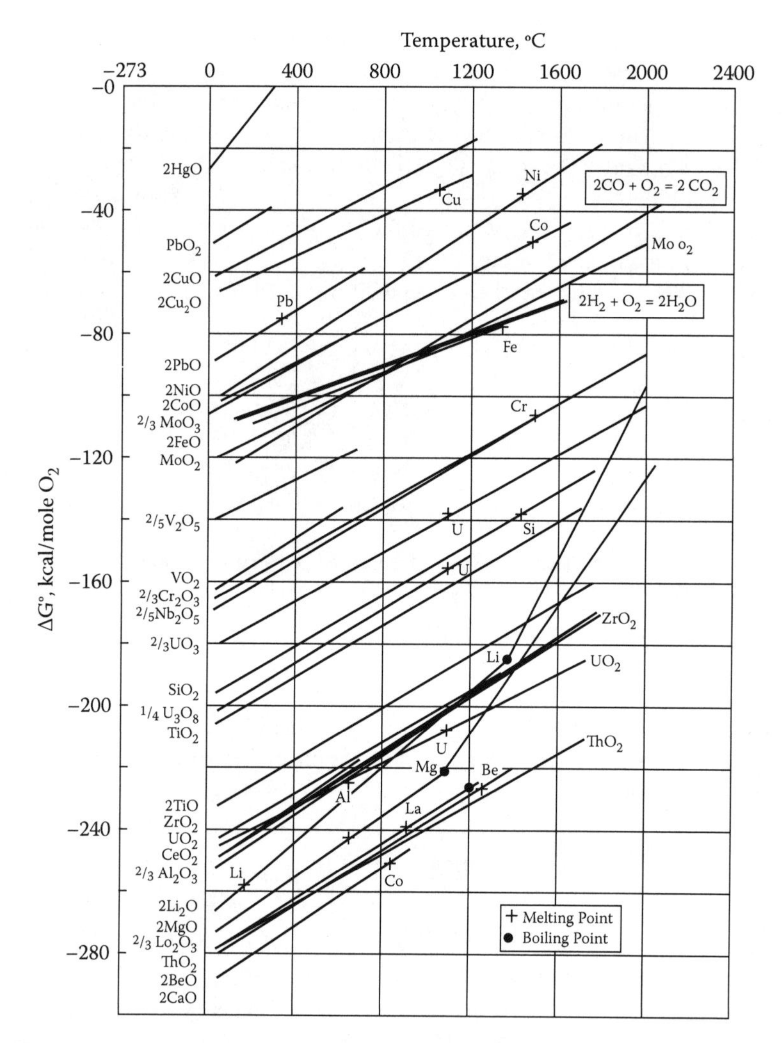

FIGURE 9.8 Ellingham diagram for the free energy of formation of oxides.

The slopes of the lines in Figure 9.8 are approximately equal, indicating that all reactions have roughly the same value of ΔS^o. The reason for this uniformity of slope is the entropy rule of thumb (Section 9.3), illustrated by Equation (9.14) for dioxides. Applying this rule to the general oxidation reaction of Equation (9.55) suggests that the slopes of the lines in Figure 9.8 should be ~ 200 J/mole-K, which is the molar entropy of one mole of O_2(g). The actual slopes in Figure 9.8 range from slightly more than 200 J/mole-K for simple oxides such as NiO and CaO to ~ 170 J/mole-K for more complex oxides such as UO_2 and ZrO_2.

The relative vertical positions of the lines in the diagram are a direct consequence of their enthalpies of formation, ΔH^o, which are the intercepts at 0 K (–273°C). The most exothermic reaction is $2Ca + O_2 = 2CaO$ and the least exothermic is $2Hg + O_2 = 2HgO$.

Several metals exhibit multiple valence states; Figure 9.8 contains lines for TiO and TiO_2, Cu_2O and CuO, and three oxides of uranium, UO_2, U_3O_8, and UO_3.* The higher the oxide (i.e., the larger the oxygen-to-metal ratio), the larger (less negative) is the free energy of formation. That is, the line for Ti/TiO_2 lies above that for Ti/TiO.

Example: Over what range of oxygen pressures is Cu_2O the stable oxide of copper?

There are two Ellingham lines for copper oxides. The lowest oxide is represented by the equilibrium reaction:

$$4Cu + O_2 = 2Cu_2O$$

for which Figure 9.8 gives $\Delta G_{Cu/Cu_2O} = -55$ kcal/mole (– 230 kJ/mole) at 400°C (673 K). At this temperature, the oxygen pressure of the Cu/Cu_2O couple is:

$$\left(p_{O_2}\right)_{Cu/Cu_2O} = \exp\left(\frac{-230\times10^3}{8.314\times673}\right) = 1.4\times10^{-18} \text{ atm}$$

Oxygen pressures less than this value permit only the metal to exist. Cu_2O is the stable oxide from the above O_2 pressure up to an O_2 pressure corresponding to the equilibrium reaction:

$$2Cu_2O + O_2 = 4CuO$$

The standard free energy of formation for this reaction is obtained by subtracting the Cu/Cu_2O equilibrium reaction from the Cu/CuO formation reaction:

$$2Cu + O_2 = 2CuO$$

From Fig. 9.8,

$$\Delta G^o_{Cu/CuO} = -48 \text{ kcal/mole } (-203 \text{ kJ/mole})$$

However, this reaction does not represent an equilibrium situation; Cu and CuO can never coexist because the lower oxide Cu_2O intervenes. Nonetheless, the standard free energy change for the Cu_2O/CuO equilibrium can be obtained from:

$$\Delta G^o_{Cu_2O/CuO} = 2\Delta G^o_{Cu/CuO} - \Delta G^o_{Cu/Cu_2O} = 2\times(-203) - (-230) = -176 \text{ kJ/mole}$$

so that:

$$\left(p_{O_2}\right)_{Cu_2O/CuO} = \exp\left(\frac{-176\times10^3}{8.314\times673}\right) = 2.2\times10^{-14} \text{ atm}$$

Therefore, at 400°C, Cu_2O is the sole oxide present in atmospheres with oxygen pressures between 1.4×10^{-18} and 2.2×10^{-14} atm. In O_2 pressures larger than the upper limit of this interval, only CuO exists.

---

* Uranium ions in UO_2 and UO_3 have valences (or oxidation states) of 4+ and 6+, respectively. The intermediate oxide U_3O_8 has a mean valence of 5.33. Since ionic charges must be integers, this oxide contains 2/3 U^{4+} and 1/3 U^{6+}. This combination is more stable than the pure 5+ oxide, U_2O_5, which does not exist.

The lines in Figure 9.8 are straight. The absence of curvature indicates that ΔH^o and ΔS^o are approximately constant. The effect of the different values of the specific heats of the products and reactants (ΔC_P) discussed in Sections 9.2 and 9.3 have either been neglected or are too small to cause appreciable nonlinearity in the Ellingham lines. This effect is included in the next subsection.

The effect of a phase change of the metal is to make ΔS^o more negative by adding entropy to the reactant side of the reaction (see the M curve in Figure 9.2). Since the slope of an Ellingham line is $-\Delta S^o$, the slopes increase discontinuously at the melting point T_M and at the boiling point T_B (i.e., when the vapor pressure equals 1 atm). The magnitude of the increase in the slope at the melting point is $\Delta h_M/T_M$. Because Δh_M is small, the change in slope at T_M is not noticeable at the plus signs on the lines in Figure 9.8.

Two metals, Li and Mg, have boiling points that fall within the temperature range of Figure 9.8. When the metals vaporize, the entropy change is $\Delta h_{vap}/T_B$. Since Δh_{vap} is typically an order of magnitude larger than Δh_M, the slope change due to the entropy of vaporization is clearly visible on the lines for these two metals.

Figure 9.8 includes lines for the CO/CO_2 equilibrium (Equation (9.26)) and the analogous equilibrium for H_2/H_2O mixtures. These reactions differ from the metal-metal oxide lines in one important aspect. The oxygen potentials of the $M\text{-}M_aO_b$ couples are given by Equation (9.56). The oxygen potentials of the two all-gas reactions, on the other hand, also depend on the molar ratio CO_2/CO (or H_2O/H_2), as shown in the example in Section 9.6.2. The lines for these two systems are $\Delta G^o_{CO/CO_2}$ and $\Delta G^o_{H_2/H_2O}$ vs. T. The two all-gas reactions are included in the Ellingham diagram because they are often needed in conjunction with the metal-metal oxide data in practical applications.

Example: In Section 9.9.4, the example gave $K_P = 4.4 \times 10^5$ at 2000 K for the CO/CO_2 equilibrium. Compare this to the value obtained from the Ellingham diagram.

At 2000 K on the CO/CO_2 line, $\Delta G^o_{CO/CO_2} = -43$ kcal/mole; using Equation (9.52) gives $K_P = 6 \times 10^4$, which is a factor of 8 lower than the value used in the example in Section 9.9.4. The discrepancy is probably due to the linear approximation for the CO/CO_2 data in Figure 9.8.

9.11.3 Analytic Representation

Some databases give free energies of formation by equations of the type:

$$\Delta G^o = A + BT \ln T + CT \tag{9.57}$$

where A, B, and C are constants. The second term on the right-hand side originates in the difference in the heat capacities between products and reactants. The effect of phase changes is not part of Equation (9.57); these are accounted for by restricting the coefficients to temperature intervals where all species maintain the same phase. Neglecting phase changes, substituting Equations (9.8) and (9.12) into Equation (9.53) gives Equation (9.57), with the coefficients identified as:

TABLE 9.3
Selected Standard Free Energies of Formation

Reaction	A, J/mole	B, J/mole-K	C, J/mole-K	Temp. range, K
$4Ag(s) + S_2(g) = 2Ag_2S(s,\alpha)$	–187,000	—	92	298–452
$4Ag(s) + S_2(g) = 2Ag_2S(s,\beta)$	–176,000	—	69	452–1115
$4Ag(L) + S_2(g) = 2Ag_2S(L)$	–217,000	—	102	1234–1500
$2Al(s) + Cl_2(g) = 2AlCl(g)$	–89,400	21	–320	298–933
$2Al(L) + Cl_2(g) = 2AlCl(g)$	–108,000	21	–301	933–2000
$Al(s) + 3/2Cl_2(g) = AlCl_3\ (g)$	–578,000	4	20	453–933
$Al(L) + 3/2Cl_2(g) = AlCl_3(g)$	–587,000	4	29	933–2000
$2Al(s) + 3Cl_2(g) = Al_2Cl_6(g)$	–1,274,000	9	18	453–933
$2Al(s) + 1.5O_2(g) = Al_2O_3(s)$	–1,677,000	–7	367	298–923
$2Al(L) + 1.5O_2(g) = Al_2O_3(s)$	–1,698,000	–7	386	923–1800
$2Al(s) + N_2(g) = 2AlN(s)$	–644,000	—	186	298–923
$4Al(s) + 3C(s) = Al_4C_3(s)$	–216,000	—	42	298–923
$4Al(L) + 3C(s) = Al_4C_3(s)$	–267,000	—	96	923–2000
$2Ba(s) + O_2(g) = 2BaO(s)$	–1,136,000	—	194	298–983
$2Ba(L) + O_2(g) = 2BaO(s)$	–1,163,000	—	223	983–1600
$3Ba(s) + N_2(g) = Ba_3N_2(s)$	–364,000	—	240	298–1000
$Be(s) + F_2(g) = BeF_2(s)$	–1,018,000	–28	343	298–818
$Be(s) + F_2(g) = BeF_2(L)$	–993,000	–36	364	818–1455
$Be(L) + F_2(g) = BeF_2(g)$	–775,000	–10	8	1557–2000
$2Be(s) + O_2(g) = 2BeO(s)$	–1,200,000	–6	235	298–1557
$2Be(L) + O_2(g) = 2BeO(s)$	–1,221,000	–6	248	1557–2000
$3Be(s) + N_2(g) = Be_3N_2(s)$	–564,000	—	170	298–1000
$2/3Bi(s) + Cl_2(g) = 2/3BiCl_3(s)$	–261,000	–22	279	298–503
$2/3Bi(L) + Cl_2(g) = 2/3BiCl_3(L)$	–266,000	–32	349	544–714
$2/3Bi(L) + Cl_2(g) = 2/3BiCl_3(g)$	–190,000	7	–14	714–1500
$C(s) + 2H_2(g) = CH_4(g)$	–69,000	22	–65	298–1200
$C(s) + 2Cl_2(g) = CCl_4(g)$	–110,000	–10	206	350–800
$C(s) + 1/2O_2(g) = CO(g)$	–112,000	—	–088	298–2000
$C(s) + O_2(g) = CO_2(g)$	–390,000	—	–1	298–2000
$H_2(g) + 1/2O_2(g) = H_2O(g)$	–237,000	8	–8	298–2500

$$A = \Delta H^o_{ref} - T_{ref}\Delta C_P$$

$$B = \Delta C_P \qquad (9.58)$$

$$C = -\left[\Delta S^o_{ref} - \Delta C_P(1 + \ln T_{ref})\right]$$

Properties of the reference state bear the subscript *ref*.

Note that if ΔC_P is assumed to be zero, the coefficients A and C reduce to the constant standard enthalpy and entropy changes, B vanishes, and Equation (9.58) reduces to Equation (9.53) with constant coefficients.

Table 9.3 gives an abbreviated example of a compilation based on Equation (9.57). Note that many of the reactions give no value for B. These are reactions for which the experimental data is not sufficiently accurate to detect curvature in the $\Delta G°$ vs. T plot.

Other representations similar to Table 9.3 are available. The JANAF (for Joint Army, Navy, Air Force) Thermochemical Tables contain analytic equations for the free energies of formation of thousands of compounds. This compilation can be accessed on the Internet at WWW.ipt.arc.nasa.gov/database2.html

9.11.4 Tabular Representation

The only type of thermochemical database that is free from the approximations $\Delta C_P = 0$ (Figure 9.8) or ΔC_P = constant (Table 9.3) are listings of h^o, s^o, and g^o at a series of temperatures for all chemical species for which data are available, including the elements. The inclusion of g^o in the listing is for convenience in forming ΔG^o from which equilibrium constants can be calculated. The basic thermochemical properties are h^o and s^o, with g^o determined as $h^o - Ts^o$. A typical tabular representation is given in Table 9.4, which has been excerpted from the most comprehensive of this kind (the tables for all species included in this database occupy two 3½-inch thick volumes). This compilation, augmented by data obtained since 1977, can be found in commercial software packages (e.g., TAPP, www.esn.software.com/tapp/).

The units of C_P and s^o in Table 9.4 are cal/mole-K and h^o and g^o are in kcal/mole. The only decision that had to be made was the reference temperature at which the molar enthalpy of the elements is specified (i.e., h_{298}^o). The convention adopted is to set h_{298}^o equal to zero for the normal state (i. e., phase) of the elements at 25°C. The reference for the molar entropies of condensed phases has been chosen as zero at 0 K. The following example illustrates the application of Table 9.4:

Example: Calculate the free energy of formation of Al_2O_3(s) (per mole of O_2) at 1000 K from the graphical, analytic, and tabular data sources. The formation reaction is: $4/3Al(L) + O_2(g) = 2/3\ Al_2O_3(s)$.

From Figure 9.8: $\Delta G^o = -220 \times 4.184 = -920$ kJ/mole.

From Table 9.3 and Equation (9.57):

$$\Delta G^o = 2/3(-1{,}678{,}000-7 \times 1000\ln 1000+386 \times 1000) \times 10^{-3} = -907 \text{ kJ/mole}$$

From Table 9.4:

$$\begin{aligned}\Delta G^o &= 2/3\, g^o_{Al_2O_3} - 4/3\, g^o_{Al(L)} - g^o_{O_2} \\ &= [2/3(-424.8) - 4/3(-10.2) - (-52.8)] \times 4.184 \\ &= -908 \text{ kJ/mole}\end{aligned}$$

All three sources are in good agreement.

The databases represented by Figure 9.8 and Tables 9.3 and 9.4 assume that the total pressure is low enough not to influence ΔG of a reaction. However, like other thermodynamic properties, ΔG is a function of both T and the total pressure p. The

TABLE 9.4
Thermochemical Properties of Inorganic Substances

			O_2(g)		
Phase	***T,K***	**C_p mole-K**	**h^o kcal/mole**	**s^o cal**	**g^o kcal/mole**
GAS	298	7.008	0	49.005	–14.611
	300	7.016	0.013	49.048	–14.702
	400	7.310	0.731	51.111	–19.714
	500	7.500	1.472	52.764	–24.910
	600	7.649	2.229	54.145	–30.257
	700	7.778	3.001	55.334	–35.733
	800	7.897	3.785	56.380	–41.319
	900	8.011	4.580	57.317	–47.005
	1000	8.120	5.387	58.167	–52.780
	1100	8.227	6.204	58.946	–58.636
	1200	8.332	7.032	59.666	–64.567
	1300	8.436	7.870	60.337	–70.568
	1400	8.540	8.719	60.966	–76.633
	1500	8.642	9.578	61.559	–82.760
	1600	8.744	10.448	62.120	–88.944
	1700	8.846	11.327	62.653	–95.183
	1800	8.948	12.217	63.161	–101.474
	1900	9.049	13.117	63.648	–107.814
	2000	9.150	14.027	64.115	–114.202
	2100	9.251	14.947	64.563	–120.636
	2200	9.352	15.877	64.996	–127.115
	2300	9.452	16.817	65.414	–133.635
	2400	9.553	17.767	65.818	–140.197
	2500	9.654	18.728	66.210	–146.798

			Al		
Phase	***T,K***	**C_p mole-K**	**h^o kcal/mole**	**s^o cal**	**g^o kcal/mole**
SOL	298	5.802	0	6.769	–2.018
	300	5.812	0.011	6.805	–2.031
	400	6.187	0.613	8.535	–2.801
	500	6.435	1.244	9.942	–3.727
	600	6.694	1.900	11.138	–4.782
	700	7.011	2.585	12.198	–5.950
	800	7.404	3.305	13.154	–7.218
	900	7.884	4.069	14.053	–8.579
	933	8.063	4.334	14.342	–9.051
			2.56	2.743	
LIQ	933	7.588	6.894	17.085	–9.051
	1000	7.588	7.400	17.609	–10.209
	1100	7.588	8.159	18.332	–12.006
	1200	7.588	8.918	18.993	–13.873
	1300	7.588	9.677	19.600	–15.803

continued

TABLE 9.4 (continued)
Thermochemical Properties of Inorganic Substances

			Al		
Phase	*T,K*	C_p mole-K	h^o kcal/mole	s^o cal	g^o kcal/mole
LIQ	1400	7.588	10.436	20.162	–17.792
	1500	7.588	11.194	20.686	–19.834
	1600	7.588	11.953	21.175	–21.928
	1700	7.588	12.712	21.635	–24.068
	1800	7.588	13.471	22.069	–26.254
	1900	7.588	14.230	22.479	–28.481
	2000	7.588	14.988	22.869	–30.749

			Al_2O_3		
Phase	*T,K*	C_p mole-K	h^o kcal/mole	s^o cal	g^o kcal/mole
SOL-I	298	18.871	–400.4	12.174	–404.030
	300	18.979	–400.365	12.291	–404.052
	400	22.987	–398.243	18.369	–405.590
	500	25.179	–395.826	23.754	–407.702
	600	26.656	–393.230	28.482	–410.319
	700	27.797	–390.505	32.680	–413.381
	800	28.757	–387.677	36.456	–416.841
SOL-II	800	28.755	–387.677	36.456	–416.841
	900	29.354	–384.770	39.878	–420.661
	1000	29.845	–381.809	42.997	–424.807
	1100	30.265	–378.803	45.862	–429.252
	1200	30.638	–375.758	48.512	–433.972
	1300	30.976	–372.677	50.978	–438.948
	1400	31.290	–369.563	53.285	–444.162
	1500	31.586	–366.420	55.454	–449.600
	1600	31.868	–363.247	57.501	–455.249
	1700	32.139	–360.046	59.441	–461.097
	1800	32.402	–356.819	61.286	–467.134
	1900	32.658	–353.566	63.045	–473.351
SOL-B	2000	32.909	–350.288	64.726	–479.740

I. Barin, O. Knacke, O. Kubachewski, 1977. Springer-Verlag, Berlin, Heidelberg, New York, Verlag Stahleisen m.b.H. Düsseldorf.

effect of the latter condition is small, so that ΔG is very nearly equal to ΔG^o (i.e., the value for $p = 1$ atm) even if the total pressure is very high (see Problem 9.19).

When a metal reacts with a gas to form a compound, the thermodynamic property of the reaction (ΔG^o) is dependent on the phase of the metal. Problems 9.16 and 9.22 analyze the effect on ΔG^o of changing the metal reactant from solid to liquid.

9.12 DISSOLUTION OF GASES IN METALS

The treatment of chemical equilibrium in the preceding sections of this chapter gives the impression that all that is required is searching the available database for ΔG^o

of the reaction, using this information to calculate the equilibrium constant and then applying the law of mass action. This approach may also require estimation of activity coefficients if solid or liquid solutions are involved, and specification of the total pressure if gas mixtures are part of the reaction. This equilibrium condition always needs to be supplemented by specification of the ratios of the elements involved in the reaction, irrespective of their molecular forms.

The equilibrium aspects of the above method are less obvious when the process involves a solution in which one of the components exists in a form different from its normal pure state. Such a situation occurs when gases dissolve in condensed phases. The dissolved gas species behave either like a liquid component if incorporated in a liquid solvent or as a solute in a crystal when dissolved in a solid. In either case, there is no pure condensed phase of the diatomic gas to serve as a standard state. However, the normal gaseous state can still be used as a standard state. The consequence in the equilibrium equation is an activity coefficient of the dissolved gas that is very far from unity (see below).

There are two distinct mechanisms by which gases dissolve in condensed phases: dissociative and nondissociative. The latter refers to physical dissolution without change in the molecular structure of the gas. Equilibrium of $O_2(g)$ between air and water is an example of nondissociative dissolution of a gas. The thermodynamics of this type of solution process is treated in Section 8.3.5.

9.12.1 Dissociative Dissolution and Sieverts' Law

A very important exception to purely physical (nondissociative) dissolution occurs when diatomic gases such as O_2, N_2 and H_2, enter metals. Many metals form M-O, M-N, and M-H bonds that are sufficiently strong to break the bond of the diatomic molecule and absorb the gas in atomic form.* Palladium of cold fusion fame is an example of this process. Denoting the diatomic gas molecule by A_2, the controlling equilibrium is:

$$\tfrac{1}{2}A_2(g) = A(sol'n) \tag{9.59}$$

The equilibrium condition for this reaction in terms of chemical potentials is:

$$\tfrac{1}{2}\mu_{A2(g)} = \mu_{A(sol'n)} \tag{9.60}$$

There is no difficulty in interpreting the chemical potential of the diatomic gas; it is given by Equation (7.44) with the standard state the molecular gas at 1 atm pressure and the temperature of the system. For A dissolved in the metal, the appropriate expression for the chemical potential is the combination of Equations (7.29) and (7.30). The choice of the standard state for A atoms dissolved in the metal is problematic. Pure A in the solid form, which would be the state closest to that in the

* The rupture of the A-A bond takes place on the metal surface.

metal, does not exist, and so is inappropriate as a standard state. Other choices are A(g) and A_2(g). Picking the latter converts the above equation to:

$$\tfrac{1}{2} g^o_{A_2(g)} + \tfrac{1}{2} RT \ln p_{A_2} = \tfrac{1}{2} g^o_{A_2(g)} + RT \ln(\gamma_A x_A)$$

where γ_A and x_A are the activity coefficient and mole fraction of A in the metal, respectively and p_{A_2} is the pressure (or partial pressure) of A_2 in the gas. The above equation is rewritten in a more useful form as:

$$\frac{x_A}{\sqrt{p_{A_2}}} = \frac{1}{\gamma_A} = K_{SA} \tag{9.61}$$

Most solutions of gases in condensed phases are sufficiently dilute that γ_A is not dependent on x_A, which is the region of Henry's law in nondissociative absorption (Section 8.3.3). The reason is that A atoms are surrounded by atoms of the solvent metal M, and do not interact with other A atoms. Consequently, the reciprocal of the activity coefficient is a temperature-dependent quantity, which is similar to an equilibrium constant in ordinary chemical reactions. It is denoted as K_{SA}, where the subscript "S" refers to Augustus Sieverts, who first demonstrated experimentally the distinctive dependence of the dissolved gas concentration on the square root of the pressure. This feature permits the type of solution process (dissociative or non-dissociative) to be determined experimentally. Nondissociative dissolution follows the linear relation given by Equation (8.10). Figure 9.9 illustrates Sieverts' law behavior of N_2 in liquid iron.

9.12.2 The Zirconium–Hydrogen Phase Diagram

The Sieverts' law behavior illustrated in Figure 9.9 does not increase the concentration of A indefinitely as p_{A_2} increases. There is a limit that the metal can accept without precipitating a new phase. This limit is called the *terminal solubility*. At this limit additional gas in the solid ends up forming an M-A compound called a hydride if A = H, a nitride if A = N, and an oxide of A = O. This process is shown by the zirconium–hydrogen phase diagram in Figure 9.10.

The Zr–H system exhibits three single-phase regions separated by three two-phase zones. The narrow α-Zr phase next to the left-hand axis consists of Zr with a hexagonal crystal structure in which hydrogen is incorporated in the interstitial sites. The terminal solubility of hydrogen in this metal is indicated by the lines forming the right-hand border of the α-Zr zone. These lines are the upper phase boundary of hydrogen in α-Zr or the solubility limit of hydrogen in this metal phase. At a constant temperature of 500°C, addition of hydrogen to the metal maintains the α-Zr structure until the solubility limit at an H/Zr ratio of 0.04. Added hydrogen precipitates the δ-hydride phase, which has the formula $ZrH_{1.33}$ at its lower phase boundary. As hydrogen is added, the α-Zr phase is converted to the δ-hydride until the mixture is completely converted to the latter. Further addition of hydrogen increases the H/Zr ratio of the hydride from its lower phase boundary value of 1.33 to the upper limit corresponding to ZrH_2.

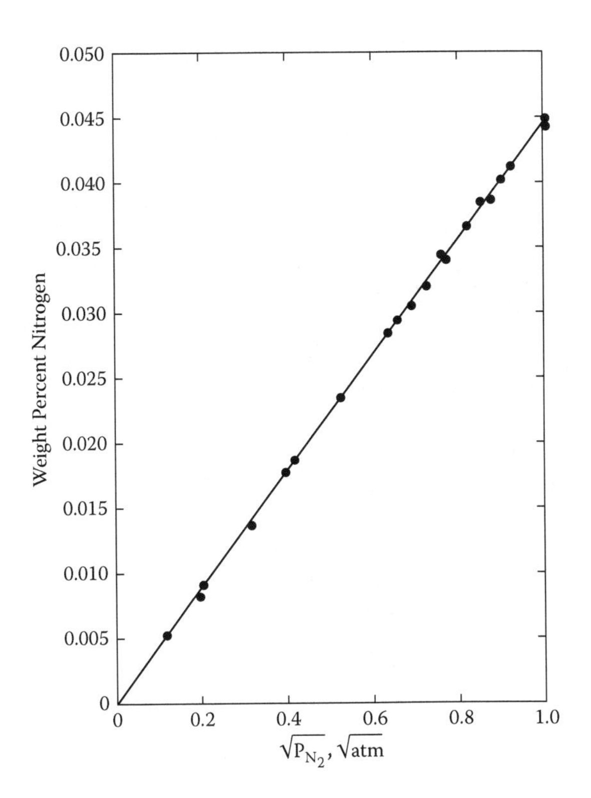

FIGURE 9.9 Solubility of nitrogen in liquid iron at 1600°C.

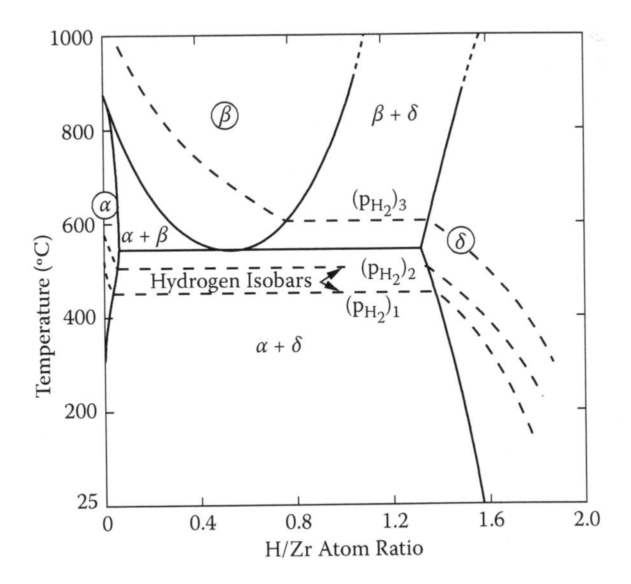

FIGURE 9.10 The hydrogen–zirconium phase diagram.

The β modification of metallic Zr is a high-temperature phase of cubic crystal structure. In hydrogen-free Zr, the α-Zr structure transforms to the β-Zr structure at 860°C. Addition of hydrogen stabilizes β-Zr at lower temperatures; at H/Zr = 0.54, β-Zr exists down to 550°C. Between 550 and 860°C, the upper phase boundary of the α-Zr phase is separated from the lower phase boundary of β-Zr by the two-phase metal zone labeled α+β in Figure 9.10. The upper phase boundary of β-Zr and the lower phase boundary of the δ-hydride are connected by the β+δ two-phase region.

Each point in Figure 9.10 has a corresponding equilibrium hydrogen pressure. This function, $p_{H_2} = f(T, \text{H/Zr})$ can be thought of as a surface in the space above the T-H/Zr plane. The dashed lines and curves in Figure 9.10 represent projections of this surface on the *T*-H/Zr plane. These curves are analogous to the elevation contours on hikers' maps of wilderness areas. In Figure 9.10, the dashed lines are called hydrogen isobars because they represent constant values of p_{H_2}.

In the three single-phase regions, the hydrogen isobars are sloped curves, indicating that p_{H_2} is a function of both *T* and H/Zr; that is, the system has two degrees of freedom. In the α-Zr and β-Zr regions, horizontal isotherms cut through the hydrogen isobars at combinations of p_{H_2} – H/Zr values that satisfy Sieverts' law, Equation (9.61) (ignoring the distinction between mole fraction and atom ratio as measures of hydrogen concentration*). Since the δ-hydride is more ceramic-like than metallic, it need not follow Sieverts' law.

In the three two-phase zones, on the other hand, the hydrogen isobars are horizontal lines, indicating that p_{H_2} is independent of H/Zr and a function of temperature only. The reason for this behavior is simple to understand: the H/Zr ratio in the two-phase regions refers to the mole-weighted average of the left- and right-hand phases (see the lever rule, Section 8.7). In moving along an isotherm in a two-phase region, the H/Zr ratios of the two phases present do not change; only the relative amounts of the two phases shift. At 500°C, for example, the hydrogen pressure in equilibrium with α-Zr with H/Zr = 0.04 is exactly the same as that in equilibrium with the δ-hydride at H/Zr = 1.33.

Problem 9.13 shows how the H_2/Zr equilibrium is applied in a laboratory device for measuring the hydrogen content of zirconium, which is important in nuclear fuels. Problem 9.14 analyzes the effect of other gas-phase reactions of a diatomic species on its dissolution in a metal.

9.13 COMPUTATIONAL THERMODYNAMICS

So far in this chapter, mainly single reactions have been analyzed. The sole exception was the two-reaction problem in Section 9.8. Mass-action laws are inherently nonlinear, and almost always require numerical solution. As the number of simultaneous reactions increases, the method of computation based on the mass-action laws becomes prohibitively complex. A new approach is needed. The literature is replete with large codes with names such as THERMOCALC and PHREEQ. All utillize equilibrium determination based on minimization of the free energy of the system, including all possible reactions and all phases.

---

* The mole fraction and the hydrogen-to-zirconium atom ratio are related by $x_H = (\text{H/Zr})/(1 + \text{H/Zr})$.

9.13.1 Method of Lagrange Multipliers

In order to understand the new computational method, a mathematical detour into the theory of Lagrange multipliers is necessary.

Consider a function $F(n_1, n_2, \ldots)$. The values of $n_1, n_2, \ldots$ at which F is a minimum are to be determined. The system is subject to the following constraints:

$$V(n_1, n_2, \ldots) = 0 \qquad \text{and} \qquad W(n_1, n_2, \ldots) = 0 \tag{9.62}$$

where F, V, and W are specified functions of the numbers, n_i.

The differential of F is:

$$dF = f_{n1}dn_1 + f_{n2}dn_2 + \ldots = 0 \tag{9.63}$$

When F is a minimum, dF is set equal to zero. The coefficients of dn_i in Equation (9.63) are:

$$f_{ni} = \frac{\partial F}{\partial n_i} \tag{9.64}$$

Taking the differentials of V and W:

$$dV = v_{n1}dn_1 + v_{n2}dn_2 + \ldots = 0 \qquad \text{and} \qquad dW = w_{n1}dn_1 + w_{n2}dn_2 + \ldots = 0 \tag{9.65}$$

where

$$v_{ni} = \frac{\partial V}{\partial n_i} \qquad w_{ni} = \frac{\partial W}{\partial n_i} \tag{9.66}$$

Multiply dV by λ_v and dW by λ_w, the Lagrange multipliers, and add to dF:

$$(f_{n1} + \lambda_v v_{n1} + \lambda_w w_{n1})dn_1 + (f_{n2} + \lambda_v v_{n2} + \lambda_w w_{n2})dn_2 + \ldots = 0 \tag{9.67}$$

Since all dn_i are arbitrary, their coefficients must be zero:

$$f_{n1} + \lambda_v v_{n1} + \lambda_w w_{n1} = 0$$

$$f_{n2} + \lambda_v v_{n2} + \lambda_w w_{n2} = 0 \tag{9.68}$$

With constraints given by Equation (9.62). These equations are solved for λ_v, λ_w and $n_1, n_2 \ldots$.

9.13.2 Water Decomposition Example

The water decomposition system, in which H_2O is partially converted to H_2, O_2, and OH, was covered in Section 9.8. Mole numbers are denoted by:

$$n_1 = H_2O;\ n_2 = H_2;\ n_3 = O_2;\ n_4 = OH;\ \ldots\ n_T = n_1 + n_2 + n_3 + n_4$$

and $n_i/n_T = x_i$ (mole fraction i).

At $p(total)$ = 1 atm, the free energy of the system is:

$$\frac{G}{RT} = n_1\left[\frac{g_1^o}{RT} + \ln\left(\frac{n_1}{n_T}\right)\right] + n_2\left[\frac{g_2^o}{RT} + \ln\left(\frac{n_2}{n_T}\right)\right]$$

$$+n_3\left[\frac{g_3^o}{RT} + \ln\left(\frac{n_3}{n_T}\right)\right] + n_4\left[\frac{g_4^o}{RT} + \ln\left(\frac{n_4}{n_T}\right)\right] \tag{9.69}$$

For free-energy minimization, $d(G/RT)$ is set equal to zero, producing an equation of the same form as Equation (9.63). In order to simplify the notation, a dimensionless free energy is defined by $g_i = g_i^o / RT$.

Following the procedure outlined in the preceding section, G is to be minimized subject to the constraints of V and W in Equation (9.62) and the following total-element constraints:

$$n_O = n_1 + 2n_3 + n_4 = \text{total moles O} \qquad n_H = 2n_1 + 2n_2 + n_4 = \text{total moles H} \tag{9.70}$$

n_O replaces V in the general theory and n_H corresponds to W.

The Lagrange multipliers are λ_O and λ_H. Equation (9.68) is:

$$\frac{\partial(G/RT)}{\partial n_1} + \lambda_O\frac{\partial n_O}{\partial n_1} + \lambda_H\frac{\partial n_H}{\partial n_1} = 0 \qquad \frac{\partial(G/RT)}{\partial n_2} + \lambda_O\frac{\partial n_O}{\partial n_2} + \lambda_H\frac{\partial n_H}{\partial n_2} = 0$$

$$\frac{\partial(G/RT)}{\partial n_3} + \lambda_O\frac{\partial n_O}{\partial n_3} + \lambda_H\frac{\partial n_H}{\partial n_3} = 0 \qquad \frac{\partial(G/RT)}{\partial n_4} + \lambda_O\frac{\partial n_O}{\partial n_4} + \lambda_H\frac{\partial n_H}{\partial n_4} = 0 \tag{9.71}$$

The derivatives of the total free energy are:*

$$f_1 = \frac{\partial(G/RT)}{\partial n_1} = g_1 + \ln\left(\frac{n_1}{n_T}\right) \qquad f_2 = \frac{\partial(G/RT)}{\partial n_2} = g_2 + \ln\left(\frac{n_2}{n_T}\right)$$

$$f_3 = \frac{\partial(G/RT)}{\partial n_3} = g_3 + \ln\left(\frac{n_3}{n_T}\right) \qquad f_4 = \frac{\partial(G/RT)}{\partial n_4} = g_4 + \ln\left(\frac{n_4}{n_T}\right) \tag{9.72}$$

---

* When taking the derivative of Equation (9.69) with respect to n_1, terms in addition to the one shown in Equation (9.72) are:

$$n_1\frac{\partial}{\partial n_1}\left[\ln\left(\frac{n_1}{n_T}\right)\right] = 1 - \frac{n_1}{n_T} + n_2\frac{\partial}{\partial n_1}\left[\ln\left(\frac{n_2}{n_T}\right)\right] = -\frac{n_2}{n_T}.+ \quad \ldots \text{ These sum to zero.}$$

The same is true for the derivatives of Equation (9.69) with respect to n_2, n_3, and n_4.

In addition, from Equation (9.70):

$$\frac{\partial n_O}{\partial n_1}=1 \quad \frac{\partial n_H}{\partial n_1}=2 \quad \frac{\partial n_O}{\partial n_2}=0 \quad \frac{\partial n_H}{\partial n_2}=2$$

$$\frac{\partial n_O}{\partial n_{31}}=2 \quad \frac{\partial n_H}{\partial n_3}=0 \quad \frac{\partial n_O}{\partial n_4}=1 \quad \frac{\partial n_H}{\partial n_4}=1 \qquad (9.73)$$

Substituting Equations (9.72) and (9.73) into Equation (9.71) yields:

$$f_1 + \lambda_O + 2\lambda_H = 0 \qquad f_2 + 2\lambda_H = 0$$

$$f_3 + 2\lambda_O = 0 \qquad f_4 + \lambda_O + \lambda_H = 0 \qquad (9.74)$$

Combining the top two equations in (9.74) gives $\lambda_O = f_2 - f_1$ and $\lambda_H = -f_2$. Eliminating λ_O and λ_H from the bottom two equations gives:

$$f_3 + 2f_2 - 2f_1 = 0 \qquad \text{and} \qquad f_4 - f_1 + \tfrac{1}{2}f_2 = 0 \qquad (9.75)$$

Inserting the definitions of $f_1 \ldots f_4$ from Equation (9.72) gives:

$$g_3 + 2g_2 - 2g_1 + \ln(n_3/n_T) + 2\ln(n_2/n_T) - 2\ln(n_1/n_T) = 0 \qquad (9.76a)$$

$$g_4 - g_1 + \tfrac{1}{2}g_2 + \ln(n_4/n_T) - \ln(n_1/n_T) + \tfrac{1}{2}\ln(n_2/n_T) = 0 \qquad (9.76b)$$

The $g_i = g_i^o/RT$ are known quantities, so Equations (9.76a) and (9.76b) contain only the four mole numbers as unknowns. The system consists of these two equations and Equations (9.70), in which n_O and n_H are specified. A method of solving this set is as follows:

From Equation (9.70), express n_2 and n_3 as functions of n_1 and n_4:

$$n_2 = \tfrac{1}{2}(n_H - 2n_1 - n_4) \qquad \text{and} \qquad n_3 = \tfrac{1}{2}(n_O - n_1 - n_4) \qquad (9.77)$$

1. Guess n_1 utilizing Equation (9.77) for n_2 and n_3:
2. Vary n_4 until Equation (9.76a) is satisfied.
3. Vary n_4 until Equation (9.76b) is satisfied.
4. If n_4 from step 2 is equal to n_4 from step 3, go to step 5; otherwise, return to step 1.
5. With the current values of $n_1, \ldots n_4$, calculate: $x_i = n_i/n_T$ for $i = 1 \ldots 4$.

For the conditions used in Section 9.8 (3000 K, $n_O = 1$, $n_H = 2$), the results of this computation are:

$$x_{H_2O} = 0.704 \qquad x_{H_2} = 0.148 \qquad x_{O_2} = 0.049 \qquad x_{OH} = 0.100$$

These results are essentially the same as those computed by the standard method in Section 9.8.

The free-energy minimization method may not appear to offer any advantage over the reaction-progress-variable method for this two-reaction case, but when there are dozens of reactions, it is the only feasible solution method.

PROBLEMS

9.1 The gas-phase reaction $2CO + O_2 = 2CO_2$ is in equilibrium at a total pressure of 1 atm in a closed container. The thermochemical properties are given in Section 9.6.2.

(a) For a stoichiometric initial mixture of reactants (i.e., $CO/O_2 = 2$), calculate and tabulate the equilibrium constant and gas composition for temperatures from 2000 K to 2800 K at 200 K intervals.

(b) A stoichiometric mixture with 0.10 moles CO and 0.05 moles O_2 initially at 300 K reacts to equilibrium in an adiabatic container. Determine the final temperature, composition, and fraction of the initial CO that has reacted. The heat capacity of the gas plus the container (which heats up along with the gas) is 11 J/K. (Note: a graphical solution utilizing the results of Part (a) is required.)

9.2 A mixture of H_2 and CH_4 establishes a "carbon activity," defined as the ratio of the carbon partial pressure generated by the reaction $C(g) + 2H_2(g) = CH_4(g)$ to the vapor pressure of solid carbon (graphite). The standard free energy change for the above reaction is $\Delta G^o(g)$. For the reaction with solid carbon, $C(s) + 2H_2(g) = CH_4(g)$, the standard free energy change is $\Delta G^o(s)$. For graphite sublimation, $C(s) = C(g)$, the free energy change is $\Delta G^o(sub)$.

(a) What is the relationship between $\Delta G^o(g)$, $\Delta G^o(s)$, and $\Delta G^o(sub)$?

(b) At 1000 K and 1 atm total pressure, what hydrogen pressure gives a carbon activity of unity? At this temperature, $\Delta G^o(s) = 19.2$ kJ/mole.

(c) What happens if the H_2 pressure is lower than the value in Part (b)?

9.3 Consider the gas phase reaction: $PCl_3(g) + Cl_2(g) = PCl_5(g)$ at equilibrium at 400 K and 1 atm total pressure. The closed system is initially charged with 1 mole of PCl_3 and 2 moles of Cl_2. The standard free energy change of this reaction is –3.53 kJ/mole. What is the mole fraction of PCl_5 at equilibrium?

9.4 Uranium metal is produced commercially by reduction of the UF_4 with magnesium. A stoichiometric initial charge of the reactants at room temperature (i.e., 1 mole UF_4 and 2 moles Mg) are placed in a reaction vessel and heated slightly to initiate the reaction. The reaction proceeds and the heat of the reaction raises the temperature to 1263°C, at which temperature the system consists of a molten fluoride salt and a liquid U-Mg alloy. These two liquids are immiscible. The overall reaction is:

Magnesium

Phase	T	H	S	G
SOL	298	0	7.81	–2.329
	300	.011	7.847	–2.343
	400	.621	9.600	–3.219
	500	1.259	11.023	–4.252
	600	1.924	12.233	–5.416
	700	2.613	13.296	–6.694
	800	3.328	14.250	–8.072
	900	4.068	15.122	–9.541
	923	4.242	15.312	–9.891
		2.14	2.319	
LIQ	923	6.382	17.631	–9.891
	1000	6.967	18.240	–11.273
	1100	7.727	18.964	–13.133
	1200	8.487	19.625	–15.063
	1300	9.247	20.234	–17.057
	1378	9.840	20.676	–18.652
		30.5	22.134	
GAS	1378	40.340	42.810	–18.652
	1400	40.449	42.889	–19.595
	1500	40.946	43.231	–23.901
	1600	41.443	43.552	–28.240

Magnesium Fluoride

Phase	T	G	S	H
SOL	298	–268.5	13.68	–272.579
	300	–268.472	13.774	–272.604
	400	–266.874	18.362	–274.219
	500	–265.178	22.144	–276.250
	600	–263.419	25.348	–278.628
	700	–261.615	28.129	–281.305
	800	–259.772	30.589	–284.244
	900	–257.896	32.799	–287.415
	1000	–255.988	34.809	–290.797
	1100	–254.050	36.656	–294.371
	1200	–252.084	38.366	–298.123
	1300	–250.090	39.962	–302.041
	1400	–248.069	41.460	–306.113
	1500	–246.021	42.873	–310.330
	1536	–245.277	43.363	–311.882
		13.9	9.049	
LIQ	1536	–231.377	52.412	–311.882
	1600	–229.933	53.333	–315.266

Uranium

Phase	T	H	S	G
SOL-I	298	0	12.03	–3.587
	300	.012	12.071	–3.609
	400	.684	14.001	–4.916
	500	1.406	15.610	–6.359
	600	2.198	17.052	–8.033
	700	3.069	18.393	–9.806
	800	4.022	19.664	–11.709
	900	5.060	20.885	–13.737
	941	5.511	21.375	–14.604
		0.70	.744	
SOL-II	941	6.211	22.119	–14.604
	1000	6.801	22.727	–15.927
	1048	7.281	23.196	–17.029
		1.15	1.097	
SOL-III	1048	8.431	24.294	–17.029
	1100	8.929	24.757	–18.304
	1200	9.887	25.591	–20.822
	1300	10.845	25.358	–23.420
	1400	11.803	27.068	–26.092
	1403	11.832	27.088	–26.173
		3.0	2.138	
LIQ	1403	14.832	29.227	–26.173
	1500	15.942	29.992	–29.046
	1600	17.087	30.731	–32.082

Uranium Tetrafluoride

Phase	T	G	S	H
SOL	298	–453.7	36.3	–464.523
	300	–453.649	36.472	–464.590
	400	–450.839	44.550	–468.659
	500	–447.957	50.978	–473.446
	600	–445.004	56.360	–478.820
	700	–441.980	61.020	–484.694
	800	–438.886	65.150	–491.006
	900	–435.722	68.876	–497.711
	1000	–432.488	72.283	–504.771
	1100	–429.183	75.432	–512.159
	1200	–425.809	78.368	–519.850
	1300	–422.354	81.125	–527.826
	1309	–422.051	81.365	–528.558
		10.2	7.792	
LIQ	1309	–411.851	89.157	–528.558
	1400	–408.256	91.812	–536.793
	1500	–404.306	94.537	–546.112
	1600	–400.356	97.087	–555.695

$$2Mg(melt) + UF_4(salt) = U(melt) + 2MgF_2(salt)$$

where "salt" means the molten fluoride salt phase and "melt" means the liquid alloy.

(a) Assuming ideal solution behavior in both phases, what is the fractional conversion of UF_4 to uranium metal (this is the "reaction progress variable"). The equilibrium constant is to be calculated from the thermochemical data shown below. (Note: because all species are in the liquid state at equilibrium, the thermochemical data for each must pertain to pure liquids, even though the stable state is not liquid at the reaction temperature.)
(b) Repeat Part (a) but using the "element conservation method" instead of the reaction progress variable method.
(c) How much heat (in kcal) is transferred to or from the reaction vessel when the products reach a temperature of 1263°C?

9.5 A mixture of CO and CO_2, is passed over molybdenum dioxide (MoO_2(s)) held in a furnace at a controlled temperature.

(a) If the CO_2/CO ratio in the flowing gas is 3, at what temperature T^* does the solid oxide decompose to the metal (i.e., the metal and oxide coexist at equilibrium)?
For the following parts, draw a stability diagram like Figure 9.6 with lines for the Mo/MoO_2 equilibrium and for the CO_2/CO equilibrium. Note that the latter depends on the CO_2/CO ratio in the gas phase.
(b) If the CO_2/CO ratio is held constant and the temperature is increased above T^*, what is the stable phase?
(c) If the temperature is held constant at T^* and the CO_2/CO ratio is reduced, what phase is stable?

Use Figure 9.8 for the necessary data.

9.6 A gas mixture containing a CO_2/CO ratio of 10^{-5} is passed over TiO_2 at 1200°C. What is the stable form of titanium in the furnace (metal, monoxide or dioxide)? (Hint: compare the O_2 pressure established by the CO_2/CO gas to that for the TiO/TiO_2 equilibrium and for the Ti/TiO equilibrium.)

9.7 In the melt/slag problem treated in Section 9.10, suppose that the oxygen pressure in a gas phase contacting the two condensed phases is specified.

(a) Derive the equations for the compositions of the melt and slag (i.e., x_M^{melt} and x_M^{slag}) as functions of the equilibrium constants K_M and K_P and p_{O_2}. (Hint: for this problem, you do not have to use the reaction progress variable method.)

(b) If the oxygen pressure is 10^{-11} atm, what are the equilibrium compositions of the melt and slag at 1000 K? Use the equilibrium constants given in the example at the end of Section 9.10.

(c) From the result obtained in Part (a), over what range of oxygen pressure are solutions possible? Physically, what happens when p_{O_2} is outside of this range?

9.8 In the melt-slag problem treated in Section 9.10, the dioxide PO_2 is replaced by the monoxide PO. The initial charge is 1 mole of M and 1 mole of PO. The system is closed, so oxygen is conserved during equilibration. Using the values of K_M and K_P given in the example in Section 9.10, calculate the compositions of the melt and slag.

9.9 In the melt-slag example in Section 9.10.2, the slag phase is ideal but M and P in the melt form a regular solution with $\Omega = 5$ kcal/mole. What are the equilibrium compositions of the two phases?

9.10 At high temperatures, diatomic molecules dissociate according to $A_2(g) = 2A(g)$.

(a) At temperature T and total pressure p, derive the expression for the degree of dissociation β, defined as the fraction of the total quantity of the element that exists in atomic form, in terms of the equilibrium constant K_P.

(b) For I_2, the thermochemical parameters of the dissociation reaction are $\Delta H^o = 157$ kJ/mole and $\Delta S^o = 113$ J/mole-K. What is the degree of dissociation in pure iodine vapor at 700 K and 0.1 atm total pressure?

(c) At very high temperatures, dissociation of diatomic molecules contributes (additively) to the heat capacity of the gas: $C_P = C_{P0} + C_{Pdiss}$ where C_{P0} is the heat capacity in the absence of dissociation. How large is C_{Pdiss} for the conditions of Part (b)?

(Hint: First calculate the heat absorbed per mole of I_2 when a fraction β is dissociated into atoms.)

9.11 If methane is burned in oxygen, the principal reaction products are carbon dioxide and water (see Equation [9.1]). However, when methane is reacted with steam, the main reaction products are CO_2 and H_2.

(a) Write the equilibrium reaction (per mole of methane).

(b) The standard free energy change of the reaction (as written in Part [a]) is −1.7 kcal/mole at 900 K. The total pressure is 1 atm and the steam-to-methane ratio of the feed gas is 50. Assuming that the reaction of Part (a) is the only one occurring, what is the percentage of the input methane that is converted to carbon dioxide at equilibrium?

(c) What is the oxygen pressure in this gas at equilibrium?

(d) Is carbon monoxide a significant component of the equilibrium gas?

9.12 The reaction XO + Y = YO + X can be broken into two couples. The stability diagram for this system is sketched below.

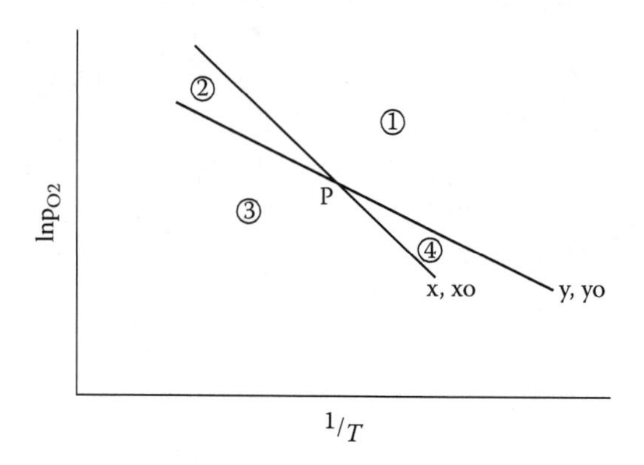

(a) What are the two couples?
(b) What equations represent the two lines?
(c) Identify the stable phases in the numbered regions and at point P.

9.13 A graduate student research project involves measuring the hydrogen content of zirconium specimens by "outgassing" at elevated temperature. The original hydrogen content is expressed as the atomic ratio (H/Zr). The specimen is contained in a vessel of known volume that is initially pumped out of all gas. When the specimen is inserted into the hot vessel, the hydrogen in the metal is partially released and causes the pressure in the apparatus to rise. The pressure rise is measured. However, not all of the hydrogen is released from the metal because at equilibrium, Sieverts' law requires that some hydrogen remain dissolved in the metal. The following information is given:

- The temperature of the specimen and the vessel, T
- The vessel volume V
- The moles of zirconium in the specimen, n_{Zr}
- The Sieverts' law constant for hydrogen dissolution in Zr, K_{SH}

Derive the equation from which the H/Zr ratio of the original specimen can be calculated from the measured pressure increase when the specimen is heated. The quantity of H remaining in the metal can be assumed to be much less than n_{Zr}.

9.14 Consider the system at temperature T and total pressure p in which the following reactions occur:

$$M(s) + A_2(g) = MA_2(g) \quad (1) \qquad \text{and} \qquad A_2(g) = 2A(g) \quad (2)$$

(a) Neglecting dissociative dissolution of A_2 in the metal M, derive the equations giving the composition (in mole fraction units) of the equilibrium gas if the equilibrium constants K_{P1} and K_{P2} are given.

(b) If the Sieverts' law constant for dissociative dissolution of A_2 in the metal is K_{SA}, how are the results of Part (a) changed? What is the mole fraction of A dissolved in the metal? The activity of the solid metal M is not affected by dissolved A and the total gas pressure is maintained at p.

9.15 When oxygen gas at 1 atm is passed over solid ruthenium metal at 2000 K, the metal reacts to form gaseous oxides. However, it is not possible to determine experimentally whether the dominant oxide is RuO(g), RuO_2(g) or RuO_3(g). Only the enthalpies of formation of these species are known. They are, respectively, 102, 37, and 30 kcal/mole of O_2. Estimate which of the three possible vapor species is dominant and provide an estimate of its equilibrium pressure under the given conditions. Assume that all gaseous species have the entropy of O_2(g) (~ 200 J/mole-K). Neglect the entropies of any solid species in the reactions.

9.16 Consider formation of Ag_2S by the reaction:
$S_2(g) + 4Ag(S \text{ or } L) = 2Ag_2S(L)$. When silver is liquid ($T > 1235$ K), the thermochemical properties of this reaction are given in Table 9.3.

(a) Calculate the equilibrium sulfur pressure over the two-phase Ag/Ag_2S liquid mixture at 1300 K.

(b) At 1200 K, Ag is solid but Ag_2S is liquid. Calculate the standard free energy change for the reaction at 1200 K. Use the enthalpy of melting of silver of 11.3 kJ/mole to correct for the change in phase of silver. What is the equilibrium sulfur pressure at this temperature?

9.17 Prove that Equation (10.25) is valid even when ΔH^o and ΔS^o are temperature dependent, but ΔC_P is constant.

9.18 At 2257 K and 1 atm total pressure, 1.77% of initially pure water is dissociated by the equilibrium reaction $2H_2O(g) = 2H_2(g) + O_2(g)$. At this temperature,

(a) What is the equilibrium constant K_P?

(b) What is the standard Gibbs free energy of the reaction?

9.19 (a) Using Figure 9.8, what is the pressure of O_2 in equilibrium with a mixture of Cr and Cr_2O_3 at 1200°C?

(b) The result of Part (a) applies to low pressures (~1 atm). How does the equilibrium O_2 pressure over this couple change if the total pressure is 1000 atm? The densities of Cr and Cr_2O_3 are 7.2 and 5.2 g/cm^3, respectively. The atomic weight of Cr is 52. (Hint: use Equation [6.15] to determine the effect of pressure on the molar free energy of the solids, assuming that their molar volumes are constant.)

9.20 One mole of an equimolar CO/CO_2 gas mixture is contacted with 2 moles of NiO in a closed container at 2000 K. At equilibrium, what is the mole fraction of CO in the gas phase and how many moles of metallic Ni are formed? At 2000 K, the free energy changes of the relevant reactions are:

$$2CO + O_2 = 2CO_2 \qquad \Delta G^\circ = -216 \text{ kJ/mole}$$

$$2Ni + O_2 = 2NiO \qquad \Delta G^\circ = -46 \text{ kJ/mole}$$

9.21 Can magnesium metal reduce UO_2 to uranium metal at 1200°C?

9.22 Two formation reactions for Al_2O_3 are given in Table 9.8, one with Al(s) and the other with Al(L).

$$2Al(s) + 1.5O_2 = Al_2O_3(s); \qquad \Delta G^\circ(s) = A_s + B_s T\ln T + C_s T$$

$$2Al(L) + 1.5O_2 = Al_2O_3(s); \qquad \Delta G^\circ(L) = A_L + B_L T\ln T + C_L T$$

where the values of A_s, ... C_L are given in Table 9.3. Assume that the specific heats of Al(s) and Al(L) are equal.

(a) Show the relation between A_L and A_s, B_L and B_s, and C_L and C_s in terms of the heat of fusion of aluminum, Δh_M and its melting point T_M.
(b) Calculate A_L from A_s, B_L from B_s, and C_L from C_s using the relations developed in Part (a) and the data: $\Delta h_M = 10.7$ kJ/mole and $T_M = 931$ K. Compare the calculated results with the numbers for A_L, B_L, and C_L given in Table 9.3.

9.23 A metal M forms a solid oxide $MO_2(s)$ and a gaseous oxide $MO_3(g)$. The standard enthalpies of the oxidation reactions are $\Delta H^o\left(MO_2(s)\right) = -350$ kJ/mole metal and $\Delta H^o\left(MO_3(g)\right) = -250$ kJ/mole metal. The metal is exposed to a gas containing O_2 at a pressure of 10^{-3} atm at 1500 K.

(a) Write the two oxidation reactions and estimate the standard entropy changes for each [i.e., $\Delta S^o\left(MO_2(s)\right)$ and $\Delta S^o\left(MO_3(g)\right)$].
(b) Does $MO_2(s)$ form?
(c) What is the partial pressure of $MO_3(g)$ in the gas exposed to the metal?

9.24 An initial mixture containing 3 moles of N_2, 1 mole of H_2 and 4 moles of NH_3 is heated to 1200 K and allowed to come into equilibrium at 1 atm total pressure.

(a) Write the reaction.
(b) Calculate the equilibrium constant for this reaction at 1200 K from the following information:

The enthalpy of formation of $NH_3(g)$ at 298K is $\Delta H^o_{298} = -136$ kJ/mole.
The heat capacities (C_P) of H_2, and N_2 are 3.5R and that of $NH_3(g)$ is 5R.
The standard entropy of the reaction at 1200 K can be estimated from the "entropy rule of thumb."

(c) What is the equilibrium composition of the gas (in terms of mole fractions)?

9.25 The $CO/CO_2/O_2$ equilibrium in a closed system at a constant pressure was analyzed in Section 9.6.2. In the present problem, the initial charge of CO and O_2 is contained in a rigid vessel at an *initial* pressure of 0.25 atm. The temperature is maintained at 2000 K. What is the equilibrium gas composition for this constant-volume condition?

9.26 A bed of 2 moles of chromium metal with some chromium oxide mixed in is held at 800°C while moist argon at 1 atm pressure flows through it. The purpose of the apparatus is to remove water vapor from the inert gas. The inlet water vapor mole fraction is 0.016.

(a) What is the mole fraction of water vapor in the exit gas?
(b) What is the heat removal/addition rate required to maintain the bed temperature?

If the argon flow rate is 0.01 moles/s, how long will the bed be able to remove water vapor?

10 Electrochemistry

10.1 ELECTROCHEMISTRY EXPLAINED

The essence of a chemical reaction is the exchange of atoms and electrons between molecular species. In most reactions, the exchange of the electrons is not manifest because it occurs in an intimate mixture of reactants and products. However, in a device known as an *electrochemical cell*, participants in a reaction are physically separated in a manner that renders the electron transfer process observable, measurable and usable for work. The study of chemical reactions with emphasis on the electron-transfer process is called *electrochemistry.*

For example, the reaction involving metals X and Y and their oxides XO and YO can be written in the conventional form:

$$XO(s) + Y(s) = YO(s) + X(s) \tag{10.1}$$

This *overall reaction* can be analyzed, as was done in Chapter 9, by breaking it up into the formation reactions of the oxides:

$$X(s) + \tfrac{1}{2}O_2 = XO(s) \qquad Y(s) + \tfrac{1}{2}O_2 = YO(s) \tag{10.2}$$

However, this manner of describing reaction (10.1) does not reveal the electron transfer process. Suppose that the initial state consists of equal numbers of moles of XO and Y. Assume also that ΔG^o is negative so that the reaction tends to proceed from left to right. What occurs in an electrochemical sense is the transfer of oxygen ions (O^{2-}) bound to X^{2+} to Y atoms to form YO. Simultaneously, Y must lose two electrons (thus becoming Y^{2+}) and X^{2+} gains two electrons to produce X. Reaction (10.1) is an example of an *oxidation-reduction reaction*: Y is *oxidized* because its *valence,* or *oxidation state* is increased from 0 to +2; X is *reduced* from the +2 oxidation state to the neutral elemental form.

Electrochemistry is divided into two branches, depending on whether or not liquid water is present. If water is absent and the reaction involves solid phases, as in reaction (10.1), the branch is called *solid-state electrochemistry.* These electrochemical reactions usually take place at high temperatures so that the kinetics are sufficiently rapid for attainment of equilibrium in a reasonable time.

Many solutes dissolved in water do so as positive and negative ions rather than molecular entities. Reactions of these ions among themselves or with dissolved gases or solid metals immersed in the water involve transfer of electrons between species. This branch of electrochemistry is called *aqueous electrochemistry.*

The electrochemical cell, whether solid-state or aqueous, provides a method for using electrical measurements for determination of free energy changes, such as ΔG^o of reaction (10.1). Alternatively, the electron flow produced by an electrochemical cell can be harnessed to perform useful work. Such a cell is a battery or a fuel cell.

10.2 THE SOLID-STATE ELECTROCHEMICAL CELL

In this section, the fundamentals of electrochemical cells are presented using the solid-state type and reaction (10.1) as a typical overall reaction. The basic theory is readily extended to aqueous cells and any type of oxidation-reduction reaction.

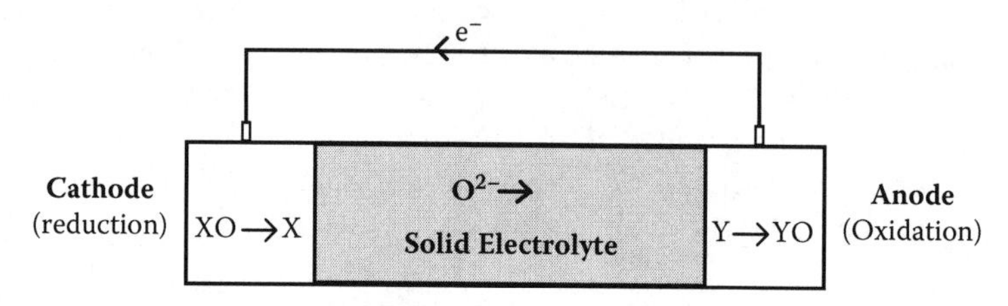

FIGURE 10.1 A short-circuited electrochemical cell for carrying out reaction (10.1).

Instead of mixing powders of the species that take part in Equation (10.1), the reaction takes place in the electrochemical cell shown in Figure 10.1. The two metals and their oxides are housed in separate compartments called *electrodes* or *half-cells*. These are joined by a bridge called a *solid electrolyte*, which is a fast conductor of oxygen ions. A typical solid electrolyte is ZrO_2 doped with CaO to enhance oxygen mobility. The compartment containing X and XO is called the *cathode* because reduction of X^{2+} (in XO) to X occurs here. The oxygen ions liberated at the cathode diffuse through the solid electrolyte. Upon reaching the opposite electrode, oxygen ions react with metal Y to form YO, releasing two electrons to the external circuit. Because Y is oxidized to Y^{2+} (in YO), this electrode is termed the *anode*. The electrons liberated at the anode supply the reduction reaction occurring in the cathode.

In the electrochemical cell, the overall reaction (10.1) is physically separated into two *half-cell reactions* that differ from formation reactions such as (10.2) by explicitly showing the oxygen ions and the electrons:

$$XO + 2e \rightarrow (O^{2-})_{cathode} + X \qquad Y + (O^{2-})_{anode} \rightarrow YO + 2e \qquad (10.3)$$

The reaction on the left is called the *cathodic reaction* and the right side is the *anodic reaction*. The sum of these half-cell reactions is the overall cell reaction of (10.1).

The short-circuited cell of Figure 10.1 serves no practical purpose other than allowing reaction (10.1) to proceed from left to right. The chemical energy released by the reaction is dissipated as heat.

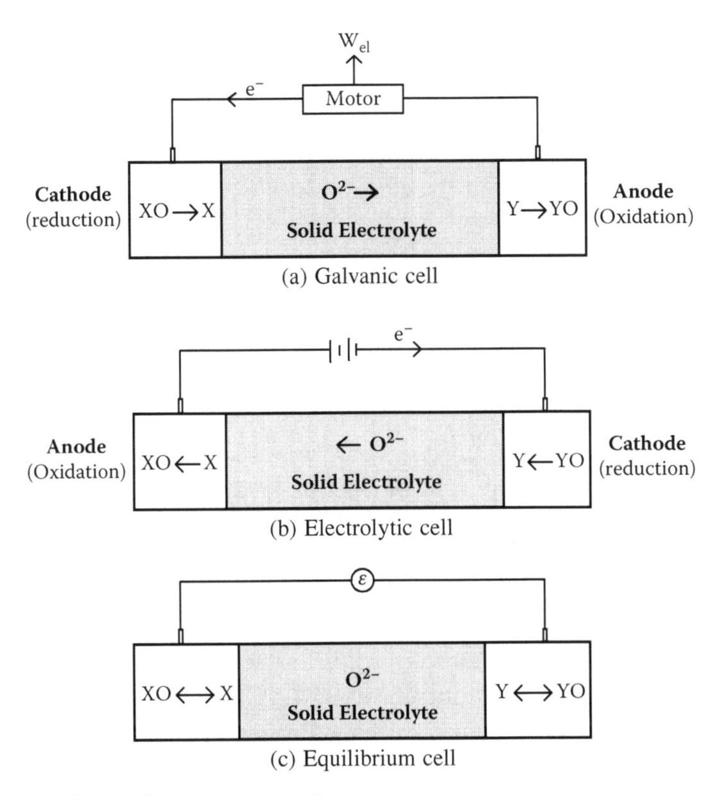

FIGURE 10.2 Three electrochemical cells.

10.2.1 Useful Electrochemical Cells

Three versions of the electrochemical cell result from inserting a device in the wire connecting the two electrodes. Figure 10.2a shows how an electrical motor permits the free-energy difference to be converted to electrical work in what is termed a *galvanic cell*. The ordinary battery and the fuel cell are examples of galvanic cells.

Figure 10.2b demonstrates how a battery or other voltage source is used to force a current through the cell in the opposite direction from the spontaneous reaction in the short-circuited cell. Reaction (10.1) now proceeds from right to left and the positions of the cathode and anode in the cell are reversed. This *electrolytic cell* mode of operation is commonly employed in practical applications of aqueous electrochemistry (e.g., to produce H_2 and O_2 by electrolysis of water).

In the third type of operation, shown in Figure 10.2c, all current flow and both half-cell reactions are stopped by placing in the line a potentiometer or an external voltage source that just balances the electric potential produced by reaction (10.1). Because no current flows, neither of the electrodes can be labeled a cathode or an anode. This *equilibrium cell* provides a method for measuring the standard free-energy changes of the overall cell reaction. To this end, the cell voltage, or *electromotive force* (EMF), denoted by ε, is measured at various cell temperatures.

Compact notation for the electrochemical cell in any of the forms shown in Figures 10.2a to 10.2c is:

$$X|XO||electrolyte||Y|YO \quad (10.4)$$

The single vertical lines separate the two species in the left- and right-hand half cells and the double vertical lines enclose the solid electrolyte that joins the half cells. In general, the specific solid electrolyte is indicated in the middle (e.g., CSZ for calcia-stabilized zirconia instead of "electrolyte").

10.3 THE CELL EMF AND THE FREE ENERGY OF THE OVERALL REACTION

The great utility of the equilibrium electrochemical cell of Figure 10.2c is that its EMF is proportional to the difference in the free energies of the mixtures in the two electrodes—that is, ε is a direct measure of ΔG of the overall cell reaction. This relationship is derived in this section.

If the z moles of electrons pass through a voltage drop ε in the external circuit of the galvanic cell in Figure 10.2a, the motor produces a quantity W_{el} of work.*

$$W_{el} = \varepsilon z e N_{Av}$$

where $e = 1.6 \times 10^{-19}$ coulombs is the electronic charge, $N_{Av} = 6 \times 10^{23}$ is Avogadro's number and the product eN_{Av} is the charge of one mole of electrons. This product is called Faraday's constant, $F = 96{,}500$ coulombs/mole. Since a coulomb is 1 Joule per volt, and there are 4.184 Joules in a calorie, a more convenient value of Faraday's constant is 96.5 kJ/mole-volt.

If the cell operates reversibly, Equation (1.20) shows that the external work is accompanied by a decrease ΔG in the free energy of the entire cell:

$$W_{ext} = -\Delta G$$

Combining the above two equations yields:

$$\Delta G = -z(eN_{Av})\varepsilon = -zF\varepsilon \quad (10.5a)$$

If all species in both half cells are in their standard states (i.e., pure liquids or solids, as in reaction [10.1]), the superscript o is appended to the quantities in Equation (10.5a) to indicate this special case:

$$\Delta G^o = -zF\varepsilon^o \quad (10.5b)$$

Equation (10.5b) applies to the equilibrium cell of Figure 10.2c because all four species in reaction (10.1) are in their standard states.

---

* This equation is the time-integrated form of the familiar electrical formula: power = voltage × current.

Example: What is the EMF of the equilibrium cell X|XO||electrolyte||Y|YO at 1000°C with X = Ni and Y = Fe?

Converting kcal to kJ in the Ellingham diagram of Figure 9.8, the free energies of formation of the oxides are:

$$\Delta G^o_{XO} = \tfrac{1}{2}(-56 \times 4.184) = -117 \text{ kJ/mole} \qquad \Delta G^o_{YO} = \tfrac{1}{2}(-88 \times 4.184) = -184 \text{ kJ/mole}$$

The factor of 1/2 accounts for the fact that the Ellingham diagram gives free energies of formation per mole of O_2 while reactions (10.2a) and (10.2b) are written for one mole of oxide. Combining the above values gives $\Delta G^o = \Delta G^o_{YO} - \Delta G^o_{XO} = -67$ kJ/mole, and Equation (10.5b) with $z = 2$ gives a cell EMF of:

$$\varepsilon^\circ = -\frac{(-67)}{2 \times 96.5} = 0.35\ V$$

Digression. Reactions (10.2) suggest an alternative interpretation of the electrochemical cell based on Reaction (10.1). Each half cell of Equation (10.4) is a MO/M couple with a characteristic oxygen pressure $(p_{O_2})_{MO/M}$ (see Digression 2 in Section 9.9.1). The electrochemical cell of Equation (10.4) could equally well be written as:

$$2O(O_2)_{XO/X} \text{ || electrolyte || } (O_2)_{YO/Y} \tag{10.4a}$$

In place of reaction (10.3), the half-cell reaction are:

$$2O^{2-} + 4e = (O_2)_{XO/X} \qquad (O_2)_{XO/X} = 2O^{2-} + 4e \tag{10.3a}$$

Instead of Equation (10.5b), the cell EMF is:

$$\varepsilon = \frac{RT}{4F}\ln\left[\frac{(p_{O_2})_{XO/X}}{(p_{O_2})_{YO/Y}}\right] \tag{10.5c}$$

That this approach and the one taken at the beginning of this section are equivalent can be shown by expressing the oxygen pressures in Equation (10.5c) in terms of the standard free energy of formation of Reaction (10.2) using the law of mass action (see Equation (9.33a)).

$$(p_{O_2})^{1/2}_{XO/X} = \exp\left[\frac{\Delta G^o_{XO/X}}{RT}\right] \qquad (p_{O_2})^{1/2}_{YO/Y} = \exp\left[\frac{\Delta G^o_{YO/Y}}{RT}\right]$$

Substituting these equations into Equation (10.5c) yields:

$$\varepsilon = \frac{1}{2F}\left(\Delta G^o_{XO/X} - \Delta G^o_{YO/Y}\right) = -\frac{\Delta G^o}{2F}$$

Which is Equation (10.5b).

The couples in the half cells are not restricted to metals and their oxides. Problem 10.7 analyzes a cell in which one electrode contains two oxides of the same metal. However, all components of the cell are pure, and are thus in standard states. Nonstandard variants of the solid-state electrochemical cell are analyzed in the following section.

10.4 NONSTANDARD SOLID-STATE ELECTROCHEMICAL CELLS

When one or more of the constituents in the half cells are not in their standard states (i.e., are *nonstandard*), Equation (10.5a) relates ΔG, the free energy change of the overall reaction, to the cell EMF. The first step in making the connection between the cell voltage and the component concentrations is to relate ΔG to the chemical potentials of the constituents in the half cells. Generalizing the analysis in Section 9.4, the version of Equation (9.16) for a nonequilibrium reaction is:

$$\Delta G = \sum_{products} \nu_i \mu_i - \sum_{reactants} \nu_i \mu_i \tag{10.6}$$

where μ_i is the chemical potential of constituent i and the sums are over the constituents in the product and reactant sides of the overall reaction, respectively. ν_i is the balancing number of constituent i in the overall reaction.

When a cell is short circuited, current flows until the compositions in the two electrodes reduce ΔG to zero. In this limit, Equation (10.6) reduces to the chemical equilibrium condition given by Equation (9.16).

The constituents in the nonstandard half cells are not restricted to pure liquids or solids, as they were in Sections 10.1 to 10.3. If the components are solid or liquid solutions, the chemical potentials are given by:

$$\mu_i = g_i^o + RT \ln a_i = g_i^o + RT \ln(\gamma_i x_i) \tag{10.7}$$

where a_i is the activity of species i in solution, or the product of the activity coefficient and the mole fraction of species i. For the special case of pure constituents (i.e., unit activity), Equation (10.7) reduces to $\mu_i = g_i^o$.

If the constituent is a gas, the chemical potential is given by:

$$\mu_i = g_i^o + RT \ln p_i \tag{10.8}$$

Substitution of Equation (10.7) and/or Equation (10.8) into Equation (10.6) gives ΔG in terms of activities and partial pressures of the constituents in the half cells. The cell EMF follows from Equation (10.5a). Several nonstandard-state electrochemical cells are described below.

10.4.1 Half Cells with Solutions

Instead of pure solids or liquids, one or more of the constituents may be in solution with a different substance. This nonstandard condition can prevail in one or both half cells. Dilution of a reactive component reduces its activity and consequently affects the cell EMF. To illustrate this effect consider the cell of Equation (10.4) modified to permit species X to be dissolved in a solvent metal Z. This cell is described by:

$$\text{X(soln in Z)|XO||electrolyte||Y|YO} \tag{10.9}$$

Using Equation (10.7) with the activities of all species except X equal to unity, Equation (10.6) gives:

$$\Delta G = \mu_{YO} + \mu_X - \mu_{XO} - \mu_Y = g^o_{YO} + (g^o_X + RT \ln a_X) - g^o_{XO} - g^o_Y = \Delta G^o + RT \ln a_X \tag{10.10}$$

where ΔG^o is the standard state free energy change of the overall cell reaction and a_X is the activity of component X in the X-Z alloy. Substituting Equation (10.10) into Equation (10.5a) gives the cell EMF:

$$\varepsilon = \varepsilon^o - \frac{RT}{2F} \ln a_X \tag{10.11}$$

The last term in this equation is the result of the dissolution of X in Z. Measuring the EMF of this cell determines the activity coefficient of X in the X/Z solution.

Example: The following is a modification of the example given in Section 10.3. Here X = Ni, Y = Fe, and Z = Pd (an inert metal). The activity of Ni in the Ni-Pd alloy is 0.5 and the temperature is 1000°C. With $\varepsilon^o = 0.35$ V from the example in Section 10.3, Equation (10.11) yields:

$$\varepsilon = 0.35 - \frac{1.986 \times 10^{-3} \times 1273}{2 \times 23.06} \ln(0.5) = 0.39\ V$$

Dilution of Ni in Pd increases the cell voltage because the tendency of NiO to be converted to Ni is enhanced by depressing the activity of Ni by alloying (Le Chatelier's principle).

Problems 10.1 and 10.6 involve cells with electrodes containing alloys that behave as nonideal solutions.

10.4.2 Cells with Gaseous Electrodes

The cell

$$O_2\ \text{||electrolyte||}\ \text{M|MO}_2 \tag{10.12}$$

contains a gas with a fixed oxygen pressure in one electrode and the M/MO_2 metal-metal oxide couple in the other. The overall cell reaction is:

$$M(s) + O_2 = MO_2(s) \tag{10.13}$$

Using Equation (10.8) for the chemical potential of O_2, the form of Equation (10.6) for this cell is:

$$\Delta G = g^o_{MO_2} - g^o_M - \left(g^o_{O_2} + RT \ln p_{O_2}\right) = \Delta G^o - RT \ln p_{O_2}$$

Applying this result to Equation (10.5a) with $z = 2$ and noting that from Equation (9.33b),

$$\Delta G^o = RT\ln\left(p_{O_2}\right)_{MO_2/M}$$

is the oxygen potential of the MO_2/M couple, gives the cell EMF:

$$\varepsilon = -\frac{RT}{2F}\left[\ln\left(p_{O_2}\right)_{MO_2/M} - \ln p_{O_2}\right] \tag{10.14}$$

Equation (10.14) provides a means of determining the free energy of formation of MO_2 by finding the oxygen pressure in the gaseous electrode, p_{O_2}, at which $\varepsilon = 0$.

Other examples of solid-state electrochemical cells with a gaseous electrode are given in Problems 10.9 and 10.10.

10.4.3 Fuel Cells

The electrochemical cell with solids as electrodes, as in Figure 10.2a, operates as a battery. That is, it produces current until one of the reactants is exhausted. The cell needs to be charged by operating in the electrolysis mode, as in Figure 10.2b.

If the active constituents of the two half cells are gaseous, however, continuous operation is possible. The preceding example dealt with a cell with one gaseous electrode and one solid electrode. In this section, a cell fueled only by gas flows is analyzed. Electrochemical cells fed by external reactant sources are called *fuel cells.*

The simplest fuel cell is one that combines hydrogen and oxygen to produce only water as a waste product. A schematic of such a cell is shown in Figure 10.3. The overall cell reaction is:

$$H_2(g) + \tfrac{1}{2}O_2(g) = H_2O(L) \tag{10.15}$$

The unit is operated at room temperature, which permits an aqueous solution to be used as the electrolyte. Pure H_2 is fed to the anode chamber where it is oxidized to H^+:

$$H_2(g) \rightarrow 2H^+ \text{ (in electrolyte)} + 2e \tag{10.16}$$

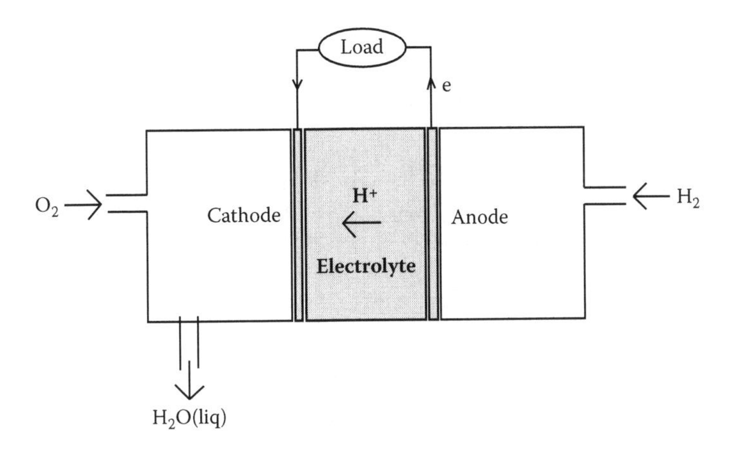

FIGURE 10.3 A hydrogen-oxygen fuel cell.

Hydrogen ions enter the electrolyte and are transported to the cathode. Simultaneously, the electrons liberated in the anodic reaction flow through the external work-producing device. At the cathode, the protons and electrons produced in the anode combine with oxygen gas fed to the cathode chamber according to the half cell reaction:

$$\tfrac{1}{2}O_2(g) + 2H^+ \text{ (from electrolyte)} + 2e \rightarrow H_2O(L) \tag{10.17}$$

The sole reaction product, pure liquid water, drains from the cell as the only waste stream.

In terms of chemical potentials, the free energy change of the overall cell reaction is:

$$\Delta G = \mu_{H_2O(L)} - \tfrac{1}{2}\mu_{O_2(g)} - \mu_{H_2(g)}$$

$$= g^o_{H_2O(L)} - \tfrac{1}{2}(g^o_{O_2} + RT \ln p_{O_2}) - (g^o_{H_2} + RT \ln p_{H_2})$$

or

$$\Delta G = \Delta G^o_{H_2O(L)} - RT \ln\left(p_{H_2}\sqrt{p_{O_2}} \right)$$

where

$$\Delta G^o_{H_2O(L)} = g^o_{H_2O(L)} - \tfrac{1}{2} g^o_{O_2} - g^o_{H_2}$$

is the standard free energy of formation of liquid water. At 25°C, $\Delta G^o_{H_2O(L)} = -239\ kJ/mole$.

Using the above equation for ΔG in Equation (10.5a) with $z = 2$ gives the cell voltage:

$$\varepsilon = \varepsilon^o + \frac{RT}{2F} \ln\left(p_{H_2}\sqrt{p_{O_2}} \right) \tag{10.18}$$

where the standard EMF of the cell is:

$$\varepsilon^o = -\frac{\Delta G^o_{H_2O(L)}}{2F} = -\frac{-239}{2 \times 96.5} = 1.23\ V$$

If the gases in the electrodes are at atmospheric pressure, the cell voltage is also 1.23 V. Operation at higher pressures would increase the available voltage but would entail construction of a more rugged unit. The actual EMF provided by the fuel cell is less than the open-circuit value given above because current is drawn to produce power. The i–V characteristics of the cell determine its performance. To obtain a usable voltage for power production, many cells must be operated in series.

10.5 AQUEOUS ELECTROCHEMISTRY

10.5.1 Cell Operational Modes

Aqueous electrochemical cells consist of half cells containing aqueous solutions of ionized species. This branch of electrochemistry exhibits an additional dimension not found in the solid-state counterpart. The added degree of complexity is illustrated in the schematic 3 × 3 matrix of Figure 10.4. The columns contain the three operational modes: galvanic, electrolytic, and equilibrium. The rows show the three types of devices: two half cells separated by a bridge; two electrodes with a common electrolyte; and electrolyte only.

In the first row, the compartments are separated by a bridge that prevents intermixing of the contents of the half cells. This is the sole mode of operation available to solid-state electrochemical devices of the sort described up until now. Combination 1(c) is particularly useful in determining the thermodynamic aspects of electrochemical systems.

In the second row labeled "whole cell," half cells no longer exist, although half cell reactions remain; two different electrodes are immersed in a common electrolyte.

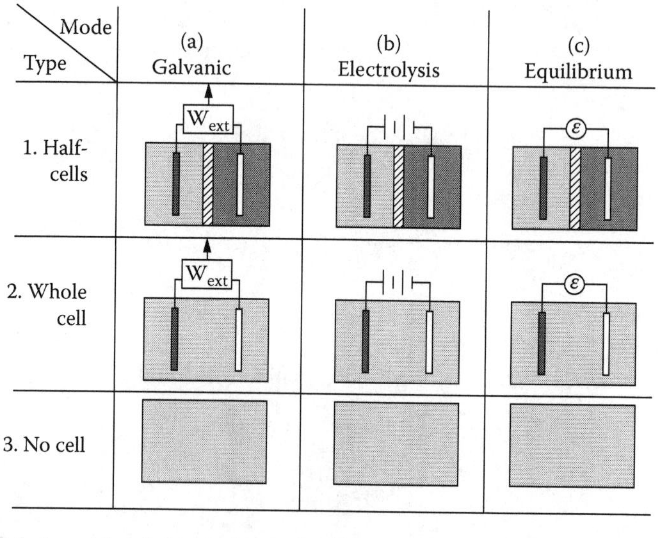

FIGURE 10.4 A matrix of electrochemical cells.

Configuration 2(b) is how nearly all practically applications of electrochemistry are conducted (e.g., electrolysis of water, plating of metals, the lead-acid battery).

In the third variant in the matrix, called "no cell," all that remains is the electrolyte and perhaps a solid that takes part in the reaction. Arguably, this mode should not even deserve to be termed electrochemistry. However, it represents ionic equilibria, and because its thermodynamic description relies on data obtained by configuration 1(c), it is properly considered as part of electrochemistry.

10.5.2 The Cell EMF and Chemical Potential of the Overall Cell Reaction

This section is a reworking of the important point that was dealt with for nonaqueous systems in Section 10.3. Here the same topic is addressed from a different direction using a typical reaction in an aqueous electrochemical cell. The objective is to demonstrate that the cell EMF is directly related to the chemical potential difference of the overall cell reaction by an equation analogous to Equation (10.5a). To this end, the cell described below is utilized.

Figure 10.5 depicts a special kind of type 1(c) cell with two metal electrodes immersed in water and different ionic species in each compartment. The cell operates in the equilibrium mode with a potentiometer measuring the voltage difference between the two electrodes. In the left-hand compartment, the electrode is a metal M and the solution contains ions of the metal M^{z+}. Positively charged ions are called *cations*. There must also be negative ions in solution (called *anions*) to maintain electrical neutrality, but these are ignored because they do not participate in the electrochemical reactions.

The left-hand half cell containing the metal M and its ion M^{z+} in solution is in equilibrium with electrons in the circuit to the left of the potentiometer. The half cell reaction describing this equilibrium is:

$$M = M^{z+} + ze \tag{10.19}$$

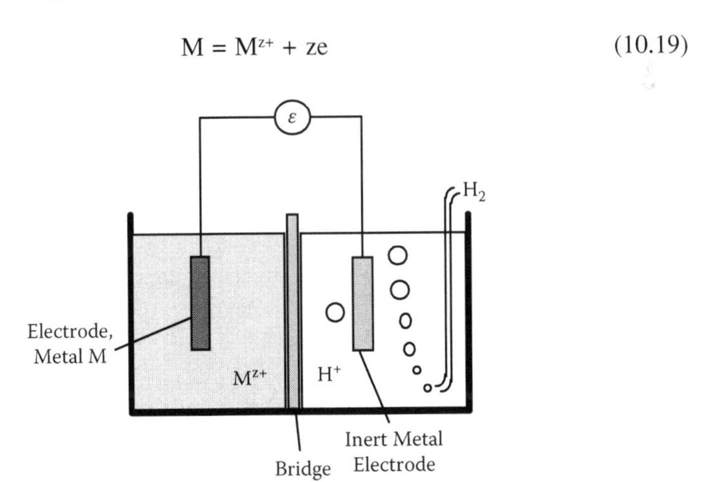

FIGURE 10.5 An aqueous electrochemical cell with an active-metal half cell and a hydrogen half cell.

The right-hand compartment comprises a hydrogen half cell. Gaseous H_2 saturates the solution with molecular hydrogen, which, by virtue of rapid reaction on the inert metal electrode, maintains equilibrium with hydrogen ions in solution and electrons in the right side of the external circuit. This half cell reaction is:

$$H_2 = 2H^+ + 2e \tag{10.20}$$

The overall-cell reaction is the difference between the two half-cell reactions (with Equation [10.20] multiplied by $z/2$ in order to cancel the electrons):

$$M + zH^+ = M^{z+} + \tfrac{1}{2}zH_2 \tag{10.21}$$

Digression. If the cell is short-circuited, the reactions in the half cells correspond to chemical attack of the metal by acid. The overall cell reaction of (10.21) proceeds from left to right. The metal half cell is the anode, where oxidation (in the sense of increasing the valence of M) occurs and reaction (10.19) proceeds from left to right. The hydrogen half cell is the cathode, where protons are reduced to elemental hydrogen, or reaction (10.20) goes from right to left.

The introduction of metal ions into the anode solution and removal of hydrogen ions from the cathode solution upsets electrical neutrality of these solutions. This charge imbalance is rectified by flow of anions (e.g., Cl^-), from right to left. This compensating charge flow must pass through the bridge in Figure 10.5. The bridge is a porous structure or a gel that serves to prevent gross mixing of the solutions in the two half cells while allowing for transport of negative ions to maintain electrical neutrality during cell operation.

In the equilibrium mode, the cell EMF is the result of the difference of the electric potentials of the electrons in the left-hand and right-hand electrodes. To determine how this EMF relates to the chemical potentials of the constituents of overall reaction (10.21), the chemical potential balance is written including the electric potentials of the electrodes. For the left-hand and right-hand cells in Figure 10.5, the total potentials are:

$$\mu_L = \mu(M^{z+}) + zF\phi_L - \mu(M) \tag{10.22L}$$

$$\mu_R = \tfrac{1}{2}z[2\mu(H^+) + 2F\phi_R - \mu(H_2)] \tag{10.22R}$$

where ϕ_L and ϕ_R are the electric potentials (in volts) in the left-hand and right-hand electrodes, respectively. They are multiplied by the Faraday constant, F = 96.5 kJ/mole-volt in order to convert the electric potentials to the same units as the chemical potentials. The $z/2$ factor in Equation (10.22R) is a consequence of the stoichiometry of overall reaction (10.21).

Equilibium is expressed by:

$$\mu_L = \mu_R \tag{10.23}$$

Substituting Equations (10.22L) and (10.22R) into Equation (10.23) yields:

$$\mu(M^{z+}) - \mu(M) - z\mu(H^+) + \tfrac{1}{2}z\mu(H_2) = -zF(\phi_L - \phi_R)$$

The term on the left is $\Delta\mu$, the chemical potential difference of overall reaction (10.21). It is the aqueous equivalent of the free energy difference ΔG used in describing nonaqueous cells. The electric potential difference on the right is the cell EMF, ε, so the equation is:

$$\Delta\mu = -zF\varepsilon \tag{10.24}$$

what remains is to express chemical potentials in terms of ion concentrations and hydrogen gas pressure.

10.5.3 Ion Standard State

The chemical potential of an ion in water is given by:

$$\mu = \mu^{\circ} + RT\ln a \tag{10.25}$$

where a, the activity of the species, is the product of its activity coefficient γ and the concentration c, in moles per liter of solution, or M, the *molarity*.*

$$a = \gamma\ c \tag{10.26}$$

μ° is the chemical potential of the ion in its standard state, and therein lies the difficulty. The standard state of a component in a nonaqueous solution is the pure substance (see Section 10.4). However, this cannot be applied to aqueous ions because a pure ion state does not exist. Instead, the standard state is chosen as the ion in an infinitely dilute solution, or, equivalently, at concentrations where Henry's law applies. This is the region where the activity is proportional to the concentration, or, the activity coefficient is independent of concentration. This condition has a solid physical basis. Ions in a dilute solution are surrounded by water molecules and are not appreciably affected by other ions in the solution.

The dilute-solution regime is chosen as the standard state and here the ion's activity coefficient is arbitrarily set equal to unity.

In this regime, the chemical potential becomes:

$$\mu = \mu^{\circ} + RT\ln c \tag{10.27}$$

This equation implies a hypothetical 1 M ideal solution standard state, which is equivalent to the real solution with an activity of unity. The former state is hypothetical because at a concentration of 1 M, the ionic species may be beyond the Henry's law region, where $\gamma \neq 1$. This standard state is best illustrated by the *concentration cell* for the H^+/H_2 couple shown at the top of Figure 10.6.** The right-hand half cell contains a fixed acid concentration c(H) in the Henry's law region—which is the reason for the designation c(H).

---

* Most treatments of electrochemistry (Atkins, 1978; Gaskell, 1981) use *molality*, which is the concentration unit expressed as moles of a species per kilogram of water. Molarity and molality differ by less than 1/2%.

** A concentration cell is one in which the two half cells contain the same solutes but at different concentrations.

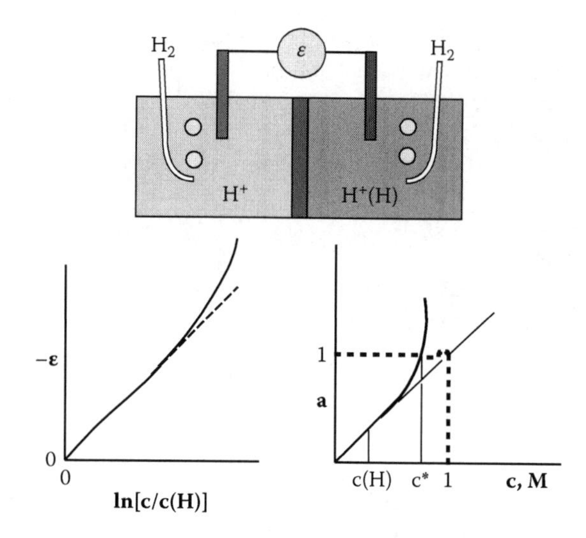

FIGURE 10.6 An H^+/H_2 concentration cell (top); variation of cell EMF with the H^+ concentration in the left-hand half cell (bottom left); activity–concentration plot derived from the cell potential measurements (bottom right).

The half-cell reaction in both sides of the cell is given by Equation (10.20). The concentration in the left-hand cell is varied and the cell EMF ε is measured. With the chemical potential of the left-hand half cell described by Equation (10.25) and (10.26), and that in the right-hand half cell by Equation (10.27), the chemical potential difference is:

$$\Delta\mu = RT\ln[\gamma c/c(H)] \tag{10.28}$$

The cell potential is related to Δμ by Equation (10.24) with $z = 2$:

$$\varepsilon = -\frac{\Delta\mu}{2F} = -\frac{RT}{2F}\left[\ln\left(\frac{c}{c(H)}\right) + \ln\gamma\right] \tag{10.29}$$

The plot on the bottom left of Figure 10.6 shows the cell potential as a function of the H^+ concentration in the left-hand half cell, which is denoted by c in Equation (10.29). The difference between the solid curve and the dashed line is due to the second term in the brackets, and so permits measurement of γ. That −ε is greater than the continuation of the Henry's law line indicates that the activity coefficient chosen for this example is greater than unity.

The values of γ as a function of c so obtained yield the activity–concentration plot shown at the bottom right of Figure 10.6. The intersection of the two dotted lines in this graph represents the hypothetical 1 M standard state of H^+ in water. To establish unit H^+ activity, however, the plot shows that the actual acid concentration must be c^*, not 1 M.

The purpose of the above analysis was to show how a unique reference half cell, the *standard hydrogen electrode*, or SHE, is constructed. The SHE is a solution with unit H^+ activity and supplied with H_2 gas at 1 atm pressure. This unique half cell is the reference with which all other half cells are paired in order to describe the thermodynamics of aqueous electrochemistry.

10.5.4 Standard Electrode Potentials

Figure 10.5 shows the M/M^{z+} half cell combined with the hydrogen half cell which, in addition, is the SHE. To establish an electrochemical database, the metal M must be pure and the M^{z+} activity must be unity. If unit activity of M^{z+} does not fall in the Henry's law region, the concentration $c^*_{M^{z+}}$ which renders $a_{M^{z+}} = 1$ is determined by the same method applied above for H^+. The M/M^{z+} half cell with these restrictions is termed a *standard electrode* and the measured EMF, ε^o in Equation (10.26) is the *standard electrode potential* of the M/M^{z+} half cell reaction. This corresponds to the overall reaction with all species in their standard states, so Equation (10.24) becomes:

$$\Delta\mu^o = -zF\varepsilon^o \quad (10.30)$$

The database for aqueous electrochemistry consists of standard electrode potentials for a large number of half cell reactions. Since most aqueous systems are at room temperature, a single table at 25°C suffices to accommodate the entire database. An abridged table of standard electrode potentials is given in Table 10.1. A more extensive tabulation can be found in *The Handbook of Chemistry and Physics* (Lide, 2006).

According to Equation (10.30), the sign of ε^o is opposite that of the standard chemical potential change of the overall cell reaction. By convention, the overall cell reaction is written with the oxidized portion of the couple written on the right-hand side, as it appears in Equation (10.21). Or, the half cell reaction for the couple is written with the electrons on the right. With this convention, couples involving a very reactive metal will tend to exhibit negative values of $\Delta\mu^o$ (because the ions are very stable in solution) and conversely, positive ε^o.

Note that the standard electrode potential for the hydrogen electrode is zero because the SHE is the reference half cell. Some of the entries in Table 10.1 involve only water and the gases oxygen and hydrogen. Another group consists of metals and their ions; a third group contains only ions of the same element which have multiple valence (or oxidation) states. The last group in the table contains half-cell reactions in which solids containing oxygen are involved.

Often the standard electrochemical potential of half-cell reactions that are not included in Table 10.1 (or similar tabulation) can be derived from reactions that appear there. In some cases, thermochemical data from sources other than the ε^o table need to be incorporated. In all cases, the method involves converting the standard electrochemical potentials to standard free energies of the overall cell reaction by Equation (10.30), combining the latter algebraically to determine $\Delta\mu^o$ of the desired half-cell reaction, and finally converting back to ε^o.

TABLE 10.1
Standard Electrode Potentials at 25°C

	Half Cell Reactionε	ε°, Volts
	Involving gases	
1	$H_2O = \frac{1}{2}O_2 + 2H^+ + 2e$	–1.229
2	$2OH^- = \frac{1}{2}O_2 + H_2O + 2e$	–0.401
3	$H_2 = 2H^+ + 2e$	0
	Involving metals	
4	$Au = Au^{3+} + 3e$	–1.498
5	$Cu = Cu^{2+} + 2e$	–0.337
6	$Ni = Ni^{2+} + 2e$	0.230
7	$Fe = Fe^{2+} + 2e$	0.440
8	$Na = Na^+ + e$	2.714
	Involving only ions	
9	$Fe^{2+} = Fe^{3+} + e$	–0.771
10	$U^{4+} + 2H_2O = UO_2^{2+} + 4H^+ + 2e$	–0.338
11	$Pu^{4+} + 2H_2O = PuO_2^{2+} + 4H^+ + 2e$	–1.043
12	$Pu^{3+} = Pu^{4+} + e$	–0.98
13	$Cu^+ = Cu^{2+} + e$	–0.16
	Involving solid oxides or hydroxides	
14	$UO_2(s) = UO_2^{2+} + 2e$	0.43
15	$Fe(s) + H_2O = FeO(s) + 2H^+ + 2e$	0.03
16	$2Cu(s) + 2OH^- = Cu_2O(s) + H_2O + 2e$	0.36
17	$Cu(s) + 2OH^- = Cu(OH)_2(s) + 2e$	0.22
18	$Ni + 2OH^- = Ni(OH)_2 + 2e$	0.66

Example: The half cell reaction $Fe = Fe^{+3} + 3e$.

This half cell reaction is the sum of reactions 7 and 9 in Table 10.1. The standard free energy changes for these reactions and for the desired one are:

$$\Delta\mu^o(Fe/Fe^{2+}) = -2F\varepsilon_7^o \qquad \Delta\mu^o(Fe^{2+}/Fe^{3+}) = -2F\varepsilon_9^o \qquad \Delta\mu^o(Fe/Fe^{3+}) = -3F\varepsilon^o$$

Summing the first two free-energy changes gives the third:

$$\Delta\mu^o(Fe/Fe^{2+}) + \Delta\mu^o(Fe^{2+}/Fe^{3+}) = \Delta\mu^o(Fe/Fe^{3+})$$

Dividing by 3 gives the standard electrode potential:

$$\varepsilon^o = \frac{2}{3}\varepsilon_7^o + \frac{1}{3}\varepsilon_9^o = \frac{2}{3}(0.44) + \frac{1}{3}(-0.77) = 0.036\ V$$

Example: The half-cell reaction: $2Fe^{2+} + 3H_2O = Fe_2O_3(s) + 6H^+ + 2e$.

This half-cell reaction can be decomposed into three simpler reactions, only two of which are found in Table 10.1:

$Fe = Fe^{2+} + 2e$ $\qquad \Delta\mu^o(Fe/Fe^{2+}) = -2 \times 96.5 \times 0.44 = -85$ kJ/mole

$\frac{4}{3}Fe + O_2 = \frac{2}{3}Fe_2O_3(s)$ $\qquad \Delta\mu^o(Fe/Fe_2O_3) = -120 \times 4.184 = 502$ kJ/mole

$2H_2O = O_2 + 4H^+ + 4e$ $\qquad \Delta\mu^o(H_2O/O_2) = -4 \times 96.5 \times (-1.23) = 475$ kJ/mole

The first and last of these half-cell reactions are numbers 7 and 1, respectively, in Table 10.1. The middle reaction does not involve either aqueous ions or electrons; its free-energy change (at 25°C) is obtained from the Ellingham diagram of Figure 9.8. The free-energy change of the desired half-cell reaction is obtained by algebraically combining the above three-component reactions:

$$\begin{aligned}\Delta\mu^o(Fe^{2+}/Fe_2O_3) &= 3/2\ \Delta\mu^o(H_2O/O_2) + 3/2\ \Delta\mu^o(Fe/Fe_2O_3) - 2\ \Delta\mu^o(Fe/Fe^{2+}) \\ &= 129 \text{ kJ/mole}\end{aligned}$$

Converting this to the half cell EMF by use of Equation (10.30) yields:

$$\varepsilon^o = -\frac{\Delta\mu^o(Fe^{2+}/Fe_2O_3)}{2F} = -\frac{129}{2 \times 96.5} = -0.67\ V$$

Problem 10.2 is another example of combining entries in Table 10.1 to obtain the standard electrode potential of a reaction not listed in the table.

10.5.5 Nernst Potential

If the ion activity in the half cell that is coupled to the SHE is not equal to the standard value of 1, the EMF can be determined by generalizing electrode reactions such as (10.19). This reaction is of the type:

$$\text{reduced form} = \text{oxidized form} + ze \tag{10.31}$$

The terms "reduced form" and "oxidized form" refer to the element that changes its valence (or oxidation state) in the half cell reaction. For example, in reaction 2 in Table 10.1, the element that changes valence is oxygen; the reduced form is OH^- and the oxidized form is O_2. The concentrations of the reduced form and the oxidized form must be low enough for the activity coefficients to be unity. The EMF of a nonstandard half cell relative to the SHE is:

$$\varepsilon_N = \varepsilon^o - \frac{RT}{zF}\ln\left(\frac{c_{oxidized}}{c_{reduced}}\right) = \varepsilon^o - \frac{0.059}{z}\log\left(\frac{c_{oxidized}}{c_{reduced}}\right) \tag{10.32}$$

where the last form employs the numerical value of RT/F for $T = 298$ K multiplied by 2.3 to convert from the natural logarithm to the base-ten logarithm. The concentrations in the argument of the logarithmic term are those of the oxidized and reduced forms of the element that changes valence. If the half cell reaction contains H^+ or OH^-, the concentrations of these species must also be included in the numerator and/or denominator of the logarithmic term. Solid species in pure form (such as metals or solid oxides or hydroxides) are not included in the logarithmic term because their activities are unity, as is that of water. Equation (10.32) is known as the *Nernst equation* and ε_N is the *Nernst potential.*

Example: Calculate the Nernst potential of half cell 10 in Table 10.1 for the following concentrations:

$$c_{H^+} = 0.01 \text{ M}, \qquad c_{UO_2^{2+}} = 1.0 \text{ M}, \qquad c_{U^{4+}} = 0.1 \text{ M}.$$

The specific form of Equation (10.32) for this half cell is:

$$\varepsilon_N = \varepsilon^o - \frac{0.059}{2}\log\left(\frac{c_{H^+}^4 c_{UO_2^{2+}}}{c_{U^{4+}}}\right) = -0.338 - \frac{0.059}{2}\log\left(\frac{0.01^4 \times 1.0}{0.1}\right) = -0.60\ V$$

10.6 NONEQUILIBRIUM AQUEOUS ELECTROCHEMICAL CELLS

The preceding section dealt with aqueous electrochemical cells operated in the equilibrium mode. The primary objective was to provide the EMF of any half cell when measured against a SHE. The purpose of the present section is to analyze electrochemical cells consisting of two arbitrary half cells, neither of which is a SHE. Such cells can be operated as galvanic cells, electrolytic cells or fuel cells.

10.6.1 BATTERY OPERATION (GALVANIC MODE)

The EMF of a cell consisting of electrodes labeled A and B can be obtained from their Nernst potentials. As shown in Figure 10.7, the Nernst potentials ε_A and ε_B are EMFs relative to the SHE.

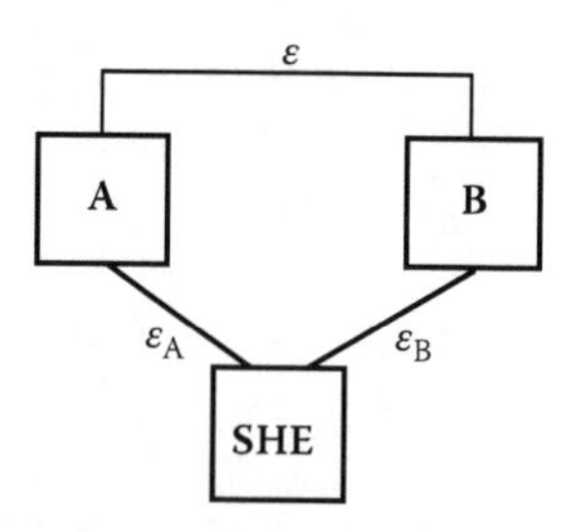

FIGURE 10.7 Relationship of the Nernst potentials of half cells A and B to the EMF of a cell comprised of these two half cells.

The half cell reactions and corresponding Nernst equations for A and B are:

$$A_{red} = A_{oxid} + z_A e \qquad \varepsilon_A = \varepsilon_A^o - \frac{0.059}{z_A}\log\left(\frac{c_{A_{oxid}}}{c_{A_{red}}}\right)$$

$$B_{red} = B_{oxid} + z_B e \qquad \varepsilon_B = \varepsilon_B^o - \frac{0.059}{z_B}\log\left(\frac{c_{B_{oxid}}}{c_{B_{red}}}\right)$$

When operated in the galvanic mode, the electrode with the most positive Nernst potential has the greater tendency to lose electrons and become oxidized. It is therefore the anode. Assume that $\varepsilon_A > \varepsilon_B$ so that half-cell A is the anode and half-cell B is the cathode. When the battery operates, the overall-cell reaction is:

$$A_{red} + B_{oxid} \rightarrow A_{oxid} + B_{red} \tag{10.33a}$$

and the battery voltage (without drawing current) is:

$$\varepsilon = \varepsilon_A - \varepsilon_B = (\varepsilon_A^o - \varepsilon_B^o) - 0.059\log\left(\frac{c_{A_{oxid}}^{z_A}\, c_{B_{red}}^{z_B}}{c_{A_{red}}^{z_A}\, c_{B_{oxid}}^{z_B}}\right) \tag{10.33b}$$

Example: Calculate the EMF of a plutonium-iron battery in which the Pu half cell contains equal concentrations of trivalent and tetravalent plutonium and the iron half cell consists of an iron metal electrode immersed in a 0.2 M Fe^{2+} solution.

The electrode reactions are numbers 7 and 12 in Table 10.1. The Nernst potentials are:

$$\varepsilon_{Pu} = \varepsilon_{Pu}^o - \frac{0.059}{1}\log\left(\frac{c_{Pu^{4+}}}{c_{Pu^{3+}}}\right) = -0.98 - 0.059\log(1) = -0.98\ V$$

$$\varepsilon_{Fe} = \varepsilon_{Fe}^o - \frac{0.059}{2}\log\left(c_{Fe^{2+}}\right) = 0.44 - \frac{0.059}{2}\log(0.2) = 0.46\ V$$

The iron electrode is the anode and the plutonium electrode is the cathode. The cell voltage is:

$$\varepsilon = \varepsilon_{Fe} - \varepsilon_{Pu} = 0.46 - (-0.98) = 1.44\ V$$

Other cell EMF calculations are given in Problems 10.4 and 10.5.

10.6.2 Electrolysis

A frequent high-school science demonstration is electrolysis of water, a schematic of which is shown in Figure 10.8. If the cell were operated in the equilibrium mode

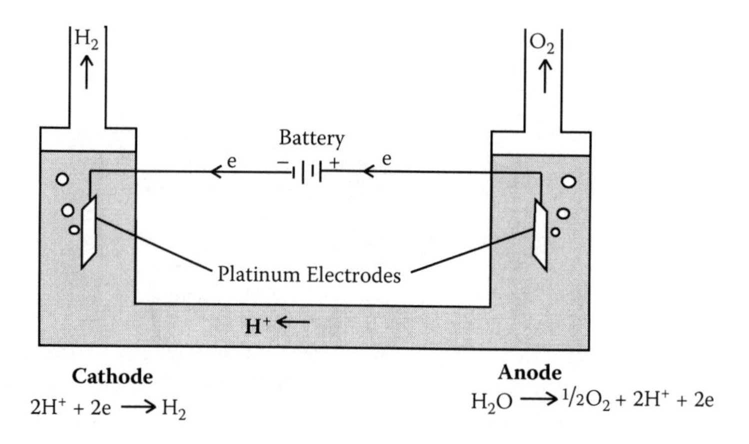

FIGURE 10.8 Cell for electrolysis of water.

(i.e., if the battery voltage just balanced the cell potential), the gases would not be produced at the two platinum electrodes. The equilibrium EMF of the cell is obtained from the Nernst equations for half cell reactions 1 and 3 in Table 10.1:

$$\varepsilon_{H_2} = 0.0 - \frac{0.059}{2}\log\left(\frac{c_{H^+}^2}{p_{H_2}}\right)$$

$$\varepsilon_{O_2} = -1.23 - \frac{0.059}{2}\log\left(c_{H^+}^2 p_{O_2}^{1/2}\right)$$

If the pressures of H_2 and O_2 in the two electrode compartments are 1 atm, the cell potential is:

$$\varepsilon = \varepsilon_{O_2} - \varepsilon_{H_2} = -1.23\ V$$

The significance of the minus sign is that the battery in Figure 10.8 must supply a voltage in excess of 1.23 V in the indicated sense in order to decompose water into hydrogen and oxygen, both at atmospheric pressure. The overall cell reaction is obtained by summing the anodic and cathodic half cell reactions shown in Figure 10.8:

$$H_2O \rightarrow \tfrac{1}{2}O_2 + H_2$$

In the anode compartment, water decomposes into (1) oxygen gas, which bubbles off and is collected; (2) hydrogen ions, which migrate towards the cathode in the bridge separating the two compartments; and (3) electrons, which enter the external circuit to supply the cathodic reaction. At the cathode, the hydrogen ions and the electrons produced in the anode are combined to produce hydrogen gas.

10.6.3 Fuel Cell Mode

The hydrogen–oxygen fuel cell has already been discussed in Section 10.4.3. Basically, the cathode and anode reactions (Equations [10.16] and [10.17]) are the reverse of those for electrolysis. The overall cell reaction is also reversed to represent the combination of hydrogen and oxygen according to Equation (10.15). The physical configuration of the fuel cell, however, is quite different from that of an electrolysis apparatus. In the former, the reacting gases must be accorded direct access to surfaces on which they decompose by producing or consuming hydrogen ions in the electrolyte and electrons for the external circuit. The EMF produced by the fuel cell is the reverse of that for electrolysis. In addition, the pressures of the reactant gases in the electrodes can be independently controlled, leading to an output voltage of:

$$\varepsilon = \varepsilon_{H_2} - \varepsilon_{O_2} = 1.23 + \frac{0.059}{2}\log\left[\left(p_{O_2}\right)_{cathode}^{1/2}\left(p_{H_2}\right)_{anode}\right]$$

10.7 EQUILIBRIUM CONSTANTS OF AQUEOUS IONIC REACTIONS

A chemical reaction such as the acid-metal reaction of Equation (10.21) need not take place in an electrochemical cell such as the one shown in Figure 10.5. Much more common is the row 3 case in Figure 10.4 in which a single electrolyte and the reactive metal M are present. This situation is equivalent to short-circuiting the electrochemical cell, which requires that the Nernst potentials of both half cells be equal:

$$\varepsilon(M^{z+}) = \varepsilon(H^+)$$

When this equality prevails, the ion concentrations become adjusted to accommodate this equilibrium requirement. The Nernst potentials for the two half cell reactions are:

$$\varepsilon(M^{z+}) = \varepsilon^o(M^{z+}) - \frac{0.059}{z}\log\left(c_{M^{z+}}\right)$$

and

$$\varepsilon(H^+) = 0 - \frac{0.059}{2}\log\left(\frac{c_{H^+}^2}{p_{H_2}}\right)$$

In general, neither of the ion concentrations are at the standard-state value of 1 M, nor need the hydrogen pressure be 1 atm. Combining the preceding equations and solving for the combination of concentrations that constitutes the law of mass action* yields:

$$K = \frac{c_{M^{z+}}\,p_{H_2}^{z/2}}{c_{H^+}^z} = 10^{z\varepsilon^o(M^{z+})/0.059}$$

* See Section 9.5 for a discussion of the law of mass action.

where K is the equilibrium constant of Equation (10.21). This equation is the aqueous analog of the equilibrium equations discussed in Chapter 9 (e.g., Equation [9.49] for the slag-melt example). Using the physical symbols for the numerical constant 0.059 in the above equation permits the equilibrium constant to be expressed in a more familiar form:

$$K = 10^{zF\varepsilon^o(M^{z+})/2.3RT} = \exp\left(\frac{zF\varepsilon^o(M^{z+})}{RT}\right) = \exp\left(-\frac{\Delta\mu^o(M^{z+})}{RT}\right)$$

The last form of K in this equation is equivalent to that employed for nonaqueous chemical equilibria, with replacement of ΔG^o in the latter by $\Delta\mu^o$.

The standard electrode potentials in Table 10.1 can be combined to yield equilibrium constants for a wide variety of aqueous ionic reactions, including those that involve solids and gases. For the general case in which the half cells are denoted by A and B, neither of which is a hydrogen half cell, the equilibrium constant for the overall reaction is:

$$K = 10^{z(\varepsilon_A^o - \varepsilon_B^o)/0.059} \tag{10.34}$$

To preserve the correct sign convention, the oxidized form of the species in half-cell A must appear on the right-hand side of the overall reaction, as in reaction (10.33a).

Example: In reprocessing of spent nuclear fuel plutonium in aqueous solution in the IV oxidation state is reduced by adding a solution of ferrous ions. In a solvent extraction step, Pu^{4+} extracts along with U^{6+}. Fe^{2+} reduces Pu^{4+} but not the U^{6+}. The reduced form of Pu is not extractable, and so separation of U and Pu is possible. If 1 liter each of 1 M Fe^{2+} and 0.5 M Pu^{4+} solutions are mixed and equilibrated, what is the ratio of Pu^{4+} to Pu^{3+} ?

The overall reaction is:

$$Fe^{2+} + Pu^{4+} = Fe^{3+} + Pu^{3+}$$

This reaction can be broken up into half-cell reactions 9 and 12 in Table 10.1. However, it is not necessary to write the Nernst equations for these half cell reactions. Instead, the equilibrium constant can be calculated from Equation (10.34). Component A in this equation is iron, because its oxidized form appears on the right-hand side of the reaction as written. The result is:

$$K = \frac{c_{Fe^{3+}}c_{Pu^{3+}}}{c_{Fe^{2+}}c_{Pu^{4+}}} = 10^{\left(\varepsilon_{Fe}^o - \varepsilon_{Pu}^o\right)/0.059} = 10^{\left[-0.77-(-0.98)\right]/0.059} = 3600$$

To determine the equilibrium composition of the solution, the reaction progress variable method is used. Let ξ = moles of Pu^{3+} formed at equilibrium. The table from which the concentrations needed in the above mass action law are determined is shown below:

Species	Initial Moles	Final Moles	Final Conc. (M)
Pu^{3+}	0	x	$\xi/2$
Pu^{4+}	0.5	$0.5 - \xi$	$0.25 - \xi/2$
Fe^{2+}	1.0	$1.0 - \xi$	$0.5 - \xi/2$
Fe	0	x	$\xi/2$

The concentrations in the last column are substituted into the mass action law to yield:

$$K = \frac{\xi^2}{(0.5-\xi)(1.0-\xi)}$$

Since K is very large, inspection of the above equation shows that ξ must be very close to 0.5. The equation can be readily solved by setting $\xi = 0.5$ in the numerator and the last term of the denominator, giving:

$$\xi = \frac{1}{2}\left(\frac{K-1}{K}\right) = 0.5(1 - 1/3600) = 0.49986$$

Forming the ratio of Pu^{4+} to Pu^{3+} from the third column of the reaction progress variable table gives:

$$\frac{c_{Pu^{4+}}}{c_{Pu^{3+}}} = \frac{0.5-\xi}{\xi} = 2.8 \times 10^{-4}$$

Thus, essentially all of the initial tetravalent plutonium is reduced to the trivalent state by addition of the ferrous ion solution.

10.7.1 Dissociation Constant of Water

The reaction describing dissociation of liquid water is:

$$H_2O = H^+ + OH^- \tag{10.35}$$

The equilibrium constant for this overall reaction is obtained from half-cell reactions 1 and 2 in Table 10.1. Subtracting half-cell reaction 2 from number 1 and dividing by two gives reaction (10.35). The Nernst potentials for these two half cell reactions are:

$$\varepsilon_1 = -1.223 - \frac{0.059}{2}\log\left(c^2_{H^+}\sqrt{p_{O_2}}\right)$$

and

$$\varepsilon_2 = -0.401 - \frac{0.059}{2}\log\left(\frac{\sqrt{p_{O_2}}}{c^2_{OH^-}}\right)$$

These two potentials are equated to reflect the condition of equilibrium of reaction (10.35). The oxygen pressure in the two equations cancel and the resulting product of the H^+ and OH^- concentrations is the water dissociation constant:

$$K_W = c_{H^+} c_{OH^-} = 10^{[-1.229-(-0.401)]/0.059} = 10^{-14} \qquad (10.36)$$

This result could also have been obtained by direct application of Equation (10.34). In neutral water (i.e., neither acidic nor basic),

$$c_{H^+} = c_{OH^-} = \sqrt{K_W} = 10^{-7}\ \mathrm{M}$$

A more convenient description of acidity is the pH:

$$\mathrm{pH} = -\log c_{H^+}$$

So that the pH of pure water at 25°C is 7.

The standard electrode potentials are temperature dependent. The coolant in nuclear reactors is high-pressure water at 300°C, at which temperature the dissociation constant is 10^{-11} and neutral water has a pH of 5.5.

10.7.2 Solubility Products

Special treatment is accorded to equilibrium constants that determine the maximum concentrations in water of the cation and anion of a solid ionic compound. Alternatively, this equilibrium can be regarded as determining the maximum concentrations of the salt that can be added to water without precipitating the solid. Dissolution of an ionic solid in water is more complex than dissolution of a nondissociating species (such as sugar). In the latter case, there exists a unique solubility at a fixed temperature. For the dissociating solid, on the other hand, all that thermodynamics provides is the product of the concentrations of the cation and anion. The actual solubility of the solid, expressed as the aqueous concentration of the cation, depends on the concentration of the anion. The latter may be varied by the addition to the solution of other ionic species that have an anion in common with the solid compound.

Consider the metal hydroxide that dissolves in water according to the reaction:

$$M(OH)_z(s) = M^{z+} + zOH^-$$

The law of mass action for this reaction, assuming the solid to be pure, is:

$$K_{SP} = c_{M^+} c_{OH^-}^z$$

where the equilibrium constant K_{SP} is called the *solubility product*. Although no oxidation or reduction of any of the elements is involved in the reaction, the solubility product can be determined from the standard electrode potentials of the following half-cell reactions:

$$M = M^{z+} + ze \qquad \text{for which} \qquad \varepsilon_1 = \varepsilon_1^o - \frac{0.059}{z}\log c_{M^{z+}}$$

$$M + zOH^- = M(OH)_z(s) + ze \qquad \text{for which} \qquad \varepsilon_2 = \varepsilon_2^o - \frac{0.059}{z}\log\frac{1}{c_{OH^-}^z}$$

Equating ε_1 and ε_2 and solving for the product of the concentrations of the cation and the square of the anion yields:

$$K_{SP} = 10^{z(\varepsilon_1^o - \varepsilon_2^o)/0.059}$$

This result could also have been obtained by direct application of Equation (10.34).

Example: What is solubility of $Ni(OH)_2$ in: (1) pure water and (2) a solution adjusted to pH = 10 by addition of NaOH?

The dissolution equilibrium is: $Ni(OH)_2$ (s) = Ni^{2+} + $2OH^-$

For Ni, Table 10.1 gives $\varepsilon_1^o = 0.23\ V$ (No. 6) and $\varepsilon_2^o = 0.66\ V$ (No. 18). From these standard electrode potentials, $K_{SP} = 2.6 \times 10^{-15}$. Irrespective of the nature of the aqueous phase, the nickel ion and hydroxide concentrations must satisfy:

$$2.6 \times 10^{-15} = c_{Ni^{2+}} c_{OH^-}^2$$

1. Pure water

   The water dissociation equilibrium is:

$$10^{-14} = c_{H^+} c_{OH^-}$$

   and electrical neutrality of the solution requires that:

$$c_{H^+} + 2c_{Ni^{2+}} = c_{OH^-}$$

   Substituting the two mass action equations into the condition of electrical neutrality yields:

$$\frac{10^{-14}}{c_{OH^-}} + 2\frac{2.6 \times 10^{-15}}{c_{OH^-}^2} = c_{OH^-}$$

Because c_{OH^-} is << 1, the first term on the left-hand side of this equation is negligible, and the hydroxyl ion concentration is:

$$c_{OH^-} = (2 \times 2.6 \times 10^{-15})^{1/3} = 1.73 \times 10^{-5} \text{ M}$$

or

$$\text{pH} = -\log(10^{-14}/1.73 \times 10^{-5}) = 9.2$$

From the solubility-product mass-action relation (above), the solubility of nickel in pure water is:

$$c_{Ni^{2+}} = \frac{2.6 \times 10^{-15}}{(1.73 \times 10^{-5})^2} = 8.7 \times 10^{-6} \text{ M}$$

2. NaOH solutions with a pH = 10

For this solution, the hydroxyl ion concentration is 10^{-4} M and the solubility of nickel from solid $Ni(OH)_2$ is:

$$c_{Ni^{2+}} = \frac{2.6 \times 10^{-15}}{(10^{-4})^2} = 2.6 \times 10^{-7} \text{ M}$$

Nickel solubility from $Ni(OH)_2$ in pure water is 30 times larger than in alkaline water. This is an example of Le Chatelier's principle: because the product $c_{Ni^{2+}}c^2_{OH^-}$ is constant, increasing c_{OH^-} decreases $c_{Ni^{2+}}$.

10.8 CHEMICAL EQUILIBRIA IN ENVIRONMENTAL WATERS

Surface and subsurface waters often contain species (both ionic and neutral) that can have deleterious effects on animals and plants, and most importantly, on human health. The equilibrium aqueous chemistry of pollutants is one of many scientific aspects of their behavior in the environment. Potentially harmful dissolved species range from inorganic ions to organic molecules to metals. Moreover, the oxidation state of the metals greatly influences the hazards posed by these species. An iconic example is hexavalent chromium, which is much more dangerous than the lower oxidation states of this element. Lead as the elemental metal is much less likely to be transported in the environment than the divalent ion dissolved in water.

10.8.1 SPECIES FROM AIR

The two most important attributes of water that control the chemical states of dissolved foreign species are the pH and the equivalent oxygen pressure. The latter is the oxygen pressure that would be in equilibrium with the actual concentration of dissolved O_2:

$$(p_{O_2})_{eq} = K_H c_{O_2} \tag{10.37}$$

where K_H = 885 atm/M is the Henry's law constant at 298 K for the equilibrium:

$$O_2 \text{ (dissolved)} = O_2 \text{ (g)}$$

Thoroughly aerated water has an equivalent oxygen pressure of 0.21 atm. Water that has been standing in pipes may have a significantly-lower $(p_{O_2})_{eq}$.

Another airborne species that affects water chemistry is carbon dioxide. Unlike oxygen, this gas reacts with water in a way that reduces the pH. The pertinent equilibria are:

$$CO_2 \text{ (dissolved)} = CO_2 \text{ (g)} \qquad K_H = 63 \text{ atm/M.} \qquad (10.38a)$$

$$CO_2 \text{ (dissolved)} + H_2O = HCO_3^- + H^+ \qquad K = 4.4 \times 10^{-7} \text{ M} \qquad (10.38b)$$

$$HCO_3^- = CO_3^{2-} + H^+ \qquad K = 5 \times 10^{-11} \text{ M} \qquad (10.38c)$$

The equilibrium constants for these reactions are for 298 K. In all, carbon dioxide is present in water in the following forms: the dissolved gas,* CO_2 (dissolved); the bicarbonate ion, HCO_3^-; and the carbonate ion, CO_3^{2-}. Air, which consists of 0.035% CO_2, is the primary source of dissolved CO_2. In the absence of a CO_2 bearing gas phase, soluble salts such as sodium bicarbonate (bicarbonate of soda) can provide all of the forms of dissolved CO_2.

10.8.2 Ionic Equilibria of Pollutants

Table 10.2 gives the electrode reactions and standard electrochemical potentials (i.e., the half cell versus SHE, all ion concentrations = 1 M, all gases at 1 atm) of five common water-impurity species and three dissolved species that influence the stability of the impurities.

TABLE 10.2
Electrochemical Properties of Environmental Pollutants

ID	Species	Half-Cell Reaction	ε^o, V
A	Ethanol	$C_2H_5OH + 3H_2O = 2CO_2 + 12H^+ + 12e^-$	–0.90
B	Dichloroethane	$C_2H_4Cl_2 + 4H_2O = 2CO_2 + 12H^+ + 2Cl^- + 10e^-$	—
C	Dimethylamine	$C_2H_7N + 7H_2O = NO_3^- + 2CO_2 + 21H^+ + 20e^-$	—
D	Arsenous acid	$H_3AsO_3 + H_2O = H_3AsO_4 + 2H^+ + 2e^-$	–0.394
E	Lead	$Pb = Pb^{2+} + 2e^-$	0.126
1	Nitrate ion	$N_2 + 6H_2O = 2NO_3^- + 12H^+ + 10e^-$	–1.241
2	Sulfate ion	$HS^- + 4H_2O = SO_4^{2-} + 9H^+ + 8e^-$	–0.248
3	Oxygen	$2OH = \frac{1}{2}O_2 + H_2O + 2e^-$	–1.229

* A small concentration of carbonic acid is formed by the reation $HCO_3^- + H^+ = H_2CO_3$.

Example: Compare the effectiveness of nitrate ion to that of oxygen for removing ethanol from water.

Nitrate ion: The overall reaction is:

$$5C_2H_5OH + 12\,NO_3^- + 12H^+ = 10CO_2 + 6N_2 + 21H_2O \tag{O1}$$

This reaction is the following combination of half cell reactions A and 1:

$$(O1) = 5 \times (A) - 6 \times (1)$$

The standard chemical potential changes for (A) and (1) are:

$$\Delta\mu_A^o = -12F\varepsilon_A^o = -12 \times 96.5 \times (-0.90) = 1042 \text{ kJ/mole}$$

$$\Delta\mu_1^o = -10F\varepsilon_1^o = -10 \times 96.5 \times (-1.241) = 1197 \text{ kJ/mole}$$

Combining:

$$\Delta\mu_{O1}^o = 5\Delta\mu_A^o - 6\Delta\mu_1^o = -1972 \text{ kJ/mole}$$

The equilibrium constant for the overall reaction is:

$$K_{O1} = \exp\left(-\frac{\Delta\mu_{O1}^o}{RT}\right) = \exp\left(-\frac{-1972 \times 10^3}{8.314 \times 298}\right) = 5 \times 10^{345}$$

The mass-action law for reaction (O1) is:

$$K_{O1} = \frac{p_{CO_2}^{10}\, p_{N_2}^{6}}{C_{ethanol}^{5}\, C_{NO_3^-}^{12}\, C_{H^+}^{12}}$$

Solving:

$$\left(C_{ethanol}\right)_{NO_3^-} = \frac{p_{CO_2}^{2}\, p_{N_2}^{1.2}}{K_{O1}^{0.2}\, C_{NO_3^-}^{2.4}\, C_{H^+}^{2.4}}$$

Dissolved oxygen: The overall reaction is:

$$C_2H_5OH + 3O_2 = 2CO_2 + 3\,H_2O \tag{O2}$$

This reaction is the following combination of half cell reactions A and 3:

$$(O2) = (A) - 6 \times (3)$$

The standard chemical potential change for (3) is:

$$\Delta\mu_3^o = -2F\varepsilon_3^o = -2 \times 96.5 \times (-1.229) = 237 \text{ kJ/mole}$$

Combining:

$$\Delta\mu^o_{O2} = \Delta\mu^o_A - 6\Delta\mu^o_3 = -380 \text{ kJ/mole}$$

The equilibrium constant for the overall reaction is:

$$K_{O2} = \exp\left(-\frac{\Delta\mu^o_{O2}}{RT}\right) = \exp\left(-\frac{-380\times10^3}{8.314\times298}\right) = 4\times10^{66}$$

The mass-action law for reaction (O2) is:

$$K_{O2} = \frac{p^2_{CO_2}}{C_{ethanol}\times p^3_{O_2}}$$

Solving:

$$\left(C_{ethanol}\right)_{O_2} = \frac{p^2_{CO_2}}{K_{O2}\times p^3_{O_2}}$$

Comparison of the two oxidants:

Assumed conditions:

$$p_{CO_2} = 3.5\times10^{-4} \text{ atm} \qquad p_{O_2} = 0.21 \text{ atm} \qquad p_{N_2} = 0.79 \text{ atm}$$

$$C_{H^+} = 10^{-7} \text{ M} \qquad C_{NO_3^-} = 0.2 \text{ M}$$

Substituting into the ethanol concentration equations yields:

$$\left(C_{ethanol}\right)_{NO_3^-} = 2\times10^{-58} \text{ M} \qquad \left(C_{ethanol}\right)_{O_2} = 3\times10^{-72} \text{ M}$$

Both oxidizing agents are very effective (thermodynamically) in breaking down ethanol in water. Removal of this particular pollutant is limited by kinetic factors.

10.9 SUMMARY

Figure 10.9 summarizes the key features and equations pertinent to the thermodynamic aspects of electrochemistry. The upper seven levels of the diagram refer to the overall reaction of acid dissolution of a metal (level 1) in the type of electrochemical cell shown as 1(c) in Figure 10.4. This reaction is chosen for illustration; the procedure represented in the figure applies to any overall reaction. Level 2 gives the chemical potential difference between the product and reactant sides of the overall reaction. The two half cells are not in equilibrium unless $\Delta\mu = 0$. Using the relation in the ellipse, $\Delta\mu$ is converted to the cell electric potential ε, or EMF, shown in level 3. Note the distinction between $\Delta\mu$ and ε: the former refers to a chemical potential difference between the left and right sides of the overall reaction; the latter represents an electric potential difference between the left and right half cells.

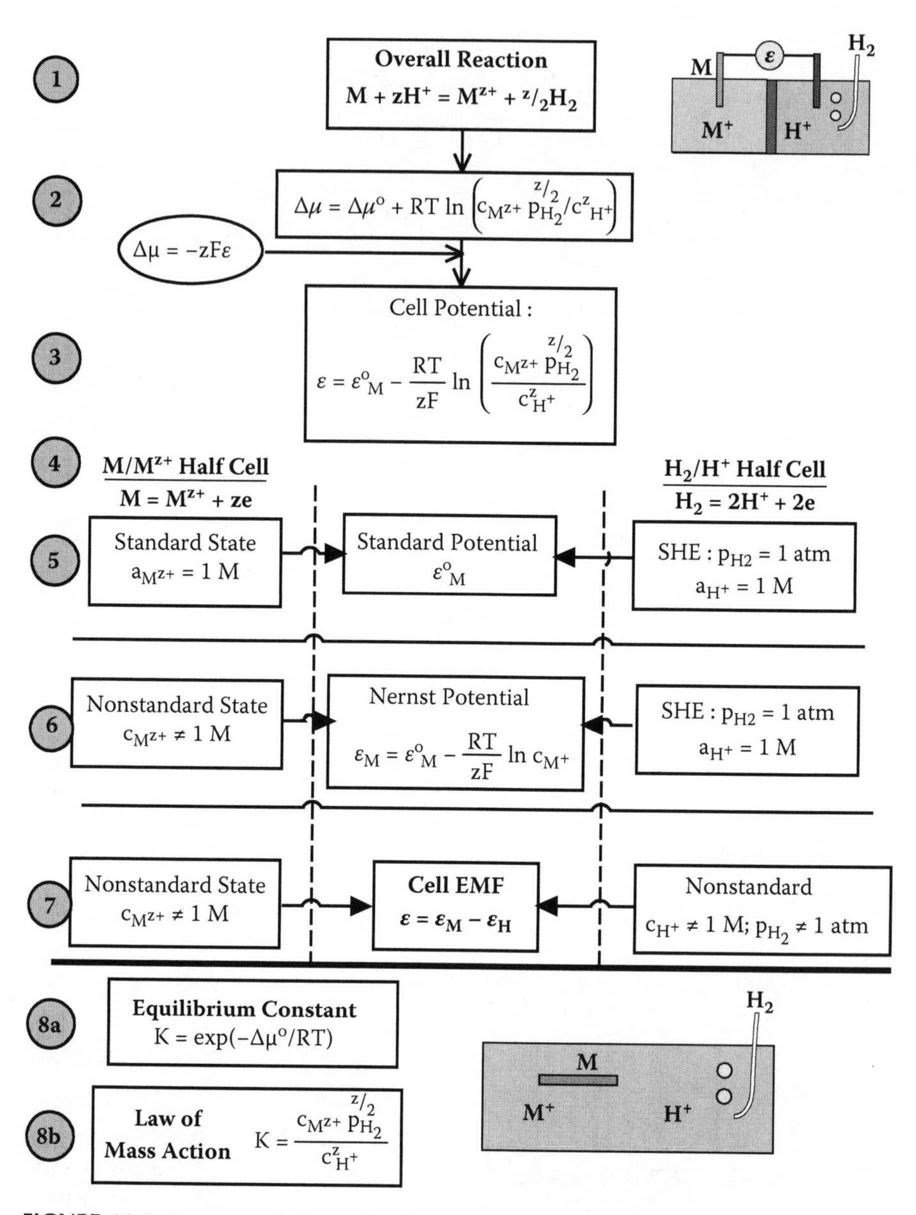

FIGURE 10.9 Summary of electrochemistry.

In level 4, the overall reaction is broken into its constituent half cell reactions. Levels 5, 6, and 7 give the cell potential difference with different combinations of conditions in the two half cells. In level 5, the species in both half cells are in their standard states (unit activity of the cation), so the cell EMF is the standard electrochemical potential of the M/M^{z+} couple. In level 6, the left half cell is no longer in its standard state, but the right half cell remains a standard hydrogen electrode (SHE).

The EMF is now the Nernst potential of the M/M^{z+} couple. In level 7, both half cells are nonstandard, and the cell delivers an EMF that is the difference between the Nernst potentials of the two half cells.

Levels 8a and 8b refer to a simple chemical reaction involving the metal M, its ion M^{z+}, the acid concentration and the H_2 pressure in equilibrium with the solution. In this case $\Delta\mu$ and ε are both zero and the equilibrium constant K is given by level 8a, where $\Delta\mu^o = -zF\left(\varepsilon_M^o - \varepsilon_H^o\right) = -zF\varepsilon_M^o$. The concentrations in solution are related to the equilibrium constant via the law of mass action shown in level 8b.

PROBLEMS

10.1 The solid-state electrochemical cell: $NbO\,|\,Nb(Ru)\,|\,|\,electrolyte\,|\,|\,Ta_2\,O_5\,|\,Ta$ consists of a Ta/Ta_2O_5 couple that produces a fixed electrode potential and a half cell containing a mixture of NbO and Nb dissolved in ruthenium. The latter is inert electrochemically and serves only to dilute the active niobium metal component, thereby changing its activity. The cell operates at 1000 K with various mole fractions of Nb dissolved in ruthenium. For the overall-cell reaction

$$\mathrm{NbO} + \frac{1}{5}\mathrm{Ta} = \mathrm{Nb} + \frac{1}{5}\mathrm{Ta_2O_5}$$

the standard-free-energy change is $\Delta G^o = 0.85$ kcal/mole.

(a) What is the cell potential if the Nb is pure (i.e., no ruthenium solvent)?
(b) When the Nb is dissolved in ruthenium so that $x_{Nb} = 0.6$, the observed cell voltage is 61 mV. What is the activity coefficient of Nb in the alloy?

10.2 Using Table 10.1, determine the standard electrode potential for the following half-cell reaction:

$$Cu_2O + 2H^+ = 2Cu^{2+} + H_2O + 2e \qquad \text{(A)}$$

10.3 A mixture of powdered Fe and FeO is stirred into a solution of uranyl nitrate maintained at a pH of 5. At equilibrium, what is the ratio of the UO_2^{2+} and U^{4+} concentrations?

10.4 Determine the EMF of the cell $Fe|Fe^{2+}||electrolyte||Fe^{2+}|Fe^{3+}$ when the ferrous ion concentration in the left-hand electrode is 0.5 M and the ratio of Fe^{3+} to Fe^{2+} concentrations in the right-hand half cell is 0.1.

10.5 The sketch shows the internals of a common lead-acid battery. The abbreviated designation is: $Pb|PbSO_4||H_2SO_4||PbO_2|PbSO_4$

The unit is constructed of alternating plates of "spongy" lead and packed lead dioxide. The lead electrodes are spongy to provide a high surface area to contact the electrolyte, which is 6 M sulfuric acid. The lead dioxide is packed into plates to form the second electrode. Operation of the battery depletes both electrodes by converting both to lead sulfate, which is insoluble in the electrolyte.

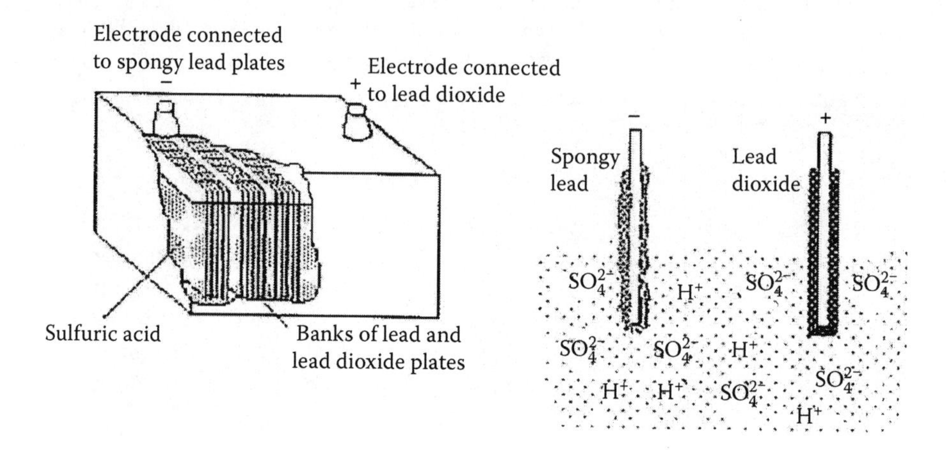

(a) Using the table below, select half cell reactions that, when combined, produce the overall reaction that characterizes battery operation.
(b) What is the voltage generated by a pair of plates?
(c) As the battery operates, sulfuric acid is consumed. What is the voltage when the acid concentration has decreased to 3 M?
(d) How are the electrodes configured to produce 12 volts output?

Species	ΔG°, kJ/mole
$PbSO_4(s)$	–813
$H_2O(L)$	–237
Pb(s)	0
$PbO_2(s)$	–217
H_2SO_4*	–745

*Δμ°, as $H^+ + HSO_4^-$

10.6 What is the EMF at 1273 K of the solid-state cell *(Fe,Ni)₁|FeO||electrolyte|| (Fe,Ni)₂|NiO* in which $(Fe,Ni)_1$ is an equimolar alloy of Fe and Ni and $(Fe,Ni)_2$ is an alloy containing 30 mole percent Ni in Fe? The activity coefficient of the nickel component is given by:

$$\ln \gamma_{Ni} = -1.875 x_{Fe}^2 + 1.55 x_{Fe}^3$$

10.7 What is the EMF produced by the cell Ni|NiO||electrolyte||TiO|TiO_2 at 800°C? Note that the electrode on the right is a two-phase mixture of two oxides of titanium, both of which appear on the Ellingham diagram (Figure 9.8).

10.8 An aqueous electrochemical cell consists of half cells denoted by A and B. The Nernst potential of half cell A is more positive than that of half cell B.

(a) If the cell were short circuited, which electrode would be the cathode and which the anode?
(b) If electrode A consists of metal A and ion A^+ and electrode B consists of metal B and ion B^{2+}, write the half cell reactions and the overall cell reaction.
(c) If the standard electrochemical potentials are ε_A^o and ε_B^o, what is the relationship between the concentrations of A^+ and B^{2+} in the two compartments of the cell when the cell EMF is ε and no current flows?

10.9 Consider the cell $Cr|Cr_2O_3||electrolyte||H_2O|H_2$ at 1000°C. If the H_2O/H_2 ratio in the gaseous electrode is 0.01, what is the cell EMF?

10.10 Consider the cell $O_2(air)||electrolyte||Pb|PbO(SiO_2)$ operating at 700°C. The anode consists of metallic lead in contact with lead oxide, which is dissolved in inert silica. The activity of PbO in the liquid oxide is 0.1. What is the EMF of this cell?

10.11 The overall reaction $A^+ + 2B^{4+} = A^{3+} + 2B^{3+}$ achieves equilibrium following addition of 1 mole of AX and 2 moles of B_2Y_3 to 1 l of water. These compounds dissociate completely but only the cations react.

(a) Write the equilibrium equation (mass-action law) and the element and charge conservation equations that apply to this system.
(b) Combine the equations in Part (a) into a single equation in the concentration of A^+.
(c) Write the half cell reactions (A) and (B) that combine to give the overall reaction.
The standard electrode potentials relevant to the overall reaction are given in the table:

ID	Half Cell Reaction	Standard Potential, V
1	$A(s) = A^+ + e$	0.52
2	$A^{2+} = A^{3+} + e$	–0.11
3	$A(s) = A^{2+} + 2e$	–1.02
B	$B^{3+} = B^{4+} + e$	–0.98

(d) What is the standard potential for half cell reaction (A) obtained in Part (c)?
(e) What is the numerical value of the equilibrium constant K for the overall reaction?
(f) Solve Part (c) for the concentrations of the four species. (Hint: if K is very different from unity, some terms in the mass-action law may be neglected.)

10.12 The electrochemical cell shown below converts hydrogen and oxygen to water and drives a motor in the process. The electrode conditions are:

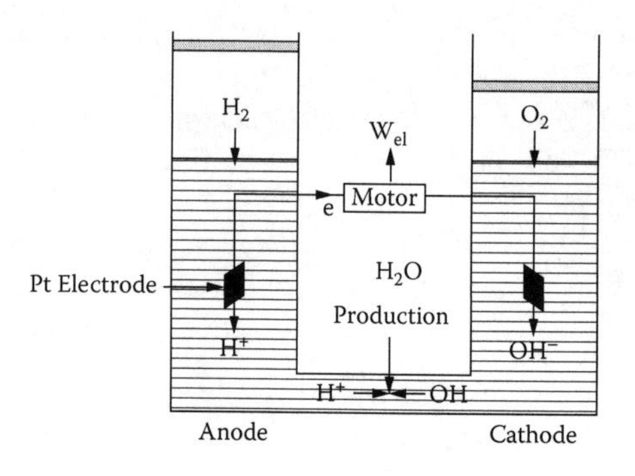

anode: pH =5; p_{H_2} = 1 atm; cathode: pH = 4; p_{O_2} = 1 atm.

(a) Write the half-cell reactions for the anode and cathode and the corresponding Nernst equations.
(b) What is the cell voltage?
(c) How much work is done by the motor per mole of water produced?
(d) How much heat is released per mole of water produced?

10.13 (a) Calculate the standard free energy of formation of liquid water at 300 K from:

1. The thermochemical properties table below.
2. The table of standard electrode potentials.

(b) What is the enthalpy of formation of liquid water?
(c) Derive the equation for the temperature dependence of the standard electrode potential. What is the standard electrode potential for water formation at 350 K?

T,K	h, kcal/mole	s, cal/mole-K	g, kcal/mole
H_2(g)			
300	.013	31.25	–9.36
400	.70	33.23	–12.59
O_2(g)			
300	.013	49.05	–14.70
400	.73	51.11	–19.71
H_2O(L)			
300	–68.29	16.83	–73.3
400	–66.48	22.03	–75.2

10.14 The following table gives standard electrode potentials of half cell reactions involving sulfur. This particular compilation uses H (atomic hydrogen) instead of H_2 or H^+. However, H can be eliminated by using the last entry in the table.

ID	Half Cell Reaction	Standard Potential, V
1	$H + HS^- = H_2S + e$	0.41
2	$2H + S^{2-} = H_2S + 2e$	0.41
3	$H + SO_4^{2-} = HSO_4^- + e$	0.07
4	$H + HSO_4^- = H_2SO_4 + e$	–0.18
5	$S^{2-} = S + 2e$	0.45
6	$S + 4H_2O = SO_4^{2-} + 8H^+ + 6e$	–0.36
7	$H = H^+ + e$	0.10

By combining the equations in the table and others previously treated, find:

(a) The standard chemical potential change for the formation reaction of sulfuric acid in water:

$$S + 2O_2 + H_2 = HSO_4^{2-} + H^+ \quad \text{(A)}$$

The standard free energy of formation of water (from H_2 and O_2) at 25°C is –237 kJ/mole. Compare your result with the value given in the table of Problem 10.5.

(b) The standard electrode potential of the half cell reaction:

$$HS^- + 4H_2O = SO_4^{2-} + 9H^+ + 8e \quad \text{(B)}$$

Compare your result with the value given for this reaction in Table 10.2.

REFERENCES

Atkins, P. W. 1978. *Physical Chemistry.* W. H. Freeman. San Francisco, CA.

Gaskell, D. R. 1981. *Introduction to Metallurgical Thermodynamics*, 2nd ed. Washington, DC: Hemisphere.

Lide, D. R. ed. 2006. *Handbook of Physics and Chemistry.* Boca Raton, FL: CRC Press.

11 Biothermodynamics

11.1 INTRODUCTION

The thermodynamics of biological systems consists of two principal components: energetics and equilibrium; or in more scientific terms, enthalpy and free energy changes of reactions in living matter. *Biothermodynamics* is a specialty within the discipline of chemical thermodynamics. Coverage in this chapter is restricted to mammals, with a special focus on the human body. Even with this limitation, biothermodynamics covers a vast array of chemical transformations. Although the strictly thermodynamic aspects of this field are relatively simple, the molecular systems that are treated are extraordinarily complex. The species involved range from the proton to huge organic molecules (*macromolecules*) with molecular weights in the tens of thousands. The specialized nomenclature of biology is particularly difficult for nonbiologists to cope with. Acronyms abound. To deal with these difficulties, this chapter defines all biological terms used and where feasible, describes large molecules with structural diagrams. Rather than attempt to explain all mammalian systems for which a thermodynamic analysis is appropriate, the following sections select only a few applications and attempt to cover them thoroughly (from a thermodynamic viewpoint, not all of the biological implications).

Molecules of biological significance are mostly organic, consisting principally of carbon, hydrogen, and nitrogen. Several inorganic species are important, as can be seen on the label of any multivitamin bottle. The organic species range in size from three-carbon segments to the complex proteins, fats, and enzymes. However, the large molecules are combinations of simpler groups. Before turning to the thermodynamics of such systems, the molecular structure of several important biological species are displayed in the following section. The selection is by no means exhaustive because the number and complexity of molecular species in mammals and plants is much too large to cover in a single chapter. Also, only a few species are singled out as examples of the application of thermodynamics, or more properly, thermochemistry, to living systems.

11.2 AMINO ACIDS

Amino acids are the monomer units of polymeric macromolecules called proteins. Amino acids are chains of carbon atoms to which various groups such as hydroxyl (OH), double-bonded oxygen (O=), carboxyl (COOH), methyl (CH_3) and amino (NH_2) are attached. The appendages to the base carbon chain often consist of 5- or 6-member rings of carbon and nitrogen. The linear skeletons contain from one to five carbon atoms, and occasionally incorporate a sulfur atom. There are twenty

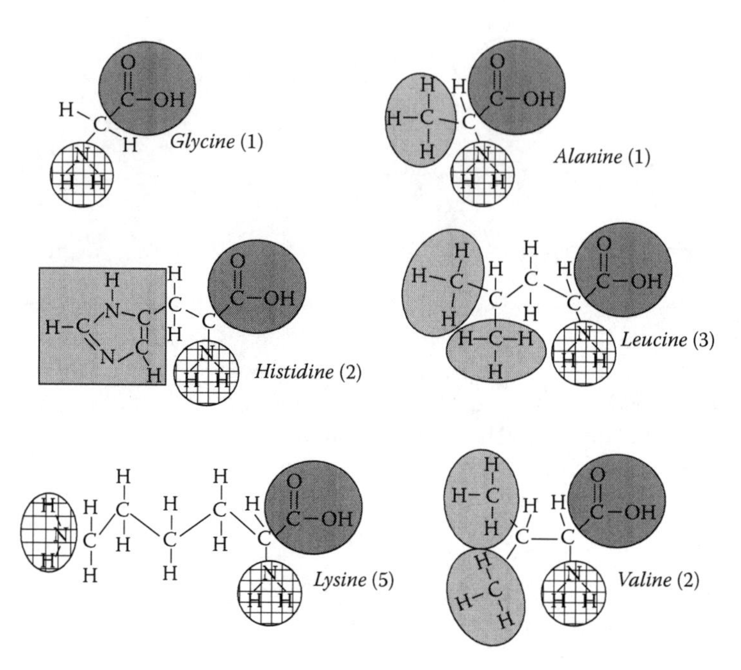

FIGURE 11.1 Amino acids.

amino acids, six of which are depicted in Figure 11.1. The lines joining the atoms represent covalent single bonds, in which an electron from each atom is shared with the adjacent atom. The names of the amino acids are shown, together with the number of carbon atoms in the basic skeleton (in parentheses). The various appendages are highlighted.

Each species has a carboxyl group and an amino group at one end. An important property of the amino acids is the ease with which the terminal H on the carboxyl group can be transferred to the neighboring amino group, creating a dipole:

$$-C\begin{matrix} \diagup COOH \\ \diagdown NH_2 \end{matrix} \longrightarrow -C\begin{matrix} \diagup COO^- \\ \diagdown NH_3^+ \end{matrix}$$

which enhances the solubility of the amino acid in water.

Reaction of the carboxyl group of one amino acid with an amino group of another results in a *peptide bond*:

$$R-C(-C(=O)-OH)(-NH-H) + HO-C(=O)-C(-NH_2)-R \longrightarrow R-C(-C(=O)-OH)-NH-C(=O)-C(-NH_2)-R + H_2O$$

Peptide Bond

A molecule of water is the other product of this reaction. "R" in the diagram is a generic designation of the remainder of the amino acid. Proteins are assembled in this manner.

11.3 GLUCOSE

In order to maintain optimum body temperature, mammals need thermal energy acquired from internal chemical reactions to make up for heat loss to the environment. Additionally, the free energy of the body's chemistry sustains functions ranging from muscle contraction to the thought processes and instinctual responses of the brain. Thermal energy and free energy, represented by ΔH and ΔG, respectively, are supplied mainly by reaction of oxygen with food-borne nutrients such as carbohydrates, proteins, and fats. The end products of these chemical conversions are carbon dioxide and water.

Glucose is produced in plants by photosynthesis, in which sunlight causes carbon dioxide and water to combine. The reaction products are glucose and oxygen:

$$6CO_2 + 6H_2O + h\nu \rightarrow C_6H_{12}O_6 + 6O_2 \tag{11.1}$$

Here hν stands for photons from the sun that provide the free energy that overcomes the highly unfavorable energy (enthalpy) difference between the products and the reactants. $C_6H_{12}O_6$ is the chemical formula for the glucose molecule, which is shown in structural representations in Figure 11.2. The oxygen product of photosynthesis is returned to the atmosphere. The radiant energy that drives this reaction is stored in the glucose molecule, to be released after vegetable matter has been consumed by humans (and other mammals) and its charge of glucose released to the body.

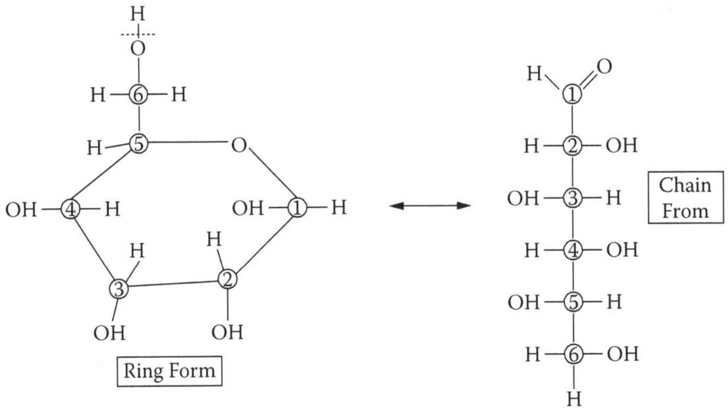

FIGURE 11.2 The ring and chain forms of glucose. The carbon atoms are numbered.

Two forms of the glucose molecule are shown in Figure 11.2: the ring and the chain. Attached to the ring of five carbon atoms plus an oxygen atom are four OH groups and an aldehyde group (CH_2OH). This form of glucose is the monomer for the polymer produced by formation of H_2O by removal of the H atom at the dashed line at the top of the ring form in Figure 11.2 from one glucose molecule and an OH group at the same location from a second glucose molecule. The two glucose molecules are then joined by a common oxygen atom. The polymeric form is called starch.

As indicated in the figure, the ring and chain forms are in equilibrium. To reach the chain form from the ring configuration, the bond between the ring oxygen and carbon atom number 5 breaks, the oxygen becomes double bonded to carbon atom number 1 and the OH group on carbon number 1 moves to carbon number 5. This transformation does not occur so simply, however. The two isomers are end states of four intermediate molecular configurations.

Numbering of the carbon atoms is helpful in following reactions of glucose. For example, if an orthophosphate ion replaces the hydroxyl group on top of the ring structure in Figure 11.2, the reaction would be written as:

$$\text{glucose} + HPO_4^{2-} = \text{glucose-6-phosphate}^{2-} + H_2O \tag{11.2}$$

indicating that the phosphate group is attached to carbon atom 6 of the glucose molecule.*

```
      O
      ||      H
      |       |
 ⁻O — P — O — C⁶ ——
      |       |
      O⁻      H
```

Adding a phosphate group to glucose is called *phosphorylation.* It is the first step in burning of glucose, a process termed *glycolysis.* The process involves a series of steps, which can be abbreviated by the following two sequential reactions:

$$C_6H_{12}O_6 + O_2 \rightarrow 2C_3H_3O_3^- + 2H^+ + 2H_2O \tag{11.3a}$$

The intermediate three-carbon ion, called *pyruvate,* has the following molecular structure:

```
    H  O
    |  ||    O⁻
H - C - C - C
    |        ⋱O
    H
```

In the presence of oxygen (in the cell), complete conversion to the final end products occurs:

$$2H^+ + 2C_3H_3O_3^- + 5O_2 \rightarrow 6CO_2 + 4H_2O \tag{11.3b}$$

The energy released in the combination of reactions (11.3a) and (11.3b) is 3000 kJ/mole glucose.

---

* Reaction (11.2) does not proceed to the right because the standard free energy change is positive (14 kJ/mole); however, ATP does this job (see Section 11.4.2).

The sum of these two partial conversions is seen to be the reverse of reaction (11.1). Reactions (11.3a) and (11.3b) (and the numerous intermediate steps) take place in the cells of the body. Oxygen is carried from the lungs to the cells by hemoglobin. In oxygen-deficient situations, such as muscle overexertion, pyruvate picks up two protons to form lactate:

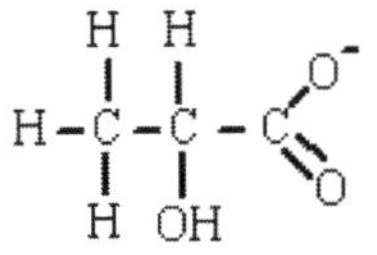

```
    H  H
    |  |     O-
H - C - C - C/
    |  |     \\O
    H  OH
```

Alternatively, the addition of three protons during fermentation of alcohol converts pyruvate to CO_2 and ethanol (CH_3-CH_2-OH).

11.4 ADENOSINE TRIPHOSPHATE (ATP)

Adenosine triphosphate (ATP) is the major provider of energy for all bodily functions. The term "energy" really means the (negative) free energy changes of the multitude of reactions that take place within the cell. These free energy changes, along with the concentrations of reactants and products via the law of mass action, provides useful work for the body's functions.

11.4.1 Synthesis

As shown in Figure 11.3, ATP consists of three components: *adenine*, composed of joined rings of carbon and nitrogen; *ribose*, which is glucose minus one carbon atom in the ring; and three orthophosphate ions. The reaction releases four water molecules and consumes two protons.

ATP is a high (free)-energy molecule, but the free energy of the reactants is even larger, so the standard free energy change of the synthesis reaction is negative. Even with the favorable thermochemistry, the reaction needs to be catalyzed by an enzyme in order to proceed at a rate sufficient for the body's needs.

11.4.2 Hydrolysis and Phosphorylation

A critical equilibrium reaction for transferring free energy to other reactions is the *hydrolysis* of ATP:

$$ATP^{4-} + H_2O = ADP^{3-} + HP^{2-} + H^+ \tag{11.4a}$$

where ADP^{3-} is adenosine diphosphate and HP^{2-} stands for the orthophosphate ion (HPO_4^{2-}). ΔG^o for this reaction is +8 kJ/mole and the equilibrium constant is:

$$K = \frac{[ADP^{3-}][HP^{2-}][H^+]}{[ATP^{4-}]} = \exp\left(-\frac{8\times 10^3}{8.314\times T}\right) \tag{11.4b}$$

At 25°C, $K = 25$, and at normal body temperature (36°C), $K = 23$.

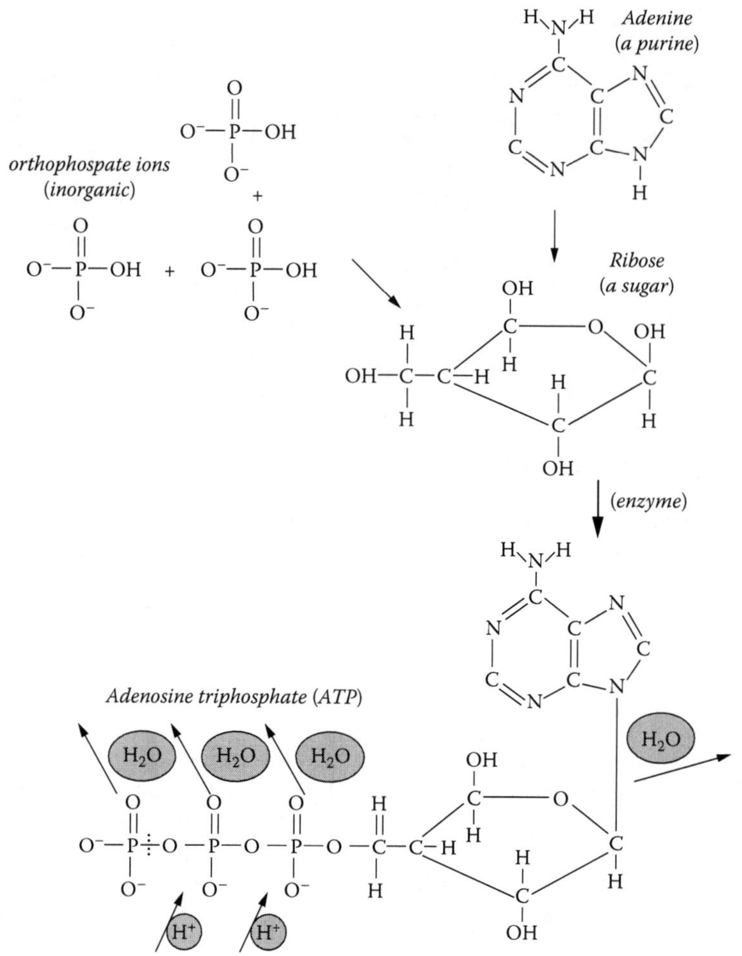

FIGURE 11.3 Synthesis of ATP.

In itself, reaction (11.4a) is not important. What is important is its effect when coupled to other reactions, such as the phosphorylation of glucose. In abbreviated notation, Equation (11.2) is:

$$G + HP^{2-} = G6P^{2-} + H_2O \qquad (11.4c)$$

for which $\Delta G^o = +14$ kJ/mole. Because of the positive standard free energy change, a phosphate group cannot be added to glucose by this reaction alone.

However, in the presence of ATP, reaction (11.4c) is "coupled" to reaction (11.4a), leading to the equilibrium reaction:

$$ATP^{4-} + G = ADP^{3-} + G6P^{2-} + H^+ \qquad (11.4d)$$

This reaction has a standard free energy of +22 kJ/mole, corresponding, at room temperature, to an equilibrium constant of $K = 1.4 \times 10^{-4}$. The equilibrium ratio of phosphorylated glucose to bare glucose is:

$$\frac{[G6P^{2-}]}{[G]} = 1.4 \times 10^{-4} \frac{[ATP^{4-}]}{[ADP^{3-}][H^+]}$$

At pH = 7, even when 95% of the ATP has been converted to ADP, nearly 99% of the glucose has been phosphorylated.

Mechanistically, reaction (11.4d) proceeds by breaking the O-H bond in the hydroxyl group attached to the number 6 carbon atom in the ring form of glucose (dotted line in Figure 11.2) and removing the proton. At the same time, the terminal PO_3^{2-} group on the ATP molecule is separated from the rest of the molecule at the dotted line in Figure 11.3 and becomes attached to the glucose molecule where the proton was removed.

Aside from the favorable thermochemistry, an electric feature of the glucose-ATP reaction is very important in cell metabolism. Glucose enters a cell via a carrier protein. Reaction (11.4d) takes place inside the cell with ATP already there. However, the membrane will not pass $G6P^{2-}$ because of its charge. This species is thus trapped in the cell and available for further decomposition, ultimately leading to pyruvate and energy release.

11.4.3 Regeneration of ATP by Oxidative Phosphorylation

If ATP is consumed by reaction (11.4d), how then does the cell regenerate it? It does so by a process termed *oxidative phosphorylation,* a reaction that adds a phosphate ion to ADP:

$$ADP^{3-} + HPO_4^{2-} + H^+ = ATP^{4-} + H_2O \tag{11.4e}$$

The equilibrium constant for this reaction yields an ATP/ADP ratio of:

$$\frac{[ATP^{4-}]}{[ADP^{3-}]} = K[HPO_4^{2-}][H^+] \tag{11.5}$$

would not normally favor the product side; despite the standard free energy change of –8 kJ/mole (K = 25), an H^+ concentration of 10^{-7} M and an orthophosphate ion concentration of ~ 10^{-3} M maintains a very small ATP/ADP ratio. However, the reaction does not occur in a manner amenable to simple thermochemical analysis. Rather, it occurs in a particular component of mammalian cells called the *mitochondrion.* Figure 11.4 illustrates the complexity of a typical cell, with the mitochondrion just one of many similar bodies, all with different functions. A typical mitochondrian is shaped like a loaf of French bread, with a length of about 7 μm and a diameter of 1 μm.

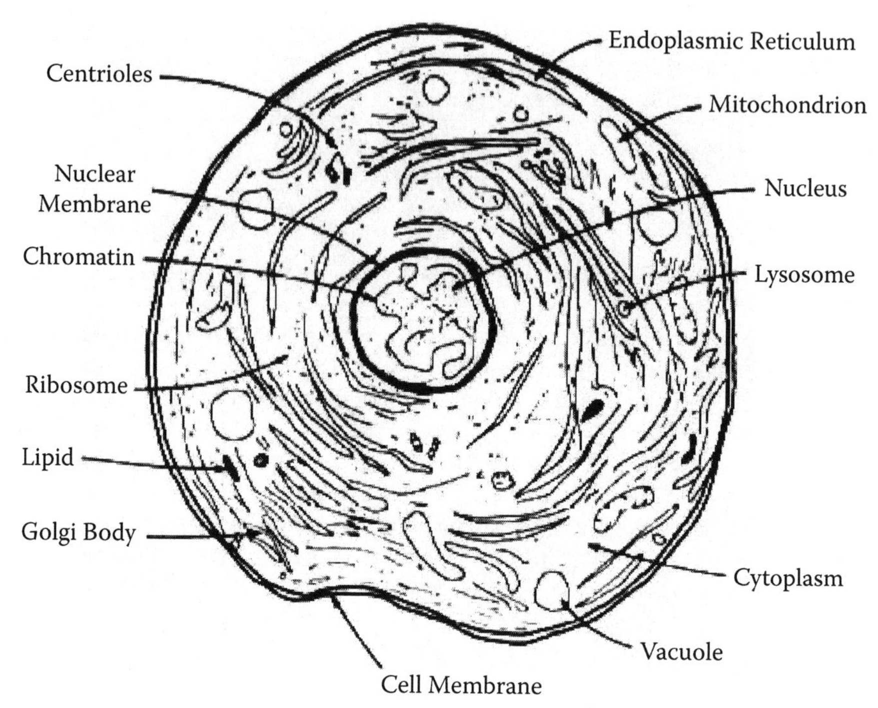

FIGURE 11.4 Components of a typical cell. (From Yang, W. 1989. *Biothermal-Fluid Sciences*. Washington, DC: Hemisphere Publishing. With permission.)

As shown in Figure 11.5, this unit is comprised of a central *matrix* separated from an *intermembrane space* by an *inner membrane*. Several enzymes are embedded in the inner membrane. Those on the right labeled I, III, and IV pump hydrogen ions against their concentration gradient, from the matrix into the intermembrane space. The mechanism by which protons produced along with the glucose-6-phosphate by reaction (11.4d) are pumped out of the matrix is too complex to describe here, but a description of it is given in Lehninger, Nelson, and Cox (1993). How this proton pump alters the unfavorable equilibrium thermochemistry is explained by the following quasi-thermodynamic model.

The "oxidative" portion of the overall process takes place in or at these enzymes. Oxygen reacts with many nucleotides, the most important of which is NADH*:

$$NADH + \tfrac{1}{2}O_2 + H^+ = NAD^+ + H_2O \tag{11.6}$$

The standard free-energy change for this reaction is –260 kJ/mole.

This reaction (and others like it), uses its large negative ΔG^o to "pump" protons from the matrix through the enzymes in the inner membrane and into the intermembrane space. This process creates a higher H^+ concentration in the intermembrane space than in the matrix, thereby setting up both an electric field and a proton driving

* NADH is the acronym for the reduced form of nicotinamide adenine dinucleotide, the structure of which is shown in Figure 13.16 of Lehninger, Nelson, and Cox (1993).

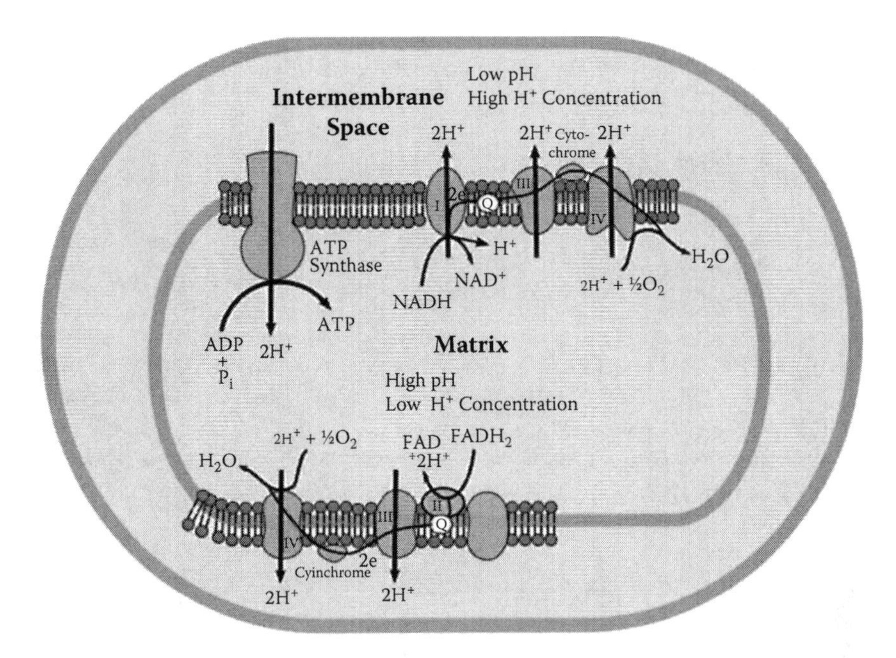

FIGURE 11.5 The mitochondrion component of the cell.

force across the inner membrane. The combination produces a reservoir of free energy given by:

$$\Delta G = F\Delta\phi + RT\ln\left(\frac{[H^+]_{IM}}{[H^+]_{MA}}\right) = F\Delta\phi + 2.3RT\Delta(pH) \tag{11.7}$$

where F is the Faraday and R is the gas constant. The subscripts IM and MA refer to the intermembrane space and the matrix, respectively. Δ(pH) is the difference in pH across the membrane, which is estimated to be about 1. The electric potential difference is approximately 0.2 V. As the protons return to the matrix via the ATP synthase enzyme* shown in the upper left in Figure 11.4b, these two effects provide an additional ~20 kJ/mole to ΔG^o of reaction (11.5), effectively upping the free energy driving force to −28 kJ/mole. From an equilibrium thermodynamic point of view, this augmented ΔG^o is still insufficient to convert a significant fraction of the ADP to ATP by reaction (11.4e). The equilibrium constant is increased from 25 to approximately 10^5, which still cannot overcome the low concentrations of the orthophosphate ions and H^+.

The process can be better viewed by converting the free energy changes to work according to Equation (1.20) (reversible work of any process that takes place at constant T and p equals the negative of the free energy change). Thus the work input in pumping protons from the matrix to the intermembrane space (w_H ~ 20 kJ per mole of H^+) is available for driving reaction (11.4e) as the protons return to the matrix. The work required for this process is calculated from the nonequilibrium

* *ATP synthase* is a very large enzyme consisting of roughly 23,000 atoms and 3000 amino acids.

version of Equation (11.5). With typical (measured) concentrations of the species involved in reaction (11.4e), the work necessary to drive the reaction is:*

$$w_{ATP} = -\Delta G^o - RT\ln\left(\frac{[ATP^{4-}]}{[ADP^{3-}][HPO_4^{2-}][H^+]}\right)$$

$$\cong -(-8) - RT\ln\left(\frac{(2\times10^{-3})}{(2\times10^{-4})(1.5\times10^{-3})(10^{-7})}\right) = -54\,\frac{kJ}{mole\,H^+}$$

Since proton pumping provides only 20 kJ of reversible work per mole of H^+, the passage of approximately three protons through the ATP synthase enzyme is needed to produce 1 mole of ATP.

The theory described above, called the *chemiosmotic model*, was proposed by Peter Mitchell circa 1961. It has withstood the rigors of considerable experimental testing, and won Mitchell a Nobel Prize.

11.5 PROTEIN STRUCTURES

If any class of molecules contains the essential building blocks of mammalian life, it is the protein. This category of very large organic molecules regulates, enables, and delivers energy derived from food, water, and oxygen to bodily parts such as cells, muscles, nerves, components of organs, and all fluids. Proteins are polymers of amino acids, which are joined by peptide bonds (Section 11.2). The sequence of amino acids in a protein molecule is genetically coded for a specific function. Egg white, technically called hen lysome, is an animal protein consisting of about 100 amino acids. The largest protein molecule contains about 27,000 amino acids. However, the monomer units in proteins are chosen from only twenty different amino acids, and not all are used in any particular protein.

Proteins perform their functions in part by "folding" into a tight compact mass not unlike a ball of yarn. Under certain conditions, a folded protein can "unfold" into a flexible one-dimensional structure that resembles a vine. The *folded* state is also called *native*, and the *unfolded* state is often termed *denatured.* Only the folded state is biologically active. Figure 11.6 is a highly-simplified sketch of the structure of the two states of a generic protein molecule. The unfolded state is a floppy chain of regularly repeating amino acids. In the folded configuration, the interwoven mass of the amino acid chains are held together by a variety of interactions, including hydrogen bonds** and ionic bonds.

The protein in an aqueous solution is in one of two states, which are in equilibrium. This situation, with the protein designated by P, is described by:

---

* W_{ATP} is negative because it represents work done on the system (the ATP synthase enzyme), not by the system.

** The type of bond responsible for many of the properties of water. The bond joins the O atom of one water molecule to the H atom of another. The bond energy is due to electrostatic attraction of H and O atoms rather than that of a true covalent bond.

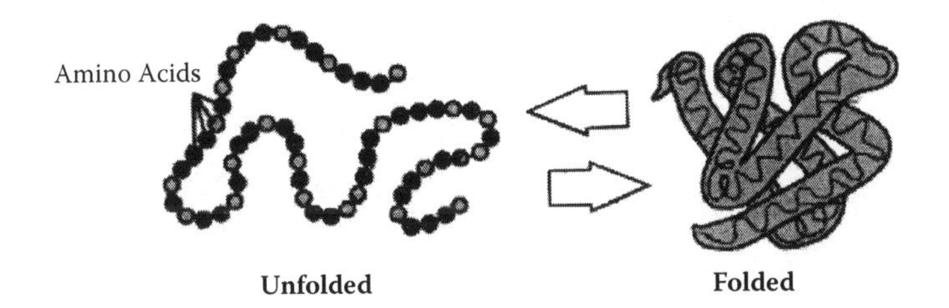

FIGURE 11.6 Protein folding–unfolding equilibrium.

$$P_f = P_U \tag{11.9}$$

with f and u meaning folded and unfolded, respectively. The equilibrium constant is

$$K_U = [P_U]/[P_f] = \exp(-\Delta G_U^o / RT) \tag{11.10}$$

Because of the huge number of atoms in protein molecules, the heat capacities of both forms are extremely large. Consequently, the ΔC_p term is important in both the enthalpy and entropy changes of the reaction. The standard free energy change of reaction (11.9) is expressed by:

$$\Delta G_U^o = \Delta H_{U,ref}^o + \Delta C_P(T - T_{ref}) - T\left[\Delta S_{U,ref}^o + \Delta C_P \ln(T / T_{ref})\right] \tag{11.11}$$

where the reference state is at 298 K. The fraction of the protein in the unfolded state is given by:

$$\frac{[P_U]}{[P_U]+[P_f]} = \frac{K_U}{1+K_U} \tag{11.12}$$

Typical values of the reference thermodynamic properties are:

$$\Delta H_{U,ref}^o = 50 \text{ kcal/mole}; \quad \Delta S_{U,ref}^o = 0.1 \text{kcal/mole-K}; \quad \Delta C_P = 1.5 \text{ kcal/mole-K}$$

The reaction is endothermic because breaking the bonds holding the folded state together requires energy. The entropy change is positive because the folded state is more highly organized than the unfolded state. The heat capacity difference is large because of the large number of atoms in the protein molecule. It is positive because the unfolded state has many more degrees of freedom (in vibration and rotation) than the tighter folded state. From the combination of Equations (11.10) to (11.12), Figure 11.7 shows the effect of ΔC_P on the variation of the unfolded fraction with temperature. The significant feature of these plots is the increasing steepness of the transition between states with increasing ΔC_P. The large values of ΔC_P causes the change of form to resemble a phase change of a single-component system, which is a step function at the transition temperature. For this reason, the folded $\rightarrow$ unfolded

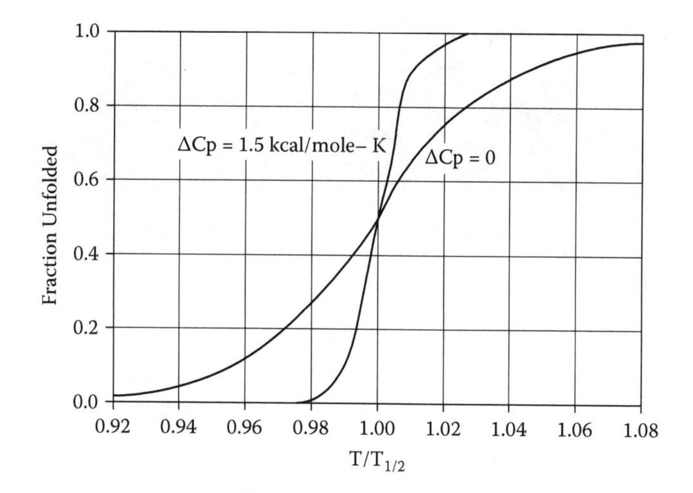

FIGURE 11.7 Effect of temperature on the folding–unfolding equilibrium of a protein in solution.

transformation is often referred to as "melting," despite the fact that the system has two components (the protein and water).

Temperature is not the only parameter affecting protein-state transition. Other components in solution may have an equally profound effect. Figure 11.8 shows the effect of urea concentration in the solution containing the protein on the intensity of the fluorescence of the solution stimulated by the light of the appropriate wavelength. The straight-line portions at either end of the curve represent fluorescence of 100% folded state at low urea concentrations and 100% unfolded state at high urea concentrations. Increasing urea concentration has the same effect on protein stability as increasing temperature.

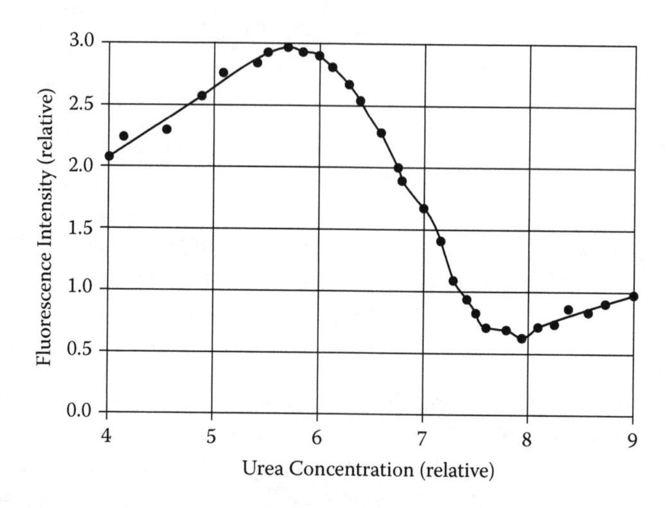

FIGURE 11.8 Effect of urea on protein unfolding as measured by fluorescence (Edsall and Gutfreund, 1983).

11.6 LIGAND BINDING TO MACROMOLECULES[6]

Small molecules called *ligands* can attach to specific *binding sites* on a large molecule (called a *macromolecule*) such as a protein, thereby greatly altering its chemical activity. Ligands range from H^+ to large organic molecules such as stearic acid ($C_{17}H_{25}COOH$). Metal ions such as Ca^{2+} can also serve as ligands. The binding sites are often metal atoms embedded in the macromolecule. Ligand binding is an equilibrium reaction treatable by modest extension of the methods of chemical thermodynamics. Binding affinity is characterized by the equilibrium (dissociation) constant for the reaction:

$$S{\cdot}L = S + L \tag{11.13}$$

where S·L denotes ligand molecule L bound to site S. S by itself connotes an empty site and L is a free ligand in solution. The empty and occupied sites are characterized by molar concentrations, just as if they were independent species in the solution. Thus, if each macromolecule possesses n binding sites and the molar concentration of the macromolecule in solution is $[mm_{tot}]$, the total concentration of binding sites in solution is $[S_o] = n \times [mm_{tot}]$. Similarly, the concentration of bound and unbound sites are [S·L] and [S], respectively. [ ] denotes molarity. Several specific cases are considered in the following section.

11.6.1 Identical and Independent Binding Sites

In this case all binding sites on the macromolecule are independent and have the same affinity for a particular ligand. The term "independent" means that the binding strength is independent of the fraction of the $[S_o]$ binding sites that are occupied. "Identical" means that the equilibrium constant K for reaction (11.13) is the same for all sites. The law of mass action for reaction (11.13) is:

$$K = \frac{[S]\times[L]}{[S\cdot L]} \tag{11.14}$$

Conservation equations for ligands and sites are:

$$[L_o] = [L] + [S{\cdot}L] \tag{11.15}$$

$$[S_o] = [S] + [S{\cdot}L] \tag{11.16}$$

$[L_o]$ is the total ligand concentration, bound plus unbound. The three equations, (11.14) to (11.16), contain three unknown concentrations. The solution is (see Problem 11.3):

$$y = \frac{[S]}{K} = \tfrac{1}{2}\left[\sqrt{(X-Y+1)^2+4Y}-(X-Y+1)\right] \tag{11.17}$$

where

$$X = \frac{[L_o]}{K} \quad \text{and} \quad Y = \frac{[S_o]}{K} \tag{11.18}$$

The fraction of the sites occupied is:

$$\frac{[S \cdot L]}{[S_o]} = \frac{[S_o]-[S]}{[S_o]} = 1 - \frac{[S]}{[S_o]} = 1 - \frac{y}{Y} \tag{11.19}$$

where y is the right-hand side of Equation (11.17).

The extent of ligand binding to a macromolecule is often measured spectroscopically. In the presence of a concentration $[S_o]$ of the binding sites, the instrument detects the concentration of unbound ligand, [L], as a function of the (known) total ligand concentration, $[L_o]$. The latter is determined by adding ligand to a solution that does not contain the macromolecule so that the detection instrument responds to the total ligand concentration. The data are plotted as $1 - [L]/[L_o]$ versus $[L_o]$ and compared to the theoretical model expressed by Equation (11.17). By combining Equations (11.15) with (11.14), the quantity obtainable from the experiment is expressed as:

$$[L_o] = [L] + [S] \times [L]/K = [L](1+y)$$

which yields the fraction of ligand bounds to sites on the macromolecule:

$$\frac{[S \cdot L]}{[L_o]} = 1 - \frac{[L]}{[L_o]} = \frac{y}{1+y} \tag{11.20}$$

Figure 11.9 is a plot of Equations (11.20) and (11.17) for various values of K and $[S_o] = 5$ μM.

11.6.1.1 Scatchard's Equation

The preceding analysis presupposes that the total concentration of sites, $[So] = n[mm_{tot}]$, is known beforehand. In most cases, however, the number of sites per macromolecule, n, is not known; only the concentration of macromolecules, $[mm_{tot}]$, can be specified. During the "titration" of mm by addition to the solution of varying concentrations $[L_o]$ of ligand, the concentration of free ligand, [L], is determined. In this instance, Figure 11.9 is useless because $[S_o]$ is unknown. To analyze such a data set, the following measure of ligand binding is defined:

$$\nu = \frac{[L_o]-[L]}{[mm_{tot}]} = \frac{\text{concentration of bound ligand}}{\text{concentration of macromolecule}} \tag{11.21}$$

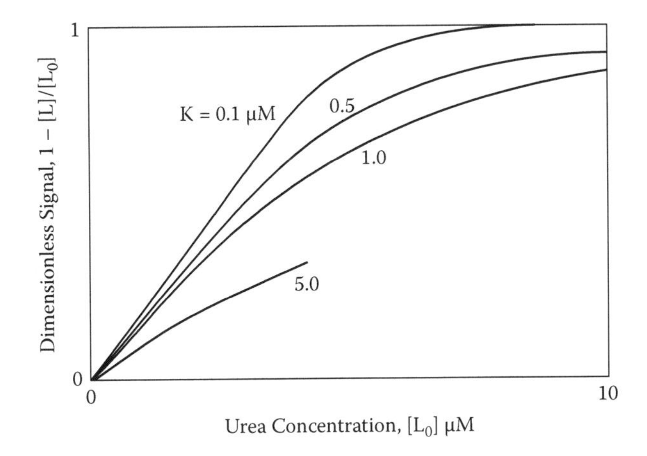

FIGURE 11.9 Saturation of binding sites for a macromolecule with identical, independent sites with various values of the ligand-binding equilibrium constant and $[S_o] = 5$ μM.

With this experimentally determinable quantity, Equations (11.14) to (11.16) can be manipulated to yield (see Problem 11.3):

$$\frac{\nu}{[L]} = \frac{n}{K} - \frac{\nu}{K} \tag{11.22}$$

In this form, a plot of ν/[L] against ν should yield a straight line with a slope of $1/K$ and an intercept of n/K. From such data treatment, both the equilibrium constant for ligand binding, K, and the number of sites per macromolecule, n, can be determined.

11.6.1.2 Site Occupancy

It is often important to know the fraction of the macromolecules that bind 0, 1, ... n ligands. This cannot be had from Scatchard's equation, but a modest extension of the model provides this information. The notation for partially saturated macromolecules is shown in Figure 11.10 for $n = 4$ (4 binding sites per macromolecule). The objective is to determine the concentrations of each of the five species, designated as $[mm{\cdot}L_k]$, where $0 \le k \le 4$. These concentrations must satisfy two conservation equations. The first is for macromolecules:

$$[mm_{tot}] = [mm] + [mm{\cdot}L] + [mm{\cdot}L_2] + [mm{\cdot}L_3] + [mm{\cdot}L_4] \tag{11.23a}$$

and the second is for bound ligands:

$$[L_o] - [L] = [mm{\cdot}L] + 2[mm{\cdot}L_2] + 3[mm{\cdot}L_3] + 4[mm{\cdot}L_4] \tag{11.24a}$$

If $f_k = [mm{\cdot}L_k]/[mm_{tot}]$ is the fraction of the macromolecules with k bound ligands, with Equation (11.21) the above equations become:

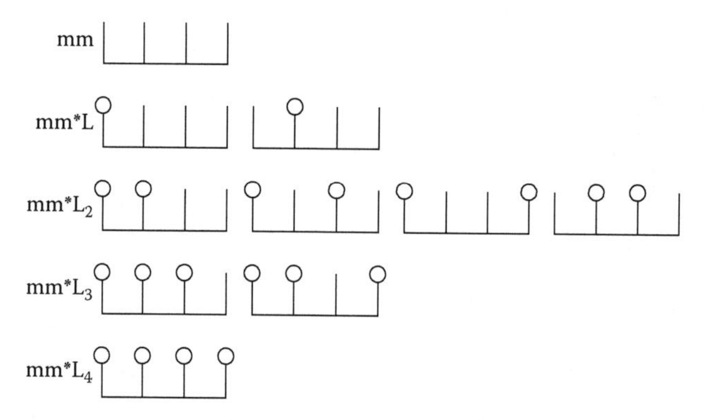

FIGURE 11.10 Combinations of ligands (circles) bound to sites on a macromolecule with four identical, independent sites.

$$f_0 + f_1 + f_2 + f_3 + f_4 = 1 \tag{11.23b}$$

$$f_1 + 2f_2 + 3f_3 + 4f_4 = \nu \tag{11.24b}$$

The most straightforward way of obtaining these fractions is by the Monte Carlo technique. Specified are *M* macromolecules with four sites on each and *N* bound ligands. These quantities are related by $\nu = N/M$ ($0 \le \nu \le 4$). The flow chart for this calculation is shown in Figure 11.11. The left-hand side (separated from the right-hand side by the vertical dashed line) is the actual Monte Carlo calculation. The heart of the computation is the site-occupancy matrix, $k_{occ}(i,m)$, which is zero if the *i*th site is empty and one if it contains a bound ligand. Ligands are "thrown" at the macromolecules until all *N* have found an empty site. The right-hand side of Figure 11.11 is the bookkeeping necessary to produce the concentrations of the macromolecules with 0, ... 4 sites occupied. This computation is repeated many times in order to obtain good statistics.

Figure 11.12 shows the result of this computation for the complete range of site saturation. The ordinates are the fractions f_0, ... f_4 multiplied by 100. Each intermediate state ($mm \cdot L_1$, $mm \cdot L_2$ and $mm \cdot L_3$) rises to a maximum and then falls to zero. As the fractional saturation approaches unity, all macromolecules have ligands bound to all four of their sites.

Note that this computation and Figure 11.12 are independent of the ligand binding equilibrium constant *K*, the ligand concentration $[L_o]$ and the macromolecule concentration $[mm_{tot}]$. These parameters determine the value of ν, however.

11.6.2 Dual Independent Binding Sites

If the macromolecule possesses two sets of independent sites, at concentrations R_o and Q_o, with n_R and n_Q per molecule, the site concentrations are related to the concentration $[mm_{tot}]$ of the macromolecule by:

$$[R_o] = n_P[mm_{tot}] \quad \text{and} \quad [Q_o] = n_Q[mm_{tot}] \tag{11.25}$$

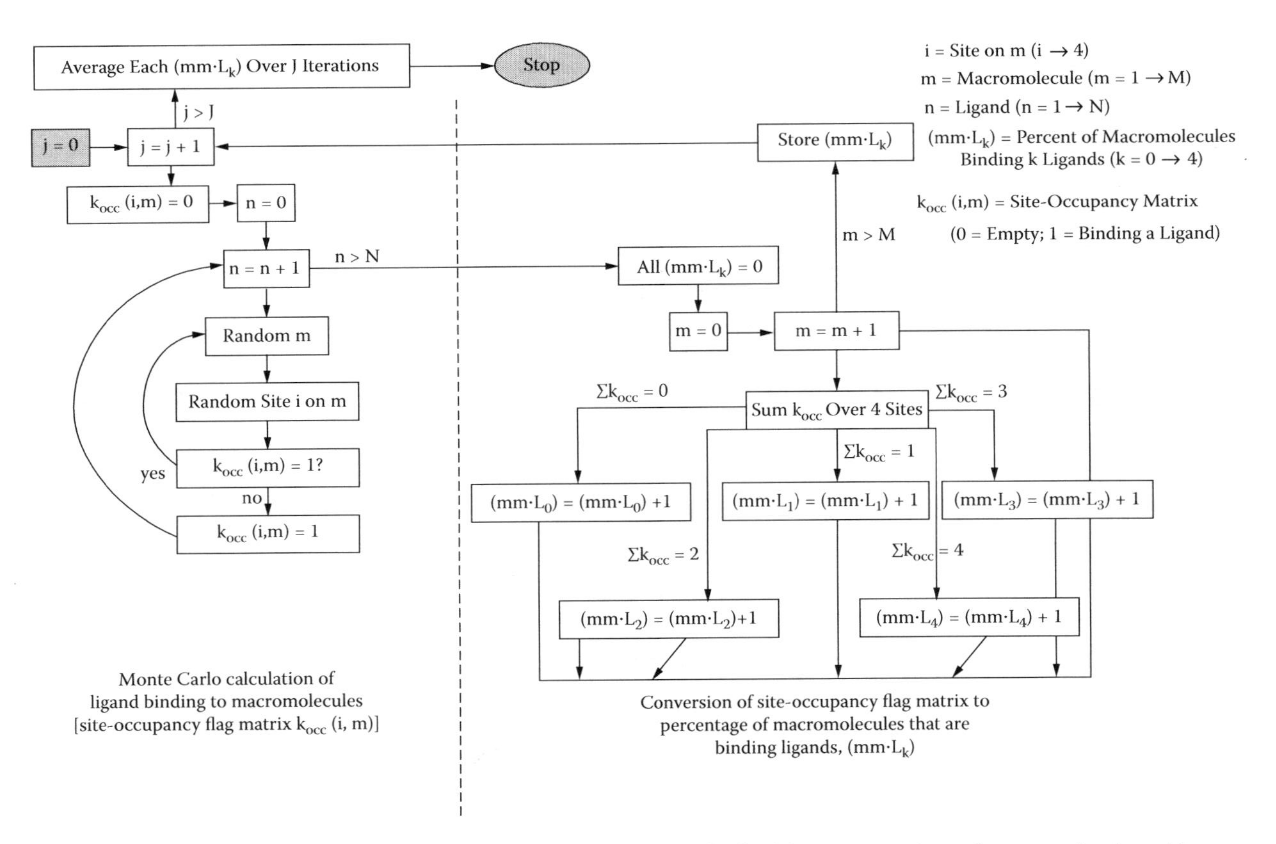

FIGURE 11.11 Flowchart for the calculation (by the Monte Carlo method) of the concentrations of macromolecules with zero to four ligands bound to identical, independent sites.

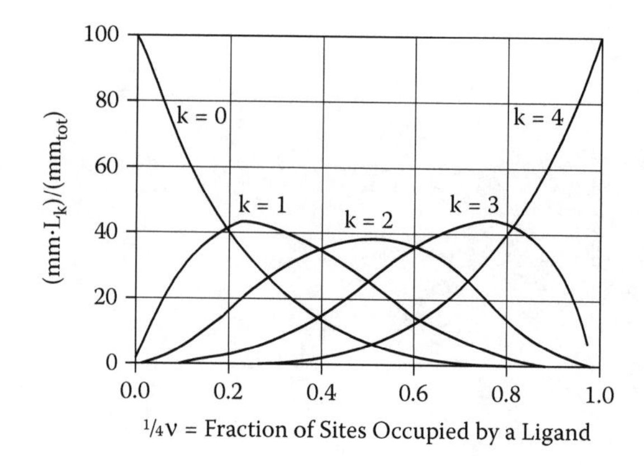

FIGURE 11.12 Percentages of total macromolecule concentration with various numbers of ligands as a function of the fraction of the sites occupied (four identical, independent binding sites per macromolecule).

There are now two equilibrium expressions, one for each site type:

$$L\cdot Q = L + Q \qquad \mathrm{K_Q} = \frac{[L][Q]}{[L\cdot Q]} \tag{11.26}$$

$$L\cdot R = L + R \qquad \mathrm{K_R} = \frac{[L][R]}{[L\cdot R]}$$

The site and ligand conservation equations are:

$$[\mathrm{R_o}] = [\mathrm{R}] + [\mathrm{L{\cdot}R}] \qquad \text{and} \qquad [\mathrm{Q_o}] = [\mathrm{Q}] + [\mathrm{L{\cdot}Q}] \tag{11.27}$$

$$[\mathrm{L_o}] = [\mathrm{L}] + [\mathrm{L{\cdot}R}] + [\mathrm{L{\cdot}Q}] \tag{11.28}$$

The four-step algebraic solution (without the math) is:

1. Eliminate [Q] and [R] in Equation (11.26) using Equation (11.27).
2. Solve for the concentrations of occupied sites and eliminate $[\mathrm{R_o}]$ and $[\mathrm{Q_o}]$ using Equation (11.25); solve for the ratios of bound sites to $[\mathrm{mm_{tot}}]$.
3. Substitute the result from step 2 into Equation (11.28); with Equation (11.21), arrive at:

$$\frac{\nu}{[L]} = \frac{n_R}{K_R + [L]} + \frac{n_Q}{K_Q + [L]} \tag{11.29a}$$

4. Multiplying by K_R + [L] and dividing by K_R yields:

$$\frac{\nu}{[L]} = \frac{n_R}{K_R} - \frac{\nu}{K_R} + \frac{n_Q}{K_R}\frac{K_R + [L]}{K_Q + [L]} \tag{11.29b}$$

which is of the same form as the single-site version Equation (11.22) with an additional term containing the effect of the second (Q) site. In order to permit plotting of the left-hand side of Equation (10.29b) against ν, the dependence of [L] on ν is needed. This is obtained by cross-multiplying Equation (11.29a) by the product of the three denominators, which yields the quadratic equation:

$$\left(\frac{n_R}{\nu} + \frac{n_Q}{\nu} - 1\right)[L]^2 + \left[\left(\frac{n_R}{\nu} - 1\right)K_Q + \left(\frac{n_Q}{\nu} - 1\right)K_R\right][L] - K_R K_Q = 0 \tag{11.30}$$

from which [L] can be calculated as a function of ν. Figure 11.13 shows the Scatchard plot for the two-independent-site model analyzed above. R is the stronger binding of the two sites. The dashed line neglects the presence of the Q sites. The effect of including these is to extend the ν/[L] plot to larger values of ν. With both sites active, the right-hand curve shows the variation of [L] with the binding ratio expressed as a combination of Equations (11.21) and (11.28):

$$\nu = \frac{[L \cdot R] + [L \cdot Q]}{[mm_{tot}]} \tag{11.31}$$

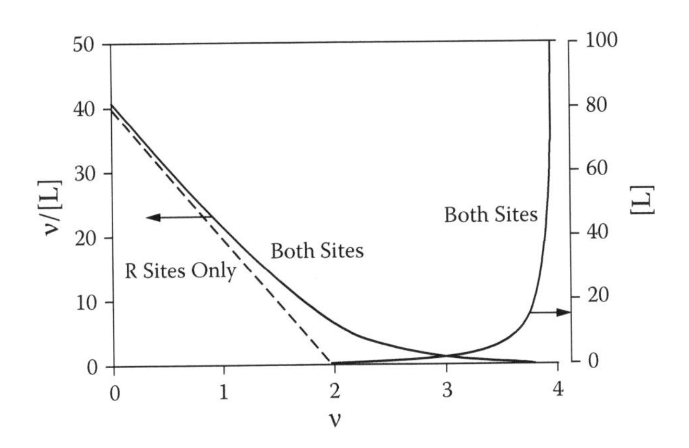

FIGURE 11.13 Scatchard plot for macromolecule troponin C and ligand Ca^{2+}. K_R = 0.05 μM, K_Q = 2.0 μM, $n_R = n_Q = 2$. (From Edsall, J. and H. Gutfreund. 1983. *Biothermodynamics.* New York: Wiley. With permission.)

Because (as in Figure 11.13) there are four sites per macromolecule, this equation shows why the saturation curve ends at $\nu = 4$. In addition, to achieve complete saturation, an infinite concentration of ligand in solution is required. This can be seen from Equation (11.26); for the free site concentrations [R] and [Q] to approach zero, [L] must approach ∞.

The two-site model contains four parameters, which must be determined by fitting data in the form of the left-hand solid curve of Figure 11.13 to the model expressed by Equations (11.29b) and (11.30). If the equilibrium constants are significantly different, the asymptote as $\nu \rightarrow 0$, in the form of the dashed line in the figure, provides information on the parameters of the stronger-binding site.

11.6.3 Cooperative Ligand Binding: Myoglobin to Hemoglobin

Ligand-binding sites may be identical when all sites in the macromolecule are empty, but when one site is occupied, the binding strength (i.e., the binding equilibrium constant) of the remaining sites changes. If the strength of binding increases with each additional ligand, this behavior is called *cooperative* binding. If subsequent additions of ligand to the same macromolecule are less strongly associated than its predecessors, the behavior is termed *anticooperative*.

Human hemoglobin is the most intensively studied of macromolecules (in this case a protein) in this regard. This protein has four cooperative sites for binding oxygen in the blood. When all sites on the protein molecule are empty, all exhibit a dissociation constant $K_1 = 25$ μM; the second, third, and fourth ligands exhibit K values of 37, 17, and 0.5 μM, respectively. Note that binding of the second site is anticooperative. The fourth site has a very low dissociation constant, meaning the O_2 is very strongly bound to the last site in hemoglobin. This general behavior provides very efficient transfer of oxygen around the body because all four sites on each protein molecule are readily saturated.

The binding sites on the hemoglobin molecule are iron atoms. An artist's rendition of a hemoglobin molecule is shown in Figure 11.14. There are four chains of amino acids in the molecule, two designated as α and two as β. The iron atoms are contained in four *heme groups*, which are shown in the sketch as small curlicues in the top center and bottom center. The molecule is designated as a tetramer $\alpha_2\beta_2$. The situation is not that simple, though. The tetramer can dissociate into αβ dimers by the equilibrium reaction $\alpha_2\beta_2 = 2(\alpha\beta)$.

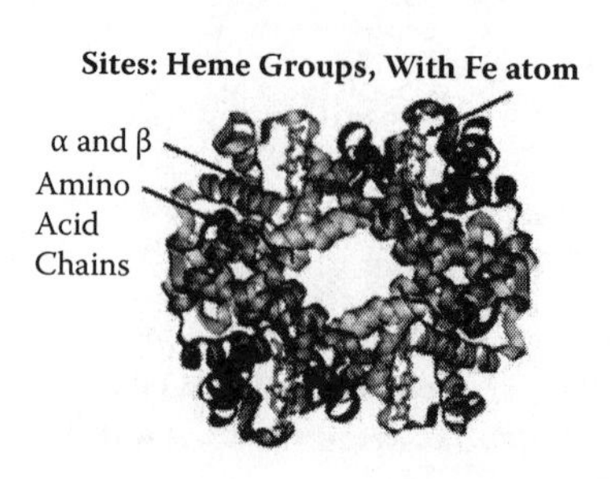

FIGURE 11.14 Depiction of the hemoglobin molecule. It has a molecular weight of about 64,000.

Myoglobin is a degenerate case of hemoglobin, consisting of only one of the four chains found in hemoglobin and possessing only a single site for binding oxygen. The dissociation constant for this site is the same as the first of hemoglobin's, namely 25 μM. The site-saturation analysis of oxygen binding to myoglobin is the same as that presented in Section 11.6.1.

The notation used in the preceding sections is retained, but the identification is H = hemoglobin and L = O_2 dissolved in the blood. L can also designate carbon monoxide, which binds to the four sites in hemoglobin even more strongly than oxygen. In the following analysis, the case of competition between CO and O_2 is not treated; the ligand is assumed to be O_2.

The liganded species are denoted by H·L, $H{\cdot}L_2$, $H{\cdot}L_3$, and $H{\cdot}L_4$. The dissociation reactions and their mass-action laws are:

$$\begin{aligned}
&H{\cdot}L = H + L && K_1 = [H][L]/[H{\cdot}L] = 25\ \mu M \\
&H{\cdot}L_2 = H{\cdot}L + L && K_2 = [H{\cdot}L][L]/[H{\cdot}L_2] = 37\ \mu M \\
&H{\cdot}L_3 = H{\cdot}L_2 + L && K_3 = [H{\cdot}L_2][L]/[H{\cdot}L_3] = 17\ \mu M \\
&H{\cdot}L_4 = H{\cdot}L_3 + L && K_4 = [H{\cdot}L_3][L]/[H{\cdot}L_4] = 0.5\ \mu M
\end{aligned} \tag{11.32}$$

The conservation equations for protein and ligands are:

$$[H_{tot}] = [H] + [H{\cdot}L] + [H{\cdot}L_2] + [H{\cdot}L_3] + [H{\cdot}L_4] \tag{11.33}$$

$$[L_o] = [L] + [H{\cdot}L] + 2[H{\cdot}L_2] + 3[H{\cdot}L_3] + 4[H{\cdot}L_4] \tag{11.34}$$

There are six equations to be solved for six unknowns. Sequentially solving the equilibrium equations yields:

$$[H{\cdot}L] = K_1^{-1}[H][L]$$

$$[H{\cdot}L_2] = K_2^{-1}[H{\cdot}L][L] = K_1^{-1}K_2^{-1}[H][L]^2$$

$$[H{\cdot}L_3] = K_3^{-1}[H{\cdot}L_2][L] = K_1^{-1}K_2^{-1}K_3^{-1}[H][L]^3$$

$$[H{\cdot}L_4] = K_4^{-1}[H{\cdot}L_3][L] = K_1^{-1}K_2^{-1}K_3^{-1}K_4^{-1}[H][L]^4$$

Substituting these into Equations (11.33) and (11.34) gives:

$$[H_{tot}]/[H] = 1 + K_1^{-1}[L] + K_1^{-1}K_2^{-1}[L]^2 + K_1^{-1}K_2^{-1}K_3^{-1}[L]^3 + K_1^{-1}K_2^{-1}K_3^{-1}K_4^{-1}[L]^4$$

$$([L_o]-[L\,H]])/[H] = K_1^{-1}[L] + 2K_1^{-1}K_2^{-1}[L]^2 + 3K_1^{-1}K_2^{-1}K_3^{-1}[L]^3 + 4K_1^{-1}K_2^{-1}K_3^{-1}K_4^{-1}[L]^4$$

The fraction of the heme sites on the hemoglobin molecule that are occupied by ligands is:

$$\frac{[L_o]-[L]}{4[H_{tot}]}=\frac{1}{4}\frac{K_1^{-1}[L]+2K_1^{-1}K_2^{-1}[L]^2+3K_1^{-1}K_2^{-1}K_3^{-1}[L]^3+4K_1^{-1}K_2^{-1}K_3^{-1}K_{41}^{-1}[L]^4}{1+K_1^{-1}[L]+K_1^{-1}K_2^{-1}[L]^2+K_1^{-1}K_2^{-1}K_3^{-1}[L]^3+K_1^{-1}K_2^{-1}K_3^{-1}K_{41}^{-1}[L]^4} \tag{11.35}$$

This is called the *Adair equation*, after its discoverer in 1925. Using the dissociation constants in Equation (11.32), Figure 11.15 is a plot of Equation (11.35) as a function of ligand concentration, L = O_2. This system is a classic example of *cooperative binding*, wherein the site-ligand bond becomes stronger for every successive site occupied (i.e., $K_{i+1} < K_i$). The importance of cooperativity in oxygen binding to hemoglobin compared to myoglobin is shown in this figure. Hemoglobin displays an S-shaped curve which rises from weak binding initially to maximal binding over a small oxygen concentration range. This provides for efficient loading and unloading of oxygen from the molecule. Myoglobin, on the other hand, requires very large oxygen concentrations to saturate the sites on the molecule.

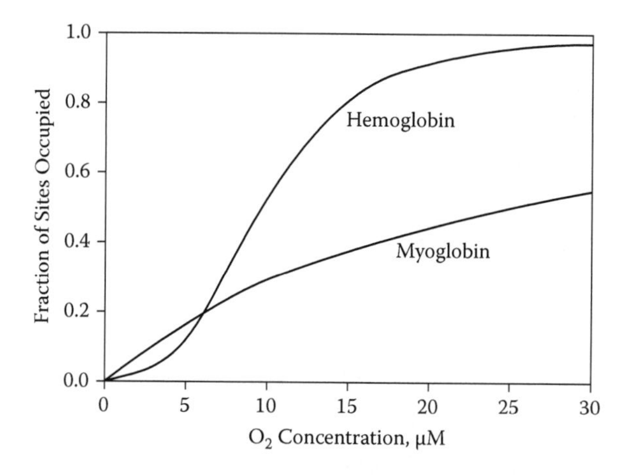

FIGURE 11.15 Adair plot for hemoglobin and myoglobin. The single equilibrium constant for myoglobin is the first-site equibrium constant for hemoglobin.

11.6.4 Competition for Sites—CO and O_2 on Hemoglobin

Binding of oxygen in the blood to hemoglobin is affected by a number of molecules that compete with O_2 for the heme sites. In particular, carbon monoxide binding to the heme sites is one to two orders of magnitude greater than oxygen binding. Analysis of a hypothetical two-site hemoglobin with both O_2 and CO dissolved in the blood is given below.

To simplify the notation, hemoglobin is designated as Hb, O_2 by L and CO by M. The bound-ligand dissociation reactions, their mass-action laws and their equilibrium constants are:

$$Hb{\cdot}L = Hb + L \qquad K_{1L} = [Hb][L]/[Hb{\cdot}L] = 25\ \mu M$$

$$Hb{\cdot}L_2 = Hb{\cdot}L + L \qquad K_{2L} = [Hb{\cdot}L][L]/[Hb{\cdot}L_2] = 15\ \mu M$$

$$Hb{\cdot}M = Hb + M \qquad K_{1M} = [Hb][M]/[Hb{\cdot}M] = 0.25\ \mu M$$

$$Hb{\cdot}M_2 = Hb{\cdot}M + M \qquad K_{2M} = [Hb{\cdot}M][M]/[Hb{\cdot}M_2] = 0.15\ \mu M$$

$$Hb{\cdot}ML = Hb{\cdot}M + L \qquad K_{ML} = [Hb{\cdot}M][L]/Hb{\cdot}ML] = 1.5\ \mu M$$

$$Hb{\cdot}ML = Hb{\cdot}L + M \qquad K_{LM} = [Hb{\cdot}L][M]/Hb{\cdot}ML] = \ldots\ \mu M$$

Except for K_{1L} and the two-order-of-magnitude factor between it and K_{1M}, these equilibrium constants have been chosen arbitrarily. The only restriction was that the dissociation constants of the di-liganded hemoglobin be smaller than the mono-ligand value in order to represent cooperative binding. The mixed di-ligand equilibrium constant, K_{ML}, was assumed to be the geometric mean of K_{2M} and K_{2L}. As shown below the K_{LM} is related to the preceding equilibrium constants.

Possible combinations of bound ligands are shown below. The sites on the hemoglobin molecule are drawn as vertical line segments.

Conservation equations for hemoglobin and the ligands O_2 and CO are:

$$[Hb_{tot}] = [Hb] + [Hb{\cdot}L] + [Hb{\cdot}M] + [Hb{\cdot}L_2] + [Hb{\cdot}M_2] + [Hb{\cdot}ML]$$

$$[L_o] = [L] + [Hb{\cdot}L] + 2[Hb{\cdot}L_2] + [Hb{\cdot}ML]$$

$$[M_o] = [M] + [Hb{\cdot}M] + 2[Hb{\cdot}M_2] + [Hb{\cdot}ML]$$

The bound-ligand concentrations are expressed in terms of the free-ligand concentrations:

$$[Hb{\cdot}L] = K_{1L}^{-1}[Hb][L]$$

$$[Hb{\cdot}L_2] = K_{2L}^{-1}[Hb{\cdot}L][L] = K_{1L}^{-1}K_{2L}^{-1}[Hb][L]^2$$

$$[Hb{\cdot}M] = K_{1M}^{-1}[Hb][M]$$

$$[Hb{\cdot}M_2] = K_{2M}^{-1}[Hb{\cdot}M][M] = K_{1M}^{-1}K_{2M}^{-1}[Hb][M]^2$$

$$[\text{Hb}\cdot\text{LM}] = K_{ML}^{-1}K_{1m}^{-1}[\text{Hb}][\text{L}][\text{M}] = K_{LM}^{-1}K_{1L}^{-1}[\text{Hb}][\text{L}][\text{M}]$$

The last of these equations shows that $K_{LM}K_{1L} = K_{ML}K_{1M}$, so that K_{LM} is not an independent equilibrium constant.

Substituting the above equations into the conservation equations reduces the number of from nine to three unknowns:

$$[\text{Hb}_{tot}] = [\text{Hb}] + K_{1L}^{-1}[\text{Hb}][\text{L}] + K_{1M}^{-1}[\text{Hb}][\text{M}] + K_{1L}^{-1}K_{2L}^{-1}[\text{Hb}][\text{L}]^2 + K_{1M}^{-1}K_{2M}^{-1}[\text{Hb}][\text{M}]^2 + K_{ML}^{-1}K_{1M}^{-1}[\text{Hb}][\text{L}][\text{M}] \quad (11.36a)$$

$$[\text{L}_o] = [\text{L}] + K_{1L}^{-1}[\text{Hb}][\text{L}] + 2K_{1L}^{-1}K_{2L}^{-1}[\text{Hb}][\text{L}]^2 + K_{ML}^{-1}K_{1M}^{-1}[\text{Hb}][\text{L}][\text{M}] \quad (11.36b)$$

$$[\text{M}_o] = [\text{M}] + K_{1M}^{-1}[\text{Hb}][\text{M}] + 2K_{1M}^{-1}K_{2M}^{-1}[\text{Hb}][\text{M}]^2 + K_{ML}^{-1}K_{1M}^{-1}[\text{Hb}][\text{L}][\text{M}] \quad (11.36c)$$

Solution method To further simplify the calculation, the above three equations are expressed in terms of the following dimensionless quantities:

$A = K_{1L}^{-1}[\text{L}_o]$; $B = K_{1M}^{-1}[\text{M}_o]$; $C = K_{1L}^{-1}K_{2L}^{-1}[\text{L}_o]^2$;

$D = K_{1M}^{-1}K_{2M}^{-1}[\text{M}_o]^2$; $E = K_{ML}^{-1}K_{1M}^{-1}[\text{L}_o][\text{M}_o]$; $X = [\text{L}]/[\text{L}_o]$;

$Y = [\text{M}]/[\text{M}_o]$; $V = [\text{Hb}_{tot}]/[\text{M}_o]$; $W = [\text{Hb}_{tot}]/[\text{L}_o]$

The first step in the solution is to factor [Hb] from the terms on the right-hand side of the Equation (11.36a) and use the result to eliminate this variable from Equations (11.36b) and (11.36c). This results in two equations with two unknowns, X and Y:

$$1 = X \times \left[1 + W\frac{A + 2CX + EY}{1 + AX + BY + CX^2 + DY^2 + EXY}\right]$$

and

$$1 = Y \times \left[1 + V\frac{B + 2DY + EX}{1 + AX + BY + CX^2 + DY^2 + EXY}\right]$$

These two equations are solved by a Monte Carlo search: X and Y are guessed as random numbers between 0 and 1 (which they must be) and substituted into the above equations. When the right-hand sides of these equations are simultaneously between 1 and 1.001, the solution has been obtained.

A modicum of reality is injected into the calculation using realistic values of $[\text{Hb}_{tot}]$, $[\text{L}_o]$ and $[\text{M}_o]$. The concentration of hemoglobin in human blood is ~ 15 g/dl, which, with a molecular weight of 68,000, corresponds to $[\text{Hb}_{tot}]$ = 2200 μM (micromoles per liter). The concentrations of O_2 and CO are taken to be those in water determined by the partial pressures of these gases (in atm) and their Henryís law constants. The latter are obtained from Table 8.1:

$$[L_o] = \left(p_{O_2}/730\right) \times 10^6 = 0.21 \times 10^6 / 766 = 274\ \mu M$$

$$[M_o] = (p_{CO}/980) \times 10^6 = p_{CO} \times 10^3\ \mu M$$

The calculation assumes that these concentrations are established by equilibration of the liquid (blood) with air containing a variable partial pressure of carbon monoxide. Equilibrium with hemoglobin takes place out of contact with air, so that $[L_o]$ and $[M_o]$ are the fixed total concentrations of the ligands, bound and free.

The results are displayed in the two graphs in Figure 11.16, in which p_{O_2} and $[Hb_{tot}]$ are fixed and p_{co} is varied. The top graph shows the fraction of the sites on the hypothetical hemoglobin molecule (two sites per molecule) that are occupied by each of the ligands. These are given by:

$$f_{O_2} = \frac{[L_o] - [L]}{2[Hb_{tot}]} = \frac{0.5}{W}(1 - X)$$

$$f_{CO} = \frac{[M_o] - [M]}{2[Hb_{tot}]} = \frac{0.5}{V}(1 - Y)$$

The horizontal line in this graph shows that oxygen binds to hemoglobin as if CO were absent, irrespective of the CO partial pressure. The constant value of f_{O_2} = 0.068 means that $X = [L]/[L_o] << 1$, a result shown in more detail in a curve in the bottom graph.

CO binding does not start in earnest until its partial pressure is ~ 0.2 atm, where the concentration in solution is ~200 μM (top). Thereafter, binding increases until at 4000 μM (4 atm partial pressure) ~ 95% of the total sites on hemoglobin are occupied by CO molecules. At any CO partial pressure, less than 0.1% of the available CO remains in solution (bottom).

The results of this calculation should not be taken to be a realistic analysis of the carbon monoxide/oxygen competition for sites on hemoglobin. It is mainly intended as a vehicle to illustrate the complex effects of the thermochemical parameters on the binding equilibria.

11.7 OSMOTIC PRESSURE

Osmosis is an important process in all living matter. This phenomenon involves aqueous solutions separated by a membrane that is permeable to some solution components but not to others, typically macromolecules and large ions. The cell wall in Figure 11.4 is a typical semipermeable membrane. If the solution on one side of the membrane contains a macromolecule that cannot pass through the membrane and the other side does not, a pressure difference between the two solutions is created as water moves through the membrane to the side containing the macromolecule.

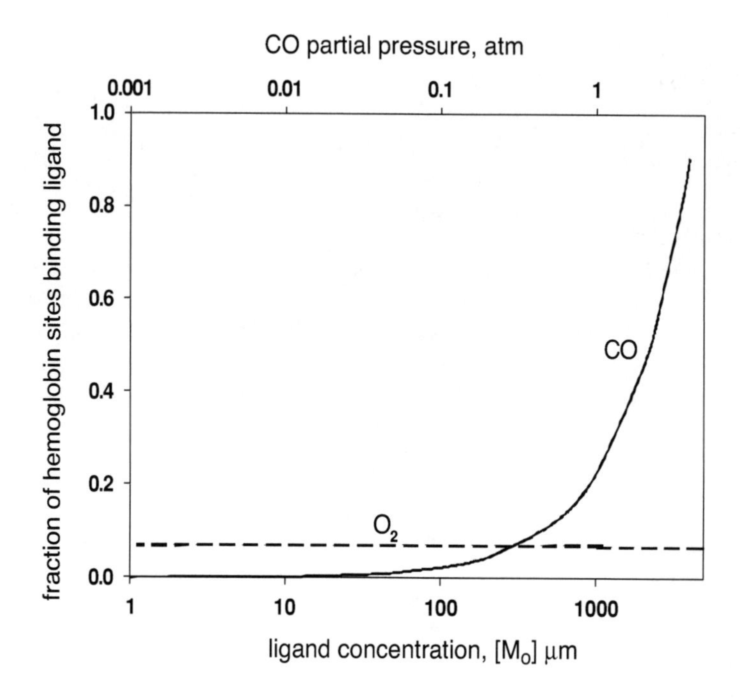

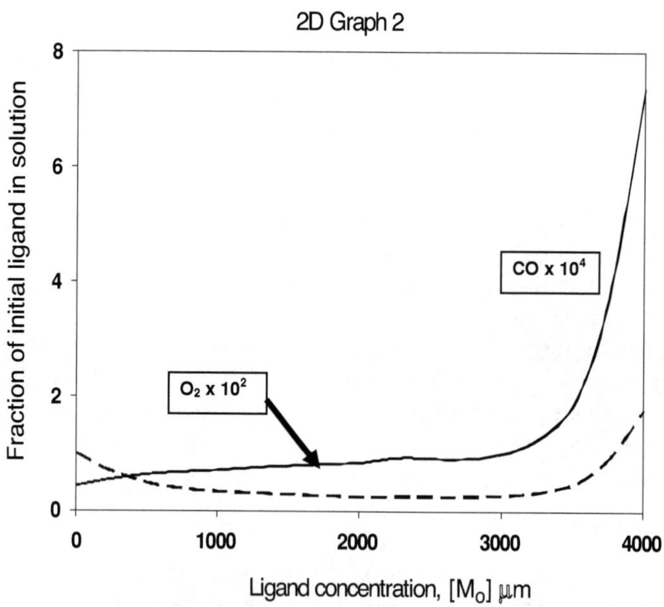

FIGURE 11.16 Binding of O_2 and CO to hemoglobin. Top: fraction of Hb sites bound to ligands; Bottom: fraction of initial ligand remaining in solution. $[Hb_{tot}]$ 2200 μM; $[L_o]$ = 274 μM.

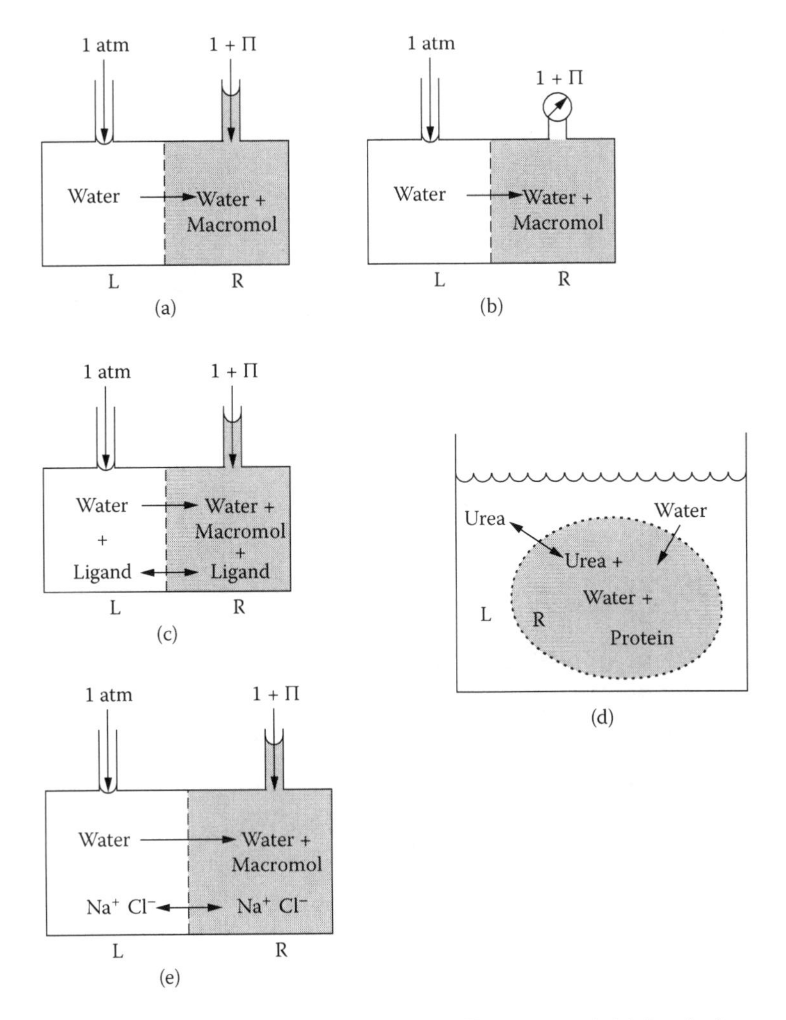

FIGURE 11.17 Devices producing osmotic pressure (Π above 1 atm). (a) Standard osmometer; (b) constant-volume osmometer; (c) with a ligand that binds to the macromolecule; (d) dialysis; (e) charge effects: Donnan equilibrium.

11.7.1 Osmometers

Figure 11.17 depicts a number of natural or man-made devices that exhibit this phenomenon, including:

- Osmometers, which are laboratory instruments intended for determination of the molecular weight of macromolecules and, at the same time, a parameter reflecting the nonideality of the macromolecule in solution. These instruments can be constructed in two ways, one in which the movement of water through the membrane is measured by the rise of the level of the macromolecule solution in a capillary tube (sketch (a) in

Figure 11.17) and a variation of (a) in which the volume of the macromolecule solution is kept constant by a rigid container and a gauge detects the rise in pressure (b).

- Osmometers containing ligands on one or both sides of the membrane and the macromolecule on one side (c). The purpose of such a device is to measure the equilibrium constant for ligand binding, as was covered in the preceding section.
- Dialysis machines, which separate urea from valuable proteins using semipermeable membranes that transmit only the former species (d).
- Osmometers which, in addition to the macromolecule, contain a completely ionized salt on both sides of the membrane (e). The salt is often NaCl, in order to mimic the saline quality of body fluids. The addition of charged species to the mix sets up an electrostatic potential between the solutions on either side of the membrane.

The analysis of osmotic pressure is prone to changes in units, so Table 11.1 summarizes them.

TABLE 11.1
Terminology Related to Macromolecules in Aqueous Solution

Property	Symbol	Units
Mass concentration	c = M[ ]	g/lit.
Molar concentration	[ ]	moles/lit.
Mole fraction	$x = v_w[\ ]$	—
Molecular wt.	M	g/mole
Molar vol. of water	v_w	lit./mole
Water	w (subscript)	—
Macromolecule	mm (subscript)	—
Protein	p (subscript	—

11.7.2 Osmotic Second Virial Coefficient

The device in Figure 11.17a functions because the macromolecule in the right-hand chamber lowers the activity of water, hence creating a driving force to move water from the left to the right through the membrane. At equilibrium, the chemical potentials of water on the two sides of the membrane are equal ($\mu_w^L = \mu_w^R$), or, in terms of the activity of water a_w^R:

$$g_w^o(1\ atm) = g_w(1+\Pi) + RT\ \ln a_w^R \tag{11.37}$$

g_w^o is the molar Gibbs free energy of liquid water at ambient temperature (usually 298 K) and the pressure rise Π is called the *osmotic pressure* of the solution containing the macromolecule. It is easily measurable by the height difference between the liquid levels in the left and right compartments (a) or directly by the pressure rise (b). a_w^R is the activity of water in the macromolecule solution on the right side of the osmometer. g_w^o is the standard-state free energy of water at 1 atm pressure. Because the pressure in the right-hand compartment is greater than 1 atm, $g_w(1+\Pi)$ is obtained by integrating the thermodynamic relation given by Equation (6.15) from 1 atm to 1 + Π atm. This procedure yields:

$$g_w(1+\Pi) = g_w^o(1\text{ atm}) + v_w\Pi \tag{11.38}$$

Substituting Equation (11.38) into Equation (11.35) yields:

$$\Pi v_w = -RT\ln a_w^R \tag{11.39}$$

The size and complexity of huge macromolecules enhances interactions between them this is manifest as deviations from ideal behavior (ideality means macromolecule interaction only with solvent (water) molecules). To account for nonideality, the activity coefficient of the macromolecule in solution is represented in a *virial expansion* because of its similarity to the virial equation of state for gases (Prausnitz et al., 1999):

$$\gamma_{mm} = 1 + b_{mm}[mm] + \ldots \tag{11.40}$$

Terms containing $[mm]^2$ and higher orders of [mm] are neglected. Equation (11.40) represents a first-order nonideality correction.

The activity of the water is obtained from the Gibbs-Duhem equation (Equation [7.28] with $\mu = g^o + RT\ln a$):

$$[w]d\ln a_w = -[mm]d\ln a_{mm} = -[mm]d\ln(\gamma_{mm}[mm])$$

The molar concentration of water is $[w] \cong 1/v_w$. Substituting Equation (11.40) into the above and integrating from a lower limit of pure water yields:

$$\ln a_w = -v_w \int_0^{[mm]} y\frac{d}{dy}\left\{\ln\left[(1+b_{mm}y)y\right]\right\}dy = \int_0^{[mm]} \left(\frac{1+2b_{mm}y}{1+b_{mm}y}\right)dy$$

$$\ln a_w = -v_w\left([mm] - \frac{1}{b_{mm}}\ln(1+b_{mm}[mm])\right) \tag{11.41}$$

In most applications, the concentration of the macromolecule is small enough to employ the approximation: $\ln(1+x) = x - \frac{1}{2}x^2$, which reduces Equation (11.41) to:

$$\ln a_w = -v_w([mm] + \tfrac{1}{2}b_{mm}[mm]^2) = -v_w\left(\frac{c_{mm}}{M_{mm}} + \frac{b_{mm}}{2M_{mm}^2}c_{mm}^2\right)$$

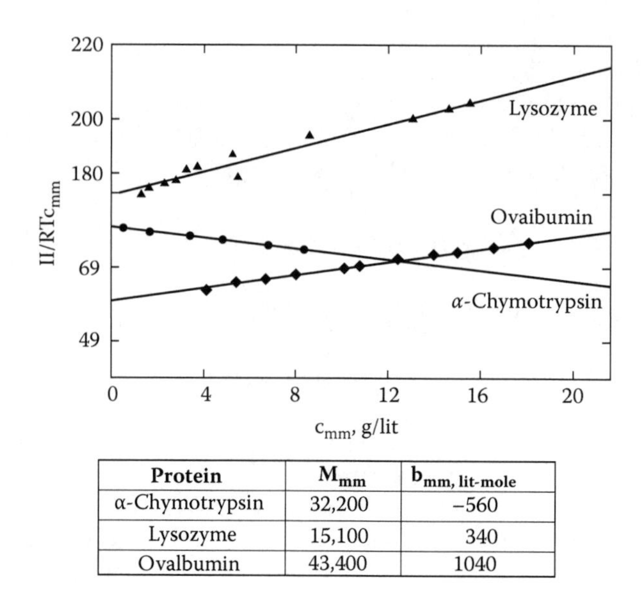

Protein	M_{mm}	b_{mm}, lit-mole
α-Chymotrypsin	32,200	−560
Lysozyme	15,100	340
Ovalbumin	43,400	1040

FIGURE 11.18 Plot for three proteins (macromolecules) according to Equation (11.42). (From Prausnitz, J. et al. 1999. *Molecular Thermodynamics of Fluid-Phase Equilibria*, 3rd ed. Englewood Cliffs, NJ: Prentice Hall. With permission.)

where c_{mm} is the mass concentration and M_{mm} is the molecular weight of the macromolecule. Substituting the above equation into Equation (11.39) yields:

$$\frac{\Pi / RT}{c_{mm}} = \frac{1}{M_{mm}} + \left(\frac{b_{mm}}{2M_{mm}^2} \right) c_{mm} \tag{11.42}$$

This very useful equation shows that plotting the left-hand side against c_{mm} should yield a straight line whose intercept is $1/M_{mm}$ and whose slope is the quantity in parentheses, which is called the *osmotic second virial coefficient*. The slope can be either positive or negative. The former indicates repulsion of two macromolecules and the latter is characteristic of attractive forces between them. As can be seen in Figure 11.18, common proteins obey Equation (11.42) very closely. The corresponding values of M_{mm} and b_{mm} are shown in the table below the graph. The osmotic second virial coefficient (more precisely, the nonideality coefficient b_{mm}) is very useful in analyzing the separation and purification of proteins (Section 11.8).

11.7.3 Osmosis in Electrolytes

Sketch (e) of Figure 11.17 illustrates an important variant of osmosis. The solutions contain low concentrations (< 0.1 M) of ionic species, either to buffer the pH or to mimic the liquid present inside cells of the human body. Typical salts (or *electrolytes)* are KCl, NaCl and $KHPO_4$. Most ions can pass through the membrane separating the two solutions. In addition, the macromolecule is invariably charged, either negatively or positively. In the following, NaCl is used as a representative electrolyte. With [ ] denoting molarity and the subscripts L and R indicating the left and right compartments of the osmometer:

$[NaCl]_o$ = initial sodium chloride in L or R
$[MCl_Z]_o$ = initial macromolecule cation with its associated chloride anion in R

Transport of Na^+ and Cl^- ions through the membrane occurs in order to equalize the chemical potentials of NaCl on the two sides of the osmometer. At equilibrium, the concentrations are:

$[M^{Z+}] = [MCl_Z]_o$ because the macromolecule cannot penetrate the membrane, its concentration remains the same
$[Na^+]_L$ = sodium ion in L
$[Cl^-]_L$ = chloride ion in L
$[Na^+]_R$ = sodium ion in R
$[Cl^-]_R$ = chloride ion in R

Assuming that the volumes of the left and right compartments are equal, these final concentrations are related by charge neutrality and species conservation in the two compartments.

Conservation of species gives:

$$[Na^+]_L + [Na^+]_R = [NaCl]_o \tag{11.43}$$

$$[Cl^-]_R + [Cl^-]_L = [NaCl]_o + Z[MCl_Z]_o \tag{11.44}$$

Electrical neutrality requires:

$$[Na^+]_L = [Cl^-]_L \tag{11.45}$$

$$[Cl^-]_R = Z[M^{Z+}] + [Na^+]_R \tag{11.46}$$

Only three of these equations are independent, as can be seen by substituting (11.45) and (11.46) into (11.44); this procedure yields Equation (11.43). Therefore, Equations (11.43), (11.45), and (11.46) constitute a sufficient summary of the above set. One additional equation is needed. This is obtained from the condition of equal chemical potentials of NaCl on either side of the membrane.

Digression. Activities of ions in solution.

Because strong electrolytes such as NaCl are completely dissociated in aqueous solution, the activity of NaCl is equal to the sum of the activities of the two ions:

$$\mu_{NaCl} = \mu^o_{Na^+} + RT \ln a_{Na^+} + \mu^o_{Cl^-} + RT \ln a_{Cl^-}$$

Alternatively, the above equation is:

$$\mu_{NaCl} = \mu^o_{NaCl} + RT \ln(a_{Na^+} a_{Cl^-}) = \mu^o_{NaCl} + RT \ln a_{\pm}$$

where the product of the activities of the two ions is denoted by $a_{\pm}$ and

$$\mu^o_{NaCl} = \mu^o_{Na^+} + \mu^o_{Cl^-}$$

$a_{\pm}$ can be divided into the usual product of an activity coefficient and a concentration:

$$a_{\pm} = \gamma_{\pm} \text{ [NaCl]}$$

where $\gamma_{\pm}$ is the *mean ionic activity coefficient.* In pure water $\gamma_{\pm}$ for NaCl drops rapidly from unity at infinite dilution to ~ 0.7 at 0.3 M, and remains essentially constant to higher concentrations. However, $\gamma_{\pm}$ is affected by even small molar concentrations of biological macromolecules in the solution.

For the present purpose, $\gamma_{\pm}$ in the L and R solutions may be assumed to be equal, thereby reducing the equilibrium condition

$$\mu^L_{NaCl} = \mu^R_{NaCl}$$

to:

$$[Na^+]_L \times [Cl^-]_L = [Na^+]_R \times [Cl^-]_R \tag{11.47}$$

This equation is known as the *Donnan equilibrium.* It is the fourth equation needed to solve for the four concentrations..

Eliminating the Cl^- concentrations in Equation (11.47) with the aid of Equations (11.45) and (11.46) and the Na^+ concentration in the left-hand compartment with Equation (11.43) gives the final result:

$$[Na^+]_R = \frac{[NaCl]_o^2}{2[NaCl]_o + Z[MCl_Z]_o} \tag{11.48}$$

from which the remaining ion concentrations follow from Equations (11.43), (11.45), and (11.46).

In order to complete the solution to this problem, the chemical potentials of water in the two compartments are set equal. However, in this case the activity of water in the left-hand compartment is not unity because of the dissolved salt so Equation (11.39) is modified to:

$$\Pi v_w = -RT(\ln a^R_w - \ln a^L_w) \tag{11.49}$$

Neglecting nonideal behavior,

$$\ln a_w^L \cong -v_w([Na^+]_L + [Cl^-]_L) \qquad \text{and} \qquad \ln a_w^R \cong -v_w([Na^+]_R + [Cl^-]_R + [M^{Z+}]_R)$$

Substituting these into Equation (11.49) gives:

$$\frac{\Pi}{RT} = -([Na^+]_L + [Cl^-]_L) \tag{11.50}$$

Eliminating the ion concentrations using Equation (11.48), (11.43), (11.45), and (11.46) yields the final result:

$$\frac{\Pi}{RT} = [MCl_Z]_o + \frac{Z^2[MCl_Z]_o^2}{2[NaCl]_o + Z[MCl_Z]_o} \tag{11.51}$$

Comparison of this result with Equation (11.42) (noting that $c_{mm}/M_{mm} = [MCL_Z]_o$) shows that the first term is the effect on the osmotic pressure of the macromolecule, ionized or not. The second term represents two effects: the added sodium chloride and the charge on the macromolecule. If NaCl is absent, Equation (11.51) reduces to:

$$\frac{\Pi}{RT} = [MCl_Z]_o(1+Z) \tag{11.51a}$$

If the macromolecule is uncharged, the second term in Equation (11.51a) vanishes. The charge effect, whether it arises from added electrolyte or from a positive charge on the macromolecule, increases the osmotic pressure.

Example: A macromolecule with a charge of +10 and a concentration of 10^{-4} M is dissolved in the solution on one side of a membrane, while a 10^{-3} M sodium chloride solution initially fills the other side. From Equation (11.51), the osmotic pressure is:

$$\Pi = 82 \times 298\left[10^{-4} + \frac{(10^{-3})^2}{2\times10^{-3} + 10\times10^{-4}}\right] = 82\times298\left[10^{-4} + 3.33\times10^{-4}\right] = 10.6 \text{ atm}$$

11.7.4 Membrane Potential Difference

Because of the imbalance in the Na^+ and Cl^- concentrations on the two sides of an osmotic cell, there exists an electric potential difference across the semipermeable membrane. The potential difference is determined by equating the chemical potentials of Na^+ on the left and right sides of the membrane:

$$\mu_{Na^+}^L = \mu_{Na^+}^o + RT\ln[Na^+]_L + F\phi_L \qquad \mu_{Na^+}^R = \mu_{Na^+}^o + RT\ln[Na^+]_R + F\phi_R \tag{11.52}$$

where F is Faraday's constant. Equating and solving for the potential difference yields:

$$\phi_L - \phi_R = \frac{RT}{F} \ln\left(\frac{[Na^+]_R}{[Na^+]_L}\right) = \frac{RT}{F} \ln\left(\frac{[NaCl]_o}{[NaCl]_o + Z[MCl_Z]_o}\right) \qquad (11.53)$$

The same result is obtained by equating the chemical potentials of Cl^- on the two sides, recognizing that because of the change in sign of the ion, the plus sign in front of the last terms in Equation (11.53) becomes a minus sign. That the same electric potential difference is obtained from either ion is simply a consequence of the Donnan equilibrium, Equation (11.47).

Example: A macromolecule with a charge of +10 is on one side of a membrane at a concentration of 10^{-4} M while a 10^{-3} M sodium chloride solution occupies the other side. The electric potential difference between the solutions on either side of the membrane is:

$$\Delta\phi = \frac{8.314 \times 298}{96,500} \ln(0.5) = -18 \text{ mV}$$

How can the existence of a potential difference across the membrane be reconciled with the condition of electrical neutrality of solutions on both sides? The answer is that the charge difference is localized at the membrane, which acts as a capacitance. The membrane potential difference has profound implications in mammalian cells because it permits species to be transferred into and out of the cell against concentration gradients.

11.8 SEPARATION OF PROTEINS BY LIQUID–LIQUID EXTRACTION

Liquid–liquid extraction [Section 8.3.4] is a chemical engineering unit operation in which solutes are equilibrated between two immiscible or partially miscible liquid phases. Its utility to biotechnology lies in the ability to separate certain proteins from a mixture of proteins and other undesirable macromolecules. Partitioning occurs by virtue of different distribution coefficients of the macromolecules between the two (usually aqueous) phases. Fermentation broths, for example, are a complex mixture of valuable proteins that require a number of steps to separate; the first is often selective extraction between two liquids.

Proteins are soluble in water, so this is the natural solvent medium. The creation of two partially miscible phases is simple; when the right pair of water-soluble polymers is dissolved in water, two phases appear, one rich in polymer A and the other rich in polymer B. When a mix of proteins is added, certain of them prefer phase I and others phase II. The efficacy of the separation of protein P is quantitatively expressed by its *distribution coefficient* between the two phases:

$$K_p = \frac{[P]_{II}}{[P]_I} \qquad (11.54)$$

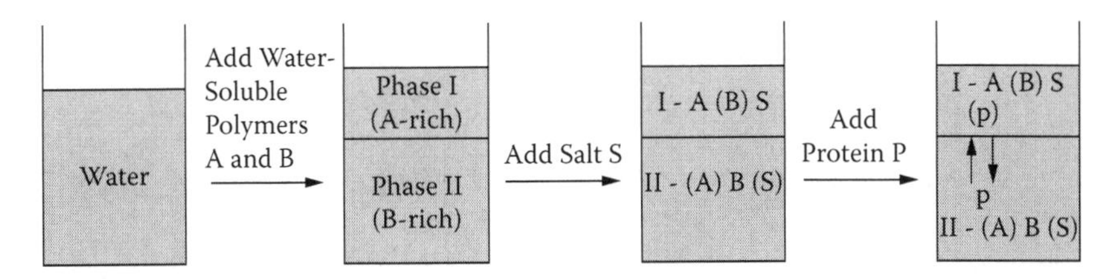

FIGURE 11.19 Protein extraction by water-soluble polymers. Letters in parentheses denote low-concentration species.

To the extent that K_p >> 1 or << 1, the separation is selective. Selectivity is enhanced by adding a completely dissociated salt to the mix. Because the activity coefficients of the salt are different in the two aqueous phases, the salt distributes unevenly, thereby setting up an electric potential difference. Most proteins in aqueous solution are charged; the electric potential difference can significantly supplement the purely "chemical" proclivity of the protein to favor one of the two phases.

The set up of the extraction system is illustrated in Figure 11.19. At equilibrium, phase I is rich in polymer A and salt S while polymer B concentrates in phase II along with the protein. Typical water-soluble polymers are dextran and polyethelyene glycol (PEG). Not shown in Figure 11.19 are additions of buffer solutions to control the pH, which affects protein distribution by its effect on the protein's charge.

11.8.1 Formation of the Two Phases

The water-soluble polymers responsible for separating water into partially miscible liquids possess osmotic second virial coefficients, which provide the following activity coefficients in water:

$$\gamma_A = 1 + b_{AA}[A] + b_{AB}[B] \quad \gamma_B = 1 + b_{BB}[B] + b_{AB}[A] \tag{11.55}$$

The second terms in the above equations account for interactions between like solute molecules and b_{AB} reflects interactions between A and B molecules in the same solution. The equilibrium distributions of the polymers are given by equations analogous to Equation (11.54):

$$K_A = \frac{[A]_{II}}{[A]_I} = \frac{\gamma_{AI}}{\gamma_{AII}} = \frac{1 + b_{AA}[A]_I + b_{AB}[B]_I}{1 + b_{AA}[A]_{II} + b_{AB}[B]_{II}} \tag{11.56a}$$

$$K_B = \frac{[B]_{II}}{[B]_I} = \frac{\gamma_{BI}}{\gamma_{BII}} = \frac{1 + b_{BB}[B]_I + b_{AB}[A]_I}{1 + b_{BB}[B]_{II} + b_{AB}[A]_{II}} \tag{11.56b}$$

These formulas are equivalent to equating the chemical potentials of A and B in the two phases.

Finally, water equilibrium between the two phases is calculated in the same manner. The activity coefficient of water in either phase is obtained from γ_A and γ_B using the Gibbs-Duhem equation for a three-component system:*

$$[A]d\ln\gamma_A + [B]d\ln\gamma_B + (1/v_w)d\ln\gamma_W = 0 \qquad (11.57)$$

In integrated form:

$$\frac{1}{v_w}\int_0^{\ln\gamma_W} d\ln\gamma_W = -\int_0^{\ln\gamma_A}[A]d\ln\gamma_A - \int_0^{\ln\gamma_B}[B]d\ln\gamma_B$$

The lower limits are all zero because this condition represents pure water. The remainder of the derivation assumes that the last two terms in Equations (11.55) are small compared to unity, so that $\ln\gamma_A \cong b_{AA}[A] + b_{AB}[B]$ and $\ln\gamma_B \cong b_{BB}[B] + b_{AB}[A]$. Substituting these into the integrals yields:

$$\frac{\ln\gamma_w}{v_w} = -b_{AA}\int_0^{[A]}[A]d[A] - b_{AB}\int_0^{[B]}[A]d[B] - b_{BB}\int_0^{[B]}[B]d[B] - b_{AB}\int_0^{[A]}[B]d[A]$$

The sum of the second and last terms on the right-hand side of this equation is the integral of d{[A][B]}, so the above equation becomes:

$$\frac{\ln\gamma_w}{v_w} = -\tfrac{1}{2}b_{AA}[A]^2 - \tfrac{1}{2}b_{BB}[B]^2 - b_{AB}[A][B] \qquad (11.58)$$

Equality of the chemical potentials of water in the two aqueous phases is expressed by:

$$\gamma_{WI}[W]_I = \gamma_{WII}[W]_{II} \quad \text{or} \quad \ln\gamma_{WI} + \ln[W]_I = \ln\gamma_{WII} + \ln[W]_{II} \qquad (11.59)$$

The molar concentration of water, [W], can be expressed in terms of mole fractions by:

$$[W] = x_w/v_w = (1 - x_A - x_B)/v_w$$

so that

$$\ln[W] = -\ln v_w + \ln(1 - x_A - x_B) \cong -\ln v_w - x_A - x_B = -\ln v_w - ([A] + [B])v_w$$

---

* Equation (11.57) is the three-component analog of Equation (7.31), with mole fractions converted to molar concentrations by dividing by v_w. Although the weight fractions of the polymers in the two phases can be as large as 0.2, their mole fractions are very much smaller because their molecular weights are in the tens of thousands.

Substituting this equation and Equation (11.58) into (11.59) yields:

$$\begin{aligned}&[A]_I + [B]_I + \tfrac{1}{2} b_{AA}[A]_I^2 + \tfrac{1}{2} b_{BB}[B]_I^2 + b_{AB}[A]_I[B]_I \\ &= [A]_{II} + [B]_{II} + \tfrac{1}{2} b_{AA}[A]_{II}^2 + \tfrac{1}{2} b_{BB}[B]_{II}^2 + b_{AB}[A]_{II}[B]_{II}\end{aligned} \tag{11.60}$$

Equations (11.56a), (11.56b) and (11.60) provide three equations with four unknowns, $[A]_I$, $[B]_I$, $[A]_{II}$, and $[B]_{II}$. The fourth equation is a statement of species conservation.

Suppose that the system consists of a total volume V of water to which n_A moles of A and n_B moles of B are added. After splitting into two phases of volumes V_I and V_{II}, the species conservation equations are:

$$n_A = [A]_I V_I + [A]_{II} V_{II} \quad \text{or} \quad [A]_{tot} = f\,[A]_I + (1-f)[A]_{II}$$

$$n_B = [B]_I V_I + [B]_{II} V_{II} \quad \text{or} \quad [B]_{tot} = f\,[B]_I + (1-f)[B]_{II}$$

where $f = V_I/V_{tot}$ is the fraction of the total water volume converted to phase I and $[A]_{tot} = n_A/V_{tot}$ and $[B]_{tot} = n_B/V_{tot}$. These are specified. Eliminating f between the two species conservation equations yields:

$$f = \frac{[A]_{tot} - [A]_{II}}{[A]_I - [A]_{II}} = \frac{[B]_{tot} - [B]_{II}}{[B]_I - [B]_{II}} \tag{11.61}$$

11.8.1.1 Solution of the Phase–Equilibrium Equations

Dimensionless forms of these equations result from the definitions:

$$X = b_{AA}[A]_I; \quad Y = b_{AA}[A]_{II}; \quad U = b_{BB}[B]_I;$$

$$V = b_{BB}[B]_{II}; \quad Z_A = b_{AA}[A]_{tot}; \quad Z_B = b_{BB}[B]_{tot}$$

$$J_A = b_{AA}/b_{AB} \quad \text{and} \quad J_B = b_{BB}/b_{AB}$$

Equations (11.56a) and (11.56b) become:

$$\frac{Y}{X} = \frac{1 + X + U/J_B}{1 + Y + V/J_B} \qquad \frac{V}{U} = \frac{1 + U + X/J_A}{1 + V + Y/J_A} \tag{11.62a,b}$$

With some rearrangement, Equation (11.60) is converted to:

$$(1+X)^2 J_B + (1+U)^2 J_A + 2XU = (1+Y)^2 J_B + (1+V)^2 J_A + 2YV \tag{11.63}$$

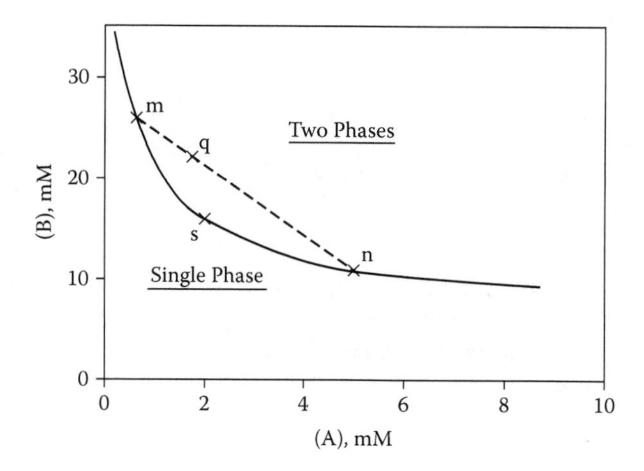

FIGURE 11.20 Calculated phase diagram for the polymers dextran T-70 and polyethylene glycol 3350 in water. The polymer molecular weights are 30,000 and 4000, respectively. The osmotic second virial coefficients used to construct the binodal were $b_{AA} = 450\ M^{-1}$ for dextran, $b_{BB} = 64\ M^{-1}$ for polyethylene glycol and $b_{AB} = 1000\ M^{-1}$.

and Equation (11.61) becomes:

$$f = \frac{Z_A - Y}{X - Y} = \frac{Z_B - V}{U - V} \tag{11.64}$$

Given Z_A, Z_B, J_A, and J_B, Equations (11.62a,b), (11.63), and (11.64) are to be solved for X, Y, U, and V.

Figure 11.20 shows a calculated phase diagram,* with coexisting, equilibrium phase concentrations [B] on the ordinate and [A] on the abscissa. The locus of equilibrium points lie along the curve, which is called the *binodal*. The region below the curve represents a single homogeneous aqueous phase containing polymers A and B (in this case dextran and polyethylene glycol, or PEG). Solutions with overall compositions lying above the binodal (e.g., point q) split into two phases, whose compositions lie on the binodal, point n for dextran-rich phase I and point m for the coexisting phase II enriched in PEG. These three points lie on a *tie line*, and are designated as

$$q = [A]_{tot}, [B]_{tot}, \qquad m = [A]_{II}, [B]_{II}, \qquad \text{and} \qquad n = [A]_{I}, [B]_{I}$$

The line m-n is one of a series of parallel tie lines. As the tie lines move down, they become shorter and ultimately disappear at point s. The two phases become one at this point, which is analogous to the critical point in the phase diagram of

* Prausnitz et al. (1999) present the model equations and a graph similar to Figure 11.20. However, they do not show how they solved the equations. The author of this book was unable to demonstrate that solution of Equations (11.62a,b), (11.63) and (11.64) results in a plot like that in Figure 11.20.

a one-component substance. All points on the binodal lying to the left at point s give the composition of B-rich phase II, and the points to the right of s characterize A-rich phase I.

The lever rule applies to the tie lines, and the volume fraction of phase I corresponding to point q is:

$$f = \frac{mq}{mn} \tag{11.65}$$

This equation gives the same value of f as either of the two ratios in Equation (11.60).

11.8.2 The Effect of Electrolyte (Salt) Addition

In the second step in Figure 11.18, an inorganic salt is added to the water–polymer A-polymer B two-phase mixture in order to enhance the distribution coefficient of the protein. Before addressing this problem, two issues are examined.

The first is the effect of the added salt on the phase diagram of Figure 11.19. In general, the salt alters the activity coefficients of polymers A and B from the salt-free values described in terms of second osmotic virial coefficients (Equations [11.55]). The dextran-PEG polymer solutions are little affected by the addition of a variety of salts up to concentrations of ~ 0.1 M (shown in Figure 3.17a of Zaslavsky, 1995).

The second is the effect of the different polymer concentrations in the two phases on the activity coefficients of the salt, or, alternatively, on the salt's distribution coefficient:

$$K_S = \frac{[S]_{II}}{[S]_I} = \frac{\gamma_{\pm I}}{\gamma_{\pm II}} \tag{11.66}$$

The effect is described empirically by the equation:

$$\ln K_S = b_S([B]_{II} - [B]_I) \tag{11.67}$$

The quantity in parentheses is the difference in the polymer-B concentration in the two coexisting phases. The coefficient b_S is a function of the electrolyte and its concentration. In dextran-PEG systems, b_S for KCl is –2.6 M^{-1} for a total salt concentration of 0.5 M (Zaslavsky, 1995, Figure 3.22). The b_S value for ammonium sulfate is approximately 20 times that for KCl. Owing to their similar chemical natures, b_S for NaCl is probably closer to that of KCl than to $(NH_4)_2SO_4$.

The difference in the electrolyte concentrations in the immiscible aqueous polymer solutions is a consequence of maintaining the same chemical potential of the salt in the two phases (this requirement leads to Equation [11.66]). In addition, the chemical potentials of the individual ions of the salt must also be the same in the two phases. This leads to a small electric potential difference between the two phases, as shown for a membrane interface in Section 11.7.4. The electric potential difference given by Equation (11.53) applies as well to immiscible phases separated by a liquid-liquid interface. For a uni-univalent salt such as NaCl, this is:

$$\phi_{II} - \phi_I = \frac{RT}{F} \ln K_S = 0.059 \times b_S \left([B]_{II} - [B]_I\right) \tag{11.68}$$

This electrostatic potential difference occurs only over a few nanometers at the interface between the two phases; the bulk phases are electrically neutral and at the same potential. Nonetheless, the interface potential difference can be measured by electrodes inserted into the bulk phases.

Example: What is the electric potential difference between the dextran-rich and PEG-rich phases for points on the tie line of the phase diagram in Figure 11.19 if the total NaCl concentration is 0.5 M?

The difference between the PEG (B) concentrations between points m and n is 15 mM. Using this value and $b_S = -2.6\ M^{-1}$ in Equation (11.68) yields an electric potential difference of –2.4 mV.

11.8.3 Protein Distribution

This section aims to calculate the distribution coefficient of a protein in a two-polymer, two-phase aqueous system with added electrolyte. The chemical potentials of the protein in the two phases are (see Equation [11.52]):

$$\mu_P^I = \mu_P^o + RT \ln a_P^I + ZF\phi_I \quad \text{and} \quad \mu_P^{II} = \mu_P^o + RT \ln a_P^{II} + ZF\phi_{II} \tag{11.69}$$

Z is the charge on the protein and is assumed to be the same in both aqueous phases. Equating these chemical potentials and replacing the activity by the product of the activity coefficient and the molar concentration leads to:

$$K_P = \frac{[P]_{II}}{[P]_I} = \left(\frac{\gamma_{PI}}{\gamma_{PII}}\right) \exp\left[-\frac{ZF}{RT}\left(\phi_{II} - \phi_I\right)\right] \tag{11.70}$$

γ_P is the “chemical” activity coefficient of the protein in the aqueous solution. It is also expressed in terms of osmotic second virial coefficients (B) reflecting the interactions between the three solutes:

$$\gamma_P = 1 + b_{PP}[P] + b_{AP}[A] + b_{BP}[B] \tag{11.71}$$

where $b = Bv_w$. The effect of the electric potential difference on K_P can be important because proteins in aqueous solution are almost always charged.

Example: If Z = 10 and the electric potential difference is –2.4 mV, the exponential term in Equation (11.70) is 1.5, which constitutes a 50% increase in K_P.

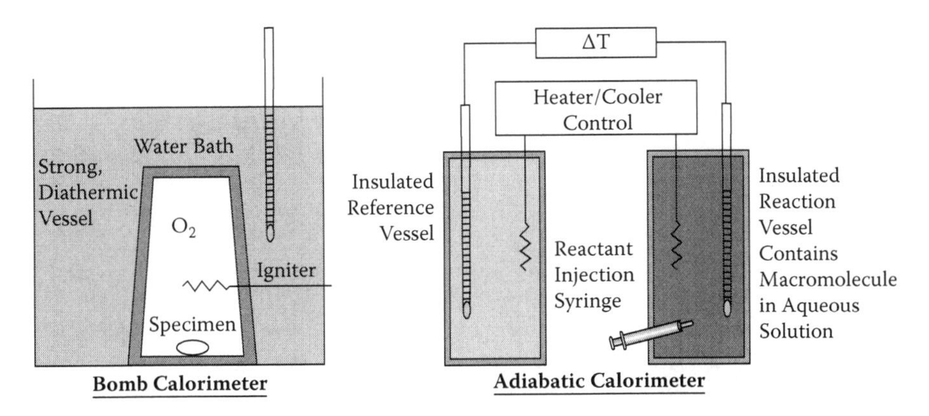

FIGURE 11.21 Calorimeters used in biological studies.

11.9 CALORIMETRY

Considerable effort has been expended in studying heat effects in biological systems. These range from measuring the calorie count of foods on supermarket shelves to the heat release from humans and animals. Laplace (ca 1800) crudely measured the heat rejected by a guinea pig in the first effort to determine animal metabolism rates. Modern evaluations of human activity yield heat release rates ranging from ~100 W while sleeping to ~200 W during a mild activity such as walking. The input energy rates are about twice these values, the difference corresponding to the various work outputs, ranging from large-scale muscle contraction to microscopic motions of cilia on mobile cell components. Energy is supplied to the body by oxidation of glucose, which produces ~3000 kJ/mole, or ~16 kJ/g. About half of this energy produces work and the remainder is rejected from the body as heat. To supply the body's energy needs requires consumption of 50 to 100 g of glucose per hour.

Scientifically oriented calorimetry involves determination of: (1) enthalpy changes of well-defined reactions, such as ligand binding to macromolecules or denaturation (unfolding) of proteins; and (2) heat releases from functioning organs such as nerve fibers or muscles.

Figure 11.21 shows the types of calorimeters used in biological studies.

11.9.1 Bomb Calorimeter

The so-called bomb calorimeter shown in Figure 11.21 is used principally to measure heat released by total combustion of organic compounds, for which a generic reaction is:

$$C_xH_yO_zN_u + (2x + \tfrac{1}{2}y - z)\,O_2 = xCO_2 + \tfrac{1}{2}yH_2O + \tfrac{1}{2}uN_2$$

This is not an equilibrium reaction, as it completely consumes the reactants and produces a mixture of products. It is therefore a standard-state reaction, and the heat evolved is the standard enthalpy of the reaction. As shown in the figure, the substance and high-pressure oxygen are contained in a strong vessel with heat-transmitting walls. Combustion is initiated by a spark, and after completion of the reaction, the temperature rise of the water bath surrounding the vessel is measured. Even though the reactant oxygen and the product gases are not pure and at 1 atm pressure, the heat released is nonetheless the standard enthalpy of the reaction ΔH^o (do you know why?). This property and the measured rise in temperature of the system (water, vessel, and reaction-product gases) are related by:

$$n\Delta H^o = (n_W C_{PW} + n_V C_{PV} + n_g C_{Pg})\Delta T$$

where n is the number of moles of the substance combusted and the subscripts W, V, and g refer to water, vessel, and gas, respectively.

11.9.2 Adiabatic Calorimetry

The pair of insulated vessels in Figure 11.21 are equipped with, in addition to temperature measuring devices (thermocouples, not thermometers as shown), provisions for heating or cooling the contents. The vessel (or cell) on the left contains a reference aqueous solution, meaning one without the reactive substance that is dissolved in the right-hand cell. The system can be operated in either of two modes.

11.9.2.1 Titration Mode

In the first mode, the reactant is periodically injected into the right-hand cell and, at a constant cooling rate, the ΔT between the left and right cells is recorded. As seen in Figure 11.22, the temperature spikes at each reactant injection (if the reaction is exothermic), and returns to the base temperature by the continuous cooling. Each temperature pulse represents the heat evolved by partial uptake of the injected reactant (often a ligand) by the reactant in the vessel (a macromolecule). As the number of injections increases, the reaction sites on the macromolecule become saturated, and the height of the spikes decreases and eventually vanishes.

What follows is a theoretical model of the "data" shown in Figure 11.22. For this analysis, the conditions of the experiment are assumed and the thermodynamic properties of the ligand-binding reaction with the sites on the macromolecule are

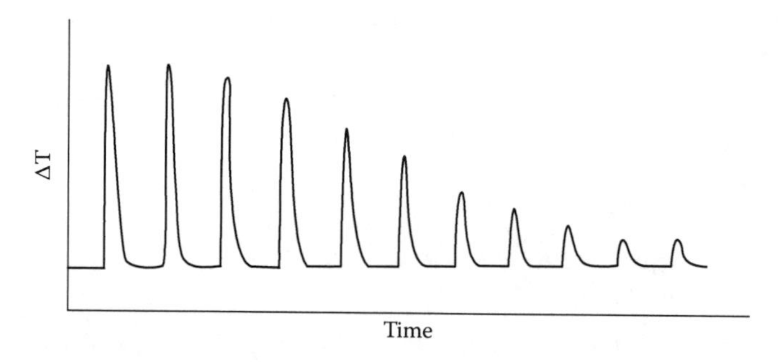

FIGURE 11.22 Temperature response of titration calorimeter.

calculated. For the purposes of illustrating the method, the simplest system consisting of identical and independent binding sites analyzed in Section 11.6.1 is treated.

The temperature spikes following each injection of ligand are due to the heat released by binding of ligands to sites on the macromolecules and the heat removed by the cooling apparatus immersed in the solution. The rate of heat removal from the solution is proportional to the departure of the temperature from the reference-cell temperature, which is also the temperature of the cooling system: $\dot{Q} = E \times \Delta T$. Here E is the product of the heat transfer coefficient between the cooling surface and the solution and the heat transfer area. The integral of the heat removal rate is the heat released during an injection pulse:

$$Q_j = E \int \Delta T \, dt$$

j is the injection number, which in Figure 11.22, runs from 1 to 11. The conditions of the experiment are:

$[L_{inj}]$ = ligand concentration in injected solution = 16 μM
V_{inj} = volume of each injection, liters = 0.4 liter
$[S_o]_0$ = concentration of binding sites in the fresh macromolecule solution = 10 μM
V_o = volume of the fresh macromolecule solution = 4 liter

After j injections, the volume of the solution is $V_o + jV_{inj}$ and a total of $jV_{inj}[L_{inj}]$ moles of ligand have been added. The total ligand concentration (corresponding to $[L_o]$ in Equation [11.15]) is:

$$[L_o]_j = \frac{jV_{inj}[L_{inj}]}{V_o + jV_{inj}}$$

After j injections of ligand, the total concentration of sites is:

$$[S_o]_j = \frac{V_o[S_o]_o}{V_o + jV_{inj}}$$

With these concentrations, and an assumed value of the binding equilibrium constant K, the quantities X and Y of Equations (11.18) are computed and the ratio of bound ligand to total ligand, $y/(1 + y)$, is obtained from Equations (11.17) and (11.20). Once done, the number of moles of bound ligand at the end of injection j is given by:

$$n_{bj} = [S.L]_j(V_o + jV_{inj}) = \frac{y}{1+y}[L_o]_j(V_o + jV_{inj})$$

and the heat released during injection j is:

$$Q_j = \Delta h_b(n_{bj} - n_{b,j-1})$$

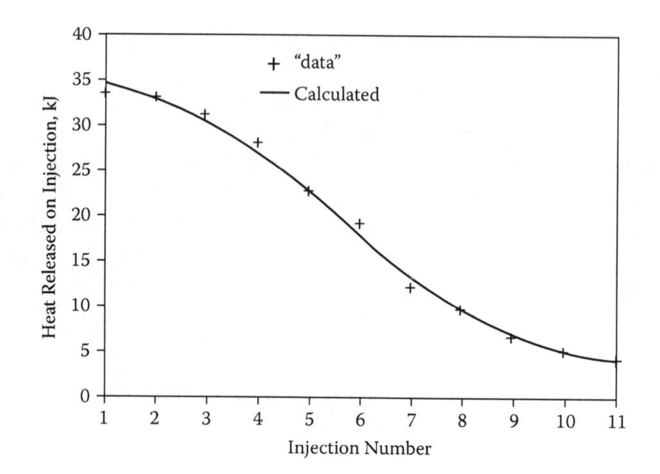

FIGURE 11.23 Comparison of injection pulse areas from Figure 11.22 with calculation.

In order to reproduce the injection pulses shown in Figure 11.22, the required binding properties are: $K = 0.72\ \mu M^{-1}$ and $\Delta h_b = 58$ kJ/mole. The comparison of the energies of each injection from the "data" in Figure 11.22 with the model predictions is shown in Figure 11.23.

The objective of the above exercise was to demonstrate that data such as those shown schematically in Figure 11.22 can be utilized to provide information not only on the energetics (Δh_b) of the biological reaction of interest, but on the equilibrium properties (K) as well.

11.9.2.2 Differential Scanning Mode

The principal application of this technique in biological research is the determination of the thermodynamic properties of denaturation (i.e., unfolding) of macromolecules, proteins in particular. The *scanning* feature indicates that the temperature of the reference cell on the left of Figure 11.21 is increased approximately linearly in time, at a rate of about 1 K/s. The temperature ramp rate needs to be slow enough for the system in the active cell to remain in equilibrium during the transient. The *differential* aspect refers to the comparison of the temperatures in the active cell on the right in Figure 11.21 with that in the reference cell. The control electronics adjusts the heating or cooling rate in the active cell to maintain a zero temperature difference between the two cells at all times.

As the temperature is raised, the control electronics records the cumulative heat inputs to the two cells. Because unfolding involves breaking many bonds within the folded protein, the process is endothermic and heat must be added to the active cell in order for its temperature to match that of the reference cell.

Consider the heat inputs in a temperature interval *dT*. Because the reference cell contains only water (and perhaps a buffer), the heat required to achieve the temperature rise is:

$$dQ_{ref} = n_W C_{PW} dT$$

where n_W is the number of moles of water in both the reference cell and the active cell. C_{PW} is the specific heat of water. In addition to n_W moles of water, the active cell contains n_p moles of the protein. In the temperature interval dT, the fraction of protein in the unfolded state increases by df, so the heat input to the active cell for the temperature interval dT is:

$$dQ = n[fC_{PU} + (1-f)C_{Pf}]dT + n\Delta H_U^o df + n_W C_{PW} dT$$

C_{PU} and C_{Pf} are the heat capacities of the unfolded and folded states of the protein, respectively, and ΔH_U^o is the enthalpy change upon unfolding. The first term in the above equation accounts for the effect of the heat capacities of the two states and their relative amounts at the current temperature. The second term is the heat that must be added to the active cell as a differential amount of unfolding (df) occurs. The third term is the heat absorbed by the water present in the active cell. The contribution of the water in the two cells is removed by combining the two heat input equations to give an *apparent heat capacity*:

$$\frac{1}{n}\frac{d(Q-Q_{ref})}{dT} = fC_{PU} + (1-f)C_{Pf} + \Delta H_U^o \frac{df}{dT}$$

The next step is to express f in terms of the thermodynamic properties of the denaturation reaction, namely ΔH_U^o, ΔS_U^o and C_{PU} and C_{Pf}. This analysis has been developed in Section11.5.1: Equation (11.10) gives the equilibrium constant, K_U, in terms of the standard Gibbs free energy of the unfolding process, ΔG_U^o; the latter is expressed in terms of the thermodynamic properties by Equation (11.11); finally, the fraction unfolded, f, is expressed in terms of the equilibrium constant by Equation (11.12). These relations, plus the Van't Hoff equation, Equation (9.25), relating the temperature derivative of K_U to the enthalpy change on unfolding, convert the apparent heat capacity to:

$$\frac{1}{n}\frac{d(Q-Q_{ref})}{dT} = C_{Pf} + \left(\frac{K_U}{1+K_U}\right)\left[(C_{PU}-C_{Pf}) + \frac{(\Delta H_U^o)^2}{(1+K_U)RT^2}\right] \tag{11.72}$$

Figure 11.24 is a plot of Equation (11.72) for the same thermodynamic properties that produced Figure 11.6. These properties are for 298 K, which is the reference temperature in Equation (11.11).

At low temperature, $K_U \to 0$ and the apparent heat capacity reduces to C_{Pf}. At the opposite limit of high temperature, $K_U \to \infty$ and the apparent heat capacity approaches C_{PU}. The bulk of the $f \to U$ conversion takes place over a narrow temperature range of approximately 15°C. It is important to note that the height and width of the peak between 370 and 380 K in Figure 11.24 depends not only upon ΔH_U^o but on ΔS_U^o and the difference in the heat capacities of the folded and unfolded states as well.

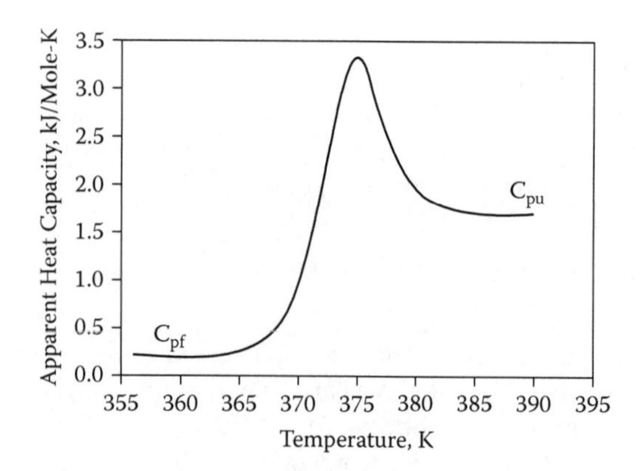

FIGURE 11.24 Response of a differential scanning calorimeter to unfolding of a protein in aqueous solution. The reaction properties at 298 K are: C_{Pf} = 0.2 kJ/mole-K; $\Delta H_U^o = 50$ kcal/mole; $\Delta S_U^o = 0.1$ kcal/mole-K; $C_{PU} - C_{Pf} = 1.5$ kcal/mole-K.

PROBLEMS

11.1 In water, glycine (an amino acid) ionizes as follows:

$$GH_2^+ = GH + H^+ \qquad \text{and} \qquad GH = G^- + H^+$$

the equilibrium constants are $K_A = 4.6 \times 10^{-3}$ M and $K_B = 2.5 \times 10^{-10}$ M, respectively. The solution contains 0.056 M of glycine and 0.044 M of NaOH. What are the pH and the concentrations of all glycine species?

11.2 Derive Equation (11.51a) starting from species conservation and change neutrality equations with $[NaCl]_o = 0$.

11.3 Starting from Equations (11.14)–(11.16), derive:

(a) Equation (11.17).
(b) Scatchard's equation.

REFERENCES

Edsall, J. and H. Gutfreund. 1983. *Biothermodynamics*. New York: Wiley.
Lehninger, A., D. Nelson and M. Cox. 1993. *Principles of Biochemistry*, 2nd ed. Worth Publishers.
Prausnitz, J. et al. 1999. *Molecular Thermodynamics of Fluid-Phase Equilibria*, 3rd ed. Englewood Cliffs, NJ: Prentice Hall.
Yang, W. 1989. *Biothermal-Fluid Sciences*. Washington, DC: Hemisphere Publishing.
Zaslavsky, B. 1995. *Aqueous Two-Phase Partitioning*. New York: Marcel Dekker.

Index

D

E

F

G

H

I

J

Q

R

S

T

U

V

W

Z